METHODS IN MOLECULAR BIOLOGY™

Series Editor
John M. Walker
School of Life Sciences
University of Hertfordshire
Hatfield, Hertfordshire, AL10 9AB, UK

For other titles published in this series, go to
www.springer.com/series/7651

Proprotein Convertases

Edited by

Majambu Mbikay

Chronic Disease Program, Ottawa Hospital Research Institute, Ottawa, ON, Canada

Nabil G. Seidah

Biochemical Neuroendocrinology Laboratory, Clinical Research Institute of Montreal (IRCM), Montreal, QC, Canada

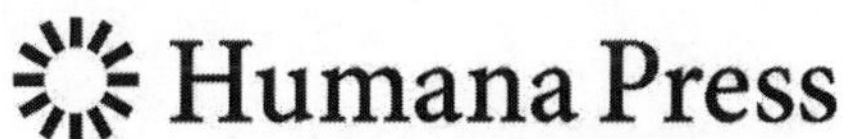

Editors
Majambu Mbikay
Chronic Disease Program
Ottawa Hospital Research Institute
K1Y 4E9 Ottawa, ON, Canada
mmbikay@ohri.ca

Nabil G. Seidah
Biochemical Neuroendocrinology Laboratory
Clinical Research Institute of Montreal (IRCM)
H2W 1R7 Montreal, QC, Canada
nabil.seidah@ircm.qc.ca

ISSN 1064-3745
e-ISSN 1940-6029
ISBN 978-1-61779-203-8
e-ISBN 978-1-61779-204-5
DOI 10.1007/978-1-61779-204-5
Springer New York Dordrecht Heidelberg London

Library of Congress Control Number: 2011933260

Printed on acid-free paper

Humana Press is part of Springer Science+Business Media (www.springer.com)

Preface

This special volume of *Methods in Molecular Biology* covers precursor endoproteolysis as a mechanism of protein activation/inactivation in the secretory pathway. Initially identified in the late 1960s as a post-translational modification leading to the production of hormonal and neural peptides, this process was found through the years to be an intervening step in the activation, and sometimes inactivation, of a wide variety of functional proteins, in nearly all cells and living organisms, from viruses to mammals. Today, activation endoproteolysis of secretory proteins is recognized as a fundamental biological mechanism of spatial and temporal regulation of protein activity as well as of diversification of protein functions. It is described at varying lengths in most cell biology textbooks written in the last two decades.

Proprotein convertases, the enzymes mediating this endoproteolysis, constitute the central theme of this volume. These endoproteinases travel through, reside within, or cycle between the various compartments of the secretory pathway. Most of them are calcium-dependent serine proteases of the subtilase subfamily, collectively designated as proprotein convertases, subtilisin/kexin type (PCSKs), but other proteases, such as cathepsin L, also appear to be able to perform similar functions, in the brain at least. The nine known PCSKs are further subdivided into seven kexin-like convertases, which cleave after basic residues, and two non-kexin-like convertases, which do not. The enzymology of kexin-like PCSKs has been extensively studied in vitro using a variety of synthetic substrates. These studies have revealed specificities, preferences, and overlaps in cleavage motif recognition. The search for specific inhibitors for these enzymes is an active field of research which should lead to novel tools for altering their expression and/or activity for experimental or therapeutic purposes. The discovery of a non-enzymatic function to PCSK9 is the latest twist in the evolving story of the proprotein convertases. Acting as a binding protein for the low-density lipoprotein receptor (LDLR), PCSK9 promotes the degradation of this receptor, thus reducing hepatic clearance of blood LDL cholesterol and causing elevation of this lipid in circulation. PCSK9 is currently the subject of intense investigation as a target for inactivation in the treatment of hypercholesterolemia and associated atherosclerosis. A better grasp of its biosynthesis and cell biology should help in the design of potent and efficacious anti-PCSK9 drugs.

From a survey of the content of this volume, it is quite apparent that, collectively, the proprotein convertases are critical players in the network of intra- and intercellular signaling events that determine normal physiology. Alterations in their expression have been associated with illnesses such as infertility, obesity, diabetes, cardiovascular diseases, and cancer. These alterations may be caused by genetic lesions, epigenetic changes, or abnormal expression of proteins that modulate their biosynthesis and enzymatic activity.

The biological relevance of the proprotein convertases has been explored mostly by studying the developmental and physiological phenotypes of mice genetically engineered not to express them. Observations from mouse studies have been corroborated by clinical cases and by genome-wide association studies in human. They have also been enriched by findings from alternative experimental models such as the zebrafish, *Caenorhabditis elegans*, and *Drosophila*. The targeted inactivation of proprotein convertase genes in mice has been either germline or somatic. Mice resulting from germline inactivation may represent useful models of inborn deficiency of a proprotein convertase in humans; those resulting from somatic inactivation, in contrast, may produce an organ- or a tissue-localized deficiency associated with a morbid phenotype mimicking human diseases caused by ageing or environmental injuries. Depending on the targeted enzyme, the phenotypes observed have ranged from developmental arrest to physical abnormalities, metabolic disturbances, and behavioral changes. The phenotyping of most of the targeted mice has been partial. Their complete and detailed characterization will undoubtedly require collaboration among many specialized fields of biology. In the meantime, comparative proteomics and peptidomics of tissues from mice expressing or not expressing the enzyme have begun to provide some insights into the nature of potential physiological substrates and the tightness of these enzymatic links, as well as the metabolic paths influenced by these enzymes. It will take refined cellular biological studies to elucidate the cascades, the cooperation, and redundancy that may be associated with the action of the proprotein convertases in the secretory pathway.

This special volume of *Methods in Molecular Biology* provides a timely assessment of the impact of activation/inactivation endoproteolysis in the secretory pathway on our current understanding of multiple physiological processes. In addition to reminiscences on the events surrounding the seminal discoveries that launched the concept in 1967, it describes the efforts that led to the elucidation in 1989 of the enzymes mediating this process as well as the evolution of the field since then. Furthermore, it offers a broader perspective on the biochemistry of the PCSKs by exploring structural and functional analogies with bacterial subtilisin and on the enzymology of endoproteolysis itself by describing the involvement in the process of non-PCSK type such as cathepsin L. Most of all, in line with the objective of the series, this volume contains a number of detailed protocols developed by prominent scientists from around the world who have been studying the biology of proprotein convertases.

This volume of *Methods in Molecular Biology* should represent an instructive and useful reference book for all scientists interested in endoproteolytic activation and/or inactivation of secretory proproteins through limited proteolysis, for experts in the field and newcomers to it as well.

Ottawa, Ontario ***Majambu Mbikay, PhD***
Montreal, Quebec ***Nabil G. Seidah, PhD***

Contents

Contributors

YOUNES ANINI • *Department of Physiology and Biophysics and Obstetrics and Gynecology, Dalhousie University, Halifax, NS, Canada B3H 1X5*

JON R. APPEL • *Torrey Pines Institute for Molecular Studies, San Diego, CA 92121, USA*

AJOY BASAK • *Chronic Disease Program, Ottawa Hospital Research Institute, Ottawa, ON, Canada K1Y 4E9; Department of Biochemistry, Microbiology and Immunology, University of Ottawa, Ottawa, ON, Canada*

MARGERY C. BEINFELD • *Department of Pharmacology and Experimental Therapeutics, Tufts University School of Medicine, Boston, MA 02111, USA*

HUGH P.J. BENNETT • *Endocrine Research Laboratory and Department of Medicine, Royal Victoria Hospital and McGill University Health Centre Research Institute, Montreal, QC, Canada H3A 1A1*

FABIEN CALVO • *INSERM, UMRS 940, Equipe AVENIR, Institut de Génétique Moléculaire, Université Paris 7, 75010 Paris, France*

ANDREW CHEN • *Chronic Disease Program, Ottawa Hospital Research Institute, Ottawa, ON, Canada K1Y 4E9*

BABYKUMARI P. CHITRAMUTHU • *Endocrine Research Laboratory and Department of Medicine, Royal Victoria Hospital and McGill University Health Centre Research Institute, Montreal, QC, Canada H3A 1A1*

HÉLÈNE CHOQUET • *CNRS-8090-Institute of Biology, Pasteur Institute, Lille, France*

MICHEL CHRÉTIEN • *Chronic Disease Program, Ottawa Hospital Research Institute, Ottawa, ON, Canada K1Y 4E9*

JENNIFER S. COLVIN • *Department of Psychiatry and Behavioral Sciences, Duke University Medical Center, Durham, NC 27710, USA*

JOHN W.M. CREEMERS • *Laboratory of Biochemical Neuroendocrinology, Center for Human Genetics, K.U. Leuven, B-3000 Leuven, Belgium*

THILINA DEWPURA • *Department of Biochemistry, Microbiology and Immunology, Ottawa Institute of Systems Biology, University of Ottawa, Ottawa, ON, Canada K1H 8M5*

LLOYD D. FRICKER • *Department of Molecular Pharmacology, Albert Einstein College of Medicine, Bronx, NY 10461, USA*

YANGXIN FU • *Departments of Obstetrics and Gynecology and Oncology, University of Alberta, Edmonton, AB, Canada*

LYDIANE FUNKELSTEIN • *Department of Neuroscience, Pharmacology, and Medicine, Skaggs School of Pharmacy and Pharmaceutical Sciences, University of California, San Diego, CA 93093, USA*

VIVIAN HOOK • *Department of Neuroscience, Pharmacology, and Medicine, Skaggs School of Pharmacy and Pharmaceutical Sciences, University of California, San Diego, CA 93093, USA*

KAI KAPPERT • *Institute of Laboratory Medicine, Clinical Chemistry and Pathobiochemistry, and Center for Cardiovascular Research (CCR), Charité – University Medicine Berlin, D-13353, Berlin, Germany*
ABDEL-MAJID KHATIB • *University of Bordeaux, INSERM, LAMC, UMR 1029, F-33400 Talence, France*
IRIS LINDBERG • *School of Medicine Anatomy and Neurobiology, University of Maryland-Baltimore, Baltimore, MD 21201, USA*
SWAPAN MAJUMDAR • *Chronic Disease Program, Ottawa Hospital Research Institute, Ottawa, ON, Canada K1Y 4E9; Department of Chemistry, Tripura University, Suryamaninagar, India*
JANICE MAYNE • *Department of Biochemistry, Microbiology and Immunology, Ottawa Institute of Systems Biology, University of Ottawa, Ottawa, ON, Canada K1H 8M5*
MAJAMBU MBIKAY • *Chronic Disease Program, Ottawa Hospital Research Institute, Ottawa, ON, Canada; Department of Biochemistry and Immunology, University of Ottawa, Ottawa, ON, Canada K1Y 4E9*
PETER METRAKOS • *Department of Surgery, McGill University, Royal Victoria Hospital, Montreal, QC, Canada H3A 1A1; College of Medicine, King Saudi University, Riyadh, Saudi Arabia*
MICHAEL G. MORASH • *Department of Physiology and Biophysics, Dalhousie University, Halifax, Canada; National Research Council of Canada, Institute for Marine Biosciences, Halifax, NS, Canada*
MARK W. NACHTIGAL • *Department of Biochemistry and Medical Genetics, Department of Obstetrics, Gynecology, and Reproductive Sciences, University of Manitoba, Winnipeg, Manitoba, Canada R3E 0J9; Manitoba Institute of Cell Biology, CancerCare Manitoba, University of Manitoba, Winnipeg, Manitoba, Canada R3E 0J9*
GUIYING NIE • *Prince Henry's Institute of Medical Research, Melbourne, VIC, Australia*
RAMONA M. RODRIGUIZ • *Department of Psychiatry and Behavioral Sciences and Mouse Behavioral and Neuroendocrine Analysis Core Facility, Duke University Medical Center, Durham, NC 27710, USA*
NATHALIE SCAMUFFA • *INSERM, UMRS 940, Equipe AVENIR, Institut de Génétique Moléculaire, Université Paris 7, 75010 Paris, France*
GUNTHER SCHMIDT • *Chronic Disease Program, Ottawa Hospital Research Institute, Ottawa, ON, Canada*
NABIL G. SEIDAH • *Biochemical Neuroendocrinology Laboratory, Clinical Research Institute of Montreal, Montreal, QC, Canada H2W 1R7*
UJWAL SHINDE • *Department of Biochemistry and Molecular Biology, Oregon Health and Science University, Portland, OR 97229, USA*
FRANCINE SIROIS • *Chronic Disease Program, Ottawa Hospital Research Institute, Ottawa, ON, Canada*
HEATHER PALMER SMITH • *Chronic Disease Program, Ottawa Hospital Research Institute, Ottawa, ON, Canada K1Y 4E9; Department of Biochemistry, Microbiology and Immunology, University of Ottawa, Ottawa, ON, Canada*
KELLY SOANES • *National Research Council of Canada, Institute for Marine Biosciences, Halifax, NS, Canada*
PHILIPP STAWOWY • *Department of Medicine/Cardiology, Deutsches Herzzentrum Berlin, D-13353 Berlin, Germany*

DONALD F. STEINER • *Departments of Biochemistry and Molecular Biology and Medicine, The University of Chicago, Chicago, IL 60637, USA*

ANDREW N. STEPHENS • *Prince Henry's Institute of Medical Research, Melbourne, VIC, Australia*

PIETER STIJNEN • *Center for Human Genetics, K.U. Leuven, Leuven, Belgium*

HAIDY TADROS • *Chronic Disease Program, Ottawa Hospital Research Institute, Ottawa, ON, Canada*

GARY THOMAS • *Vollum Institute, Oregon Health and Science University, Portland, OR 97229, USA*

JONATHAN WARDMAN • *Department of Molecular Pharmacology, Albert Einstein College of Medicine, Bronx, NY 10461, USA*

WILLIAM C. WETSEL • *Departments of Psychiatry and Behavioral Sciences, Cell Biology, and Neurobiology, Mouse Behavioral and Neuroendocrine Analysis Core Facility, Duke University Medical Center, Durham, NC 27710, USA*

Section I

Reminiscences

Chapter 1

On the Discovery of Precursor Processing

Donald F. Steiner

Abstract

Studies of the biosynthesis of insulin in a human insulinoma beginning in 1965 provided the first evidence for a precursor of insulin, the first such prohormone to be identified. Further studies with isolated rat islets then confirmed that the precursor became labeled more rapidly than insulin and later was converted to insulin by a proteolytic processing system located mainly within the secretory granules of the beta cell and was then stored or secreted. The precursor was designated "proinsulin" in 1967 and was isolated and sequenced from beef and pork sources. These structural studies confirmed that the precursor was a single polypeptide chain which began with the B chain of insulin, continued through a connecting segment of 30–35 amino acids and terminated with the A chain. Paired basic residues were identified at the sites of excision of the C-peptide. Human proinsulin and C-peptide were then similarly obtained and sequenced. The human C-peptide assay was developed and provided a useful tool for measuring insulin

Member of the National Academy of Science (USA) and Laureate of the Wolf Prize in Medicine, 1985.

M. Mbikay, N.G. Seidah (eds.), *Proprotein Convertases*, Methods in Molecular Biology 768,
DOI 10.1007/978-1-61779-204-5_1, © Springer Science+Business Media, LLC 2011

levels indirectly in diabetics treated with insulin. The discovery of other precursor proteins for a variety of peptide hormones, neuropeptides, or plasma proteins then followed, with all having mainly dibasic cleavage sites for processing. The subsequent discovery of a similar biosynthetic pathway in yeast led to the identification of eukaryotic families of specialized processing subtilisin-like endopeptidases coupled with carboxypeptidase B-like exopeptidases. Most neuroendocrine peptides are processed by two specialized members of this family – PC2 and/or PC1/3 – followed by carboxypeptidase E (CPE). This brief report concentrates mainly on the role of insulin biosynthesis in providing a useful early paradigm of precursor processing in the secretory pathway.

Key words: Proinsulin, pulse-chase labeling, proteolytic conversion, C-peptide, convertase, carboxypeptidase.

Although the subtopic headings "protein precursor" and "protein processing" did not appear in the Index Medicus until 1973 and 1983, respectively, Bayliss and Starling in 1902 (1), who discovered secretin and famously called it a *hormone*, also proposed its possible storage as a zymogen-like prohormone or "prosecretin" to possibly explain their difficulties in extracting it. Similarly, in 1916, 5 years before insulin was discovered, Edward Schäfer similarly proposed that *insuline*, the putative antidiabetic substance believed to be located in the pancreatic islets, was eluding discovery because it probably was stored in the tissue as "proinsuline" (2). Fortunately, especially for diabetics, their reasoning was faulty, as we now know that most prohormones and proneuropeptides are efficiently processed proteolytically within their cells of origin and then stored as readily releasable active forms. As soon as suitable extraction methods for peptide hormones were developed, the idea of prohormones faded out of sight and did not surface again until the mid-1960s when interest in how peptide hormones were made finally led to the discovery of the first authentic prohormone – proinsulin (3). What prompted this resurgence of interest was the development of methods for elucidating the structures, first of small peptides like oxytocin and vasopressin by Vincent DuVigneaud (4) and then of proteins (insulin) by Fred Sanger in the 1950s (5).

Determining the structure of insulin was a daunting task even though it fortunately turned out to be not so large as was first believed to be the case (36 kDa), which was due to its tendency to self-associate. Its monomeric molecular weight is actually only about 6,000 and it contains only 51 amino acids. Nonetheless it took Sanger and his colleagues almost 10 years to first work out the structures of each of its two separated chains and then to solve the enigma of the arrangement of the disulfide bonds. The complete covalent structure of beef and pig insulin appeared in 1955, and human insulin was completed by his students Nicol and Smith by 1960 (6). Insulin thereby became the "Rosetta Stone" of modern protein chemistry and earned Sanger his first Nobel Prize. As a medical student I was just completing the second quarter of my freshman biochemistry course in 1953 when Watson and Crick announced their structure for DNA (7). By 1955, when I was a junior, Sanger's results demonstrating the first species differences in the amino acid sequence of the insulin molecule had provided the first clear indications of the existence of the genetic code. The challenge posed by insulin which invited considerable speculation at that time was the mechanism by which the chains were made and then assembled into the active molecule.

In 1955 I embarked on my first biochemical research effort – finding a way to keep spleen cells alive for several days in culture so as to be able to determine whether antibodies were made de novo in response to an antigen or were derived by a template, or imprinting, mechanism from pre-existing immunoglobulins in the guiding presence of an antigen such as bovine serum albumin. To do this I first developed a workable tissue culture system to maintain a whole rabbit spleen cell suspension (ca. 1 g) for 3–4 days while the cells carried out a "secondary" response to a BSA stimulus given 2 days before splenectomy. The system worked well and by adding labeled amino acids to the medium during the active phase of antibody secretion, we could show that specific antibodies precipitable by BSA were made de novo (8). Pauling's instructive hypothesis was not confirmed. During my senior year, as the work continued, I entertained notions of using my culture system to study such problems as the origin of the Bence Jones proteins associated with multiple myeloma or perhaps of insulin in a human islet cell tumor, but there were no opportunities to do so before my graduation and internship.

Ten years later I was an assistant professor of biochemistry, back in Chicago from Seattle, where at the University of Washington, I first became involved in research on insulin. None of my experiments then involved anything other than the injection of insulin into diabetic animals or its solution in incubation media for studies on its action mechanism. However, quite suddenly in October 1965 an opportunity to study insulin biosynthesis appeared – a patient with an insulinoma was in the hospital and would be undergoing exploratory surgery for a possible insulin secreting pancreatic tumor, the next day!

I seized this opportunity and quickly prepared a pilot experiment to study insulin biosynthesis in the tumor using two tritiated amino acids on hand in my laboratory (Phe and Leu) to label batches of slices prepared from the tumor and then froze the labeled material for later extraction and work up. The ensuing analysis provided the first solid evidence (3) for a larger insulin-related molecule which indeed turned out to be proinsulin, the single chain precursor of insulin (*see* **Fig. 1.1**). (For more details see previous accounts of this discovery: 9–11.) This experiment also provided evidence that proinsulin began with the B chain, ended with the A chain, and contained an additional peptide

HUMAN PROINSULIN

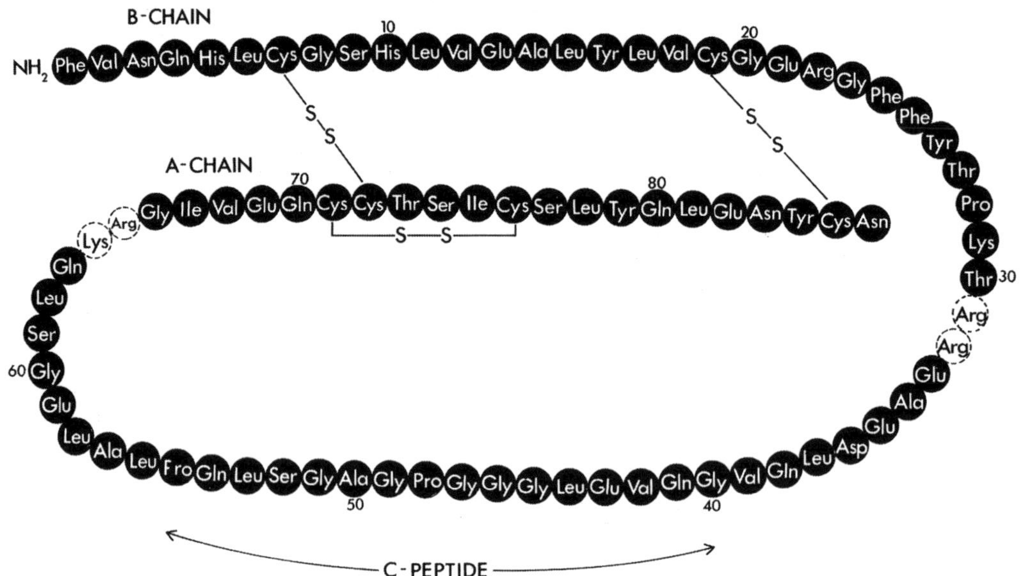

Fig. 1.1. The structure of human proinsulin predicted on the basis of the known sequence of human insulin and the deduced sequence of the connecting segment (*see* ref. 19 for details).

linking the chains and adding an additional 50% to its molecular size. We also looked for such a molecule in commercial insulin preps and soon found that it was indeed present at a level of about 1% of the total insulin (3), thus providing us with a potential source of proinsulin for further studies. With the help of generous gifts of crystalline insulin from pharmaceutical companies [Lilly (Indianapolis) and Novo (Denmark)], we were able within a year or so to develop chromatographic procedures for its purification to homogeneity. We chose to study the bovine prohormone, mainly with help from Novo, while at Lilly, Ronald Chance and coworkers isolated the porcine prohormone. Although we succeeded in purifying enough bovine proinsulin by November 1967 (12) to begin sequence studies in collaboration with protein chemists Chris Nolan and Emmanuel Margoliash in North Chicago, progress was slow due to various technical problems. Ron and coworkers at Lilly outdistanced us and published the covalent structure of porcine proinsulin in 1968 (13).

However, in the meantime we also had continued to study the biosynthesis of insulin using isolated rat islets, a new technique then. With the help of several students the precursor–product relationship of proinsulin to insulin could be demonstrated using pulse-chase labeling methods (14). In 1967, a graduate student, Jeffrey Clark, joined in that effort and did an extensive series of biochemical and biosynthetic studies in which he isolated a large supply of the rat insulins (both rats and mice have two insulin genes) from a kilogram of rat pancreas, purified the two rat insulins, isolated, and identified both proinsulins and in doing so corrected the structure of rat insulin-2 (15). In the course of his biosynthetic studies on rat islets, he identified a fast moving labeled band on PAGE which he identified as the rat C-peptide. Based on these findings we then undertook isolation of the bovine C-peptide from pancreas via a purification procedure that began with modified acid–ethanol extraction, followed by gel exclusion chromatography. On gel filtration the C-peptide comigrates with insulin, which explains why, in earlier experiments with the leucine-labeled human proinsulin fraction from the tumor studies, treatment with low amounts of trypsin resulted in an apparently seamless conversion of proinsulin to insulin (3). The simplest way to separate the C-peptide from insulin was by paper electrophoresis in 30% formic acid which (after drying) gave a sharp band with ninhydrin staining that was negative to two other protein stains (Sakaguchi or Pauli) which recognize Arg or His/Tyr, respectively. These amino acids do not occur in the bovine C-peptide, but are present in insulin. Having pure preparations of bovine pancreatic C-peptide now available, we undertook its amino acid sequence determination in order to prove unequivocally that it was identical to the connecting peptide segment of bovine proinsulin, then being sequenced by our collaborators in North Chicago. This work was done mainly by Philip Oyer and Jim Peterson, while I was writing the manuscript for the paper, but to our great surprise, our sequence differed slightly from that of bovine proinsulin. We checked our data and Chris Nolan reexamined his – in fact he had misplaced an Ala residue in a chain of several glycines, a difficult sequence to sort out. It was fortunate that we had gone the extra mile of sequencing, in addition to simply peptide mapping, the bovine C-peptide, as now we could be confident that both sequences were correct and their identity also established the origin of the pancreatic C-peptide from proinsulin (16, 17).

Biosynthetic studies carried out by Jeff Clark and Arthur Rubenstein, who had joined my laboratory group in 1968, demonstrated that the molar ratio of insulin to C-peptide both within rat islets and also in the medium after secretion was very close to unity, confirming that the C-peptide and insulin are stored together in the secretory granules after conversion and co-secreted into the medium (18). To obtain the human C-peptide sequence, Sooja Cho in our group collected pancreatic material from autopsies weekly for more than a year and extracted it. It had usually undergone some autolysis which lowered yields. However, Philip Oyer was able to secure enough pure material to determine its sequence (**Fig. 1.1**), and we could thus predict the structure of human proinsulin (19). Arthur Rubenstein also spearheaded studies showing the presence of secreted proinsulin in both the blood and the urine of human subjects (20, 21). The C-peptide radioimmunoassay was developed in 1970 (22) in order to demonstrate its presence in human serum. Arthur Rubinstein, Kenneth Polonsky, and their colleagues then accomplished a feat of translational medicine by refining and calibrating the C-peptide assay, which is now widely utilized clinically to measure endogenous insulin production under a variety of conditions (23, 24).

Other important early findings that emerged from Jeff Clark's pulse-chase studies with rat islets were (a) that glucose, strongly and rapidly stimulates the biosynthesis of insulin selectively and without a requirement for new RNA synthesis and (b) that energy is required for the transfer of newly synthesized proinsulin from the ER to a new intracellular compartment, where it comes into contact with the proteolytic enzymes that result in its efficient conversion to insulin. Thus, once conversion to insulin begins it no longer can be inhibited by antimycin A (J. Clark, PhD thesis, University of Chicago, 1969) (25). The pioneering studies of Jamieson and Palade (26) on the intracellular migration of newly synthesized secretory protein in pancreatic exocrine cells helped us to rationalize that the initial energy-requiring step was the transfer via vesicles from ER to the Golgi apparatus (long known to be associated with secretory activity) and, based on the observed time constraints for processing, it was likely that conversion might begin in the TGN, but normally it mainly occurs in the secretory granules (27). This was definitively demonstrated in 1985 by Orci et al. (28), using a monoclonal antibody specific for proinsulin that was produced in our laboratory (29). Studies of the secretion of proinsulin and insulin from rat islets indicated that proinsulin is not selectively secreted in significant amounts by non-granule pathways, consistent with its role as an intermediate in biosynthesis (30).

Another early concern was the nature of the conversion process of proinsulin to insulin within the beta cell. The earliest studies on proinsulin revealed its great sensitivity to trypsin, suggesting that basic amino acids would be involved (3). This was subsequently confirmed when the sequence of porcine proinsulin was announced. However, conversion with trypsin resulted in cleavage at Lys B29 (13). This problem was overcome by adding an excess of carboxypeptidase B with trypsin to the reaction mixture (31). This combination works so well because the carboxypeptidase quickly removes the Arg-doublet at position 31 and 32 beyond B30 which greatly reduces the susceptibility of the B29 lysine to trypsin. This efficient method of conversion is still being used today for the conversion of biosynthetic human proinsulin to insulin at the Lilly Company.

The search for the enzymes involved in the conversion of proinsulin and many other prohormones finally culminated in

1989 in the discovery of PC2 and then of PC1/3, members of a larger family of calcium-dependent subtilisin-like converting endoproteases that includes furin, PACE 4, PC4, PC5/6 A and B, and PC7 (32, 33). The discovery of yeast kexin was the key to finally solving this problem (34). The processing carboxypeptidases, CPE, and then CPD began to be identified before the mammalian endoproteases, in the mid-1980s (35). With the advent of molecular cloning in the late 1970s the structural elucidation of precursor proteins rapidly increased with many interesting and exciting revelations of larger polyprotein precursors containing multiple copies of structurally or functionally related biologically active peptides (36). Thus as time has passed the dimensions of the field of precursor processing have continually enlarged far beyond the simple secretory pathway paradigm of proinsulin and its role in insulin biosynthesis. But it was the identification of and a wealth of studies on proinsulin and insulin biosynthesis that provided a solid base – a corner stone – for this ever-enlarging field that we celebrate in this volume today (for a more detailed account on this topic see (37)).

Acknowledgments

I am indebted to many students, postdocs, and colleagues who have contributed to the work from my laboratory discussed in this review. In particular, I would like to mention Philip Oyer, Jeffrey Clark, Wolfgang Kemmler, Dennis Cunningham, Hiroyuki Sando, Franco Melani, Arthur Rubenstein, Simon Pilkis, Howard Tager, James Peterson, Shu Jin Chan, Susan Terris, Sooja Cho Nehrlich, Ray Carroll, Ole Madsen, Sture Falkmer, Stefan Emdin, Christoph Patzelt, Åke Lernmark, Cecelia Hoffman, Jon Marsh, Masakazu Haneda, Simon Kwok, Kevin Docherty, John Hutton, David Nielsen, Michael Welsh, Susumu Seino, Kishio Nanjo, Steve Smeekens, Graeme Bell, Kenneth Polonsky, Steve Duguay, Motoshige Miyano, Masahiro Nishi, Shinya Ohagi, Machi Furuta, Shinya Nagamatsu, Yasunao Yoshimasa, Jonathan Whittaker, Tadashi Hanabusa, Hisako Ohgawara, Yves Rouille, Jeremy Paul, Mohammad Pashmforoush, Louis Philipson, Grigory Lipkind, Sean Martin, Tony Oliva, An Zhou, Gene Webb, Joe Bass, Xiaorong Zhu, Takeshi Kurose, Arunangsu Dey, Jie Wang, Iris Lindberg, Per Westermark, Soo Young Park, and Gunilla Westermark. Work from my laboratory has been supported by NIH grants DK13914 and DK20595 and by the Howard Hughes Medical Institute.

References

1. Bayliss, W. M., and Starling, E. H. (1902) The mechanism of pancreatic secretion *J Physiol* **28**, 325–53.
2. Schäfer, E. A. (1916) *The Endocrine Organs-An Introduction to the Study of Internal Secretion*, p. 128. Longmans, Green and Co., London.
3. Steiner, D. F., and Oyer, P. E. (1967) The biosynthesis of insulin and a probable precursor of insulin by a human islet cell adenoma *Proc Natl Acad Sci USA* **57**, 473–80.
4. du Vigneaud, V. (1956) Hormones of the posterior pituitary gland: Oxytocin and vasopressin *Harvey Lect* **50**, 1–26.
5. Sanger, F. (1959) Chemistry of insulin *Science* **129**, 1340–4.
6. Nicol, D. S., and Smith, L. F. (1960) Amino-acid sequence of human insulin *Nature* **187**, 483–5.
7. Watson, J. D., and Crick, F. H. (1953) Molecular structure of nucleic acids; a structure for deoxyribose nucleic acid *Nature* **25**, 737–8.
8. Steiner, D. F., and Anker, H. S. (1956) On the synthesis of antibody protein in vitro *Proc Natl Acad Sci USA* **42**, 580–6.
9. Steiner, D. F. (1991) The biosynthesis of biologically active peptides: A perspective. In: *Peptide Biosynthesis and Processing* (L. D. Fricker, ed.), pp. 1–15, CRC Press, Boca Raton, FL.
10. Steiner, D. F. (2001) The prohormone convertases and precursor processing in protein biosynthesis. In: *The Enzymes*, Vol. XXII (R. E. Dalbey and D. S. Sigman, eds.), pp. 163–98, Academic, New York, NY.
11. Steiner, D. F., Clark, J. L., Nolan, C., Rubenstein, A. H., Margoliash, E., Aten, B., and Oyer, P. E. (1969) Proinsulin and the biosynthesis of insulin *Recent Prog Horm Res* **25**, 207–82.
12. Steiner, D. F., Hallund, O., Rubenstein, A. H., Cho, S., and Bayliss, C. (1968) Isolation and properties of proinsulin, intermediate forms, and other minor components from crystalline bovine insulin *Diabetes* **17**, 725–36.
13. Chance, R. E., Ellis, R. M., and Bromer, W. W. (1968) Porcine proinsulin: Characterization and amino acid sequence *Science* **161**, 165–7.
14. Steiner, D. F., Cunningham, D., Spigelman, L., and Aten, B. (1967) Insulin biosynthesis: Evidence for a precursor *Science* **157**, 697–700.
15. Clark, J. L., and Steiner, D. F. (1969) Insulin biosynthesis in the rat: Demonstration of two proinsulins *Proc Natl Acad Sci USA* **62**, 278–85.
16. Steiner, D. F., Cho, S., Oyer, P. E., Terris, S., Peterson, J. D., and Rubenstein, A. H. (1971) Isolation and characterization of proinsulin C-peptide from bovine pancreas *J Biol Chem* **246**, 1365–74.
17. Nolan, C., Margoliash, E., Peterson, J. D., and Steiner, D. F. (1971) Structure of bovine proinsulin *J Biol Chem* **246**, 2780–95.
18. Rubenstein, A. H., Clark, J. L., Melani, F., and Steiner, D. F. (1969) Secretion of proinsulin C-peptide by pancreatic B cells and its circulation in blood *Nature* **224**, 697–9.
19. Oyer, P. E., Cho, S., Peterson, J. D., and Steiner, D. F. (1971) Studies on human proinsulin: Isolation and amino acid sequence of the human pancreatic C-peptide *J Biol Chem* **246**, 1375–86.
20. Rubenstein, A. H., Cho, S., and Steiner, D. F. (1968) Evidence for proinsulin in human urine and serum *Lancet* **1**, 1353–5.
21. Melani, F., Rubenstein, A. H., and Steiner, D. F. (1970) Human serum proinsulin *J Clin Invest* **49**, 497–507.
22. Melani, F., Rubenstein, A. H., Oyer, P. E., and Steiner, D. F. (1970) Identification of proinsulin and C-peptide in human serum by a specific immunoassay *Proc Natl Acad Sci USA* **67**, 148–55.
23. Polonsky, K., and Rubenstein, A. H. (1986) Current approaches to measurement of insulin secretion *Diabetes Metab Rev* **2**, 315–29.
24. Brandenburg, D. (2008) History and diagnostic significance of C-peptide *Exp Diabetes Res*, Article ID: 576862, 1–7.
25. Steiner, D. F., Clark, J. L., Nolan, C., Rubenstein, A. H., Margoliash, E., Melani, F., and Oyer, P. E. (1970) The biosynthesis of insulin and some speculations regarding the pathogenesis of human diabetes. In: The Pathogenesis of Diabetes Mellitus. Nobel Symposium **13**, pp. 57–80. Almqvist and Wiksell, Stockholm.
26. Jamieson, J. D., and Palade, G. E. (1967) Intracellular transport of secretory proteins in pancreatic exocrine cell. I. Role of peripheral elements of golgi complex *J Cell Biol* **34**, 577–96.
27. Kemmler, W., Steiner, D. F., and Borg, J. (1973) Studies on the conversion of proinsulin to insulin. III Studies in vitro with a crude secretion granule fraction isolated from rat islets of Langerhans *J Biol Chem* **248**, 4544–51.

28. Orci, L., Ravazzola, M., Amherdt, M., Madsen, O., Vassalli, J. -D., and Perrelet, A. (1985) Direct identification of prohormone conversion site in insulin-secreting cells *Cell* **42**, 671–81.
29. Madsen, O. D., Frank, B. H., and Steiner, D. F. (1984) Human proinsulin specific antigenic determinants identified by monoclonal antibodies *Diabetes* **33**, 1012–16.
30. Sando, H., Borg, J., and Steiner, D. F. (1972) Studies on the secretion of newly synthesized proinsulin and insulin from isolated rat islets of Langerhans *J Clin Invest* **51**, 1476–85.
31. Kemmler, W., Peterson, J. D., and Steiner, D. F. (1971) Studies on the conversion of proinsulin to insulin. I. Conversion in vitro with trypsin and carboxypeptidase B *J Biol Chem* **246**, 6786–91.
32. Rouillé, Y., Duguay, S. J., Lund, K., Furuta, M., Gong, Q., Lipkind, G., Oliva, A. A., Jr., Chan, S. J., and Steiner, D. F. (1995) Proteolytic processing mechanisms in the biosynthesis of neuroendocrine peptides: The subtilisin-like proprotein convertases *Frontiers Neuroendocrinol* **16**, 322–61.
33. Zhou, A., Webb, G., Zhu, X., and Steiner, D. F. (1999) Proteolytic processing in the secretory pathway *J Biol Chem* **274**, 20745–8.
34. Julius, D., Brake, A., Blair, L., Kunisawa, R., and Thorner, J. (1984) Isolation of the putative structural gene for the lysine-arginine-cleaving endopeptidase required for processing of yeast prepro-alpha-factor *Cell* **37**, 1075–89.
35. Fricker, L. D., Evans, C. J., Esch, F. S., and Herbert, E. (1986) Cloning and sequence analysis of cDNA for bovine carboxypeptidase E *Nature* **323**, 461–4.
36. Douglass, J., Civelli, O., and Herbert, E. (1984) Polyprotein gene expression: Generation of diversity of neuroendocrine peptides *Annu Rev Biochem* **53**, 665–715.
37. Steiner, D. F. (2011) Adventures with insulin in the islets of Langerhans *J Biol Chem* **286**, 17399–421.

Chapter 2

The Prohormone Theory and the Proprotein Convertases: It Is All About Serendipity

Michel Chrétien

Abstract

When I became a physician and an endocrinologist in the early 1960s, peptide hormone sequencing was still in its infancy; it was also far removed from my immediate interests. Through chance encounters with prominent teachers and mentors, I later became increasingly convinced that elucidation of the primary sequence of peptide hormones is key to understanding their production as well as their functions in human health and disease. My interest for pituitary hormones led me to discover that the sequence of β-melanocyte-stimulating hormone was contained within that γ and β-lipotropins and could be released from the latter by limited endoproteolysis. This prohormone theory became the leitmotiv of my career as a clinician/scientist. Through serendipity and the efforts of many laboratories including mine, this theory

M. Mbikay, N.G. Seidah (eds.), *Proprotein Convertases*, Methods in Molecular Biology 768,
DOI 10.1007/978-1-61779-204-5_2, © Springer Science+Business Media, LLC 2011

has now been widely confirmed, extended to various precursor proteins and implicated in many diseases. It has led to our discovery of the proprotein convertases.

Key words: Prohormone theory, lipotropin, adrenocorticotropic hormone, endorphin, proopiomelanocortin, proprotein convertase.

"Dans les champs de l'observation, la chance sourit aux esprits préparés."
(In the field of observation, chance favors the prepared mind).
Louis Pasteur (1822–1895)

It was in the years 1955–1956 that, as a 1st-year medical student at the University of Montreal, I first became enthused by the chemistry of biology. All biological phenomena, stressed some of my teachers, will eventually be explained in terms of chemical reactions. In my 3rd year, I attended numerous lectures on various clinical sub-specialties. Endocrinology was one of my favorites. It was taught by the prestigious French professor Henri Bricaire. Dr Bricaire possessed a unique ability to explain endocrinology in a simple, yet profound way. The pathophysiology of Cushing's syndrome was his particular domain of expertise. Listening to him, I became fascinated by the signaling process through the hypothalamo-pituitary-adrenal axis and the cascade of chemical signals that emanate from the hypothalamus to induce in the pituitary other chemical signals targeted to peripheral endocrine organs.

In my first two residency years (1960–1962), I had the good fortune to be a research fellow under Dr Jacques Genest at the Hôtel-Dieu Hospital and the University of Montreal Medical School. With Dr Genest, I investigated the role of the renin–angiotensin–aldosterone system in arterial hypertension. It was a combination of bench work and clinical research. I carried out the first angiotensin II infusions in human subjects and developed, with Dr Roger Boucher, the first biochemical assay to measure the blood levels of angiotensin II. The end result was the first demonstration in human that angiotensin II is a potent aldosterone-stimulating agent (1–3). Along the way, I learned with amazement that a two-amino acid difference between angiotensin I and II increased by many folds the aldosterone-stimulating effects and that small modifications on the cyclopentanoperhydrophenanthrene ring to, as examples, a ketone, a hydroxyl, or an aldehyde group, would completely change the biological activities of steroid hormones.

In my third and fourth residency years (1963–1964), I joined Drs George Thorn and George Cahill at Harvard Medical School. There, I was exposed to a most diverse cohort of patients with diseases of the pituitary–adrenal axis, i.e., Cushing's and Nelson's syndromes, hypopituitarism, and Addison's disease. All these pathological syndromes involve different states of adrenocorticotropic hormone (ACTH) secretion. The chemistry of ACTH had just been elucidated by Dr Choh Hao Li at the Hormone Research Laboratory at University of California at Berkeley. Although I had absolutely no background in protein chemistry, I applied to join Dr Li's group. To my pleasant surprise, he accepted me with no course prerequisites.

When I arrived in his laboratory in late 1964, Dr Li handed me a little vial containing a white powder which was a putatively novel peptide hormone. He had isolated it from ovine pituitaries, had found it to exhibit some lipolytic activity, and hence had named it β-lipotropic hormone (β-LPH). This peptide was known to be composed of about 90 amino acids (aa), but the sequence of these amino acids was yet to be unraveled. The determination of this sequence was assigned to me as a research project. For my further training, Dr Li also asked me to purify on my own hundreds of milligrams of β-LPH from sheep pituitary glands. While conducting this purification, I stumbled on a side fraction which shared similar biological and chemical properties as the original powder given to me by Dr Li. I was given the task to solely complete its chemical characterization.

The only sequencing method available at the time consisted of successive analyses of the N-terminal amino acid released by Edman degradation. It was entirely manual, labor intensive and time consuming, permitting the determination of a single N-terminal residue every 3 days. After completing the sequence of the two peptides (4, 5), we realized that the peptide in the side fraction represented the first 58-aa fragment of β-LPH (**Fig. 2.1**). We named it γ-LPH. Of further interest was the fact that amino acids 41–58 in both β- and γ-LPHs represented the sequence of

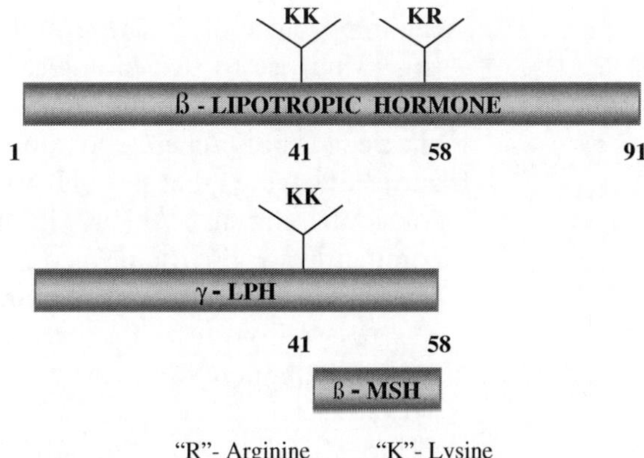

Fig. 2.1. β-Lipotropic hormone (β-LPH) is a precursor protein. Its sequence contains that of γ-LPH which contains that of α-MSH. Dilysyl basic pairs (KK) flank the subfragments of β-LPH.

the β-melanocyte-stimulating hormone (β-MSH) peptide elucidated a short time earlier by Dr Li's group. Thus, while embedded within the β-LPH molecule, β-MSH represented the C terminus of γ-LPH, as if the latter peptide resulted from the removal of a C-terminal fragment of β-LPH at a Lys-Arg pair of basic amino acid residues. This C-terminal fragment, it was later learned, is β-endorphin (β-END).

In our first report of this finding in the *Canadian Journal of Biochemistry* on 5 January 1967 (5), Dr Li and I stated: "The results reported in this paper raise the interesting possibility that the pituitary gland synthesizes de novo a number of peptides having identical sequences, or, alternatively, that the pituitary produces one large molecule that is subsequently broken down into smaller fragments."

Back in Canada at the newly opened Institut de recherches cliniques de Montréal (IRCM), I decided to explore this new paradigm. In those early years, I had the good fortune to recruit Suzanne Benjannet, a talented cell biologist who, together with Xavier Bertagna, a post-doctoral fellow in my laboratory, carried out important studies of peptide hormone biosynthesis in pituitary cells. Suzanne Benjannet has been a precious collaborator ever since and is still making important contributions to the field of endoproteolysis. In the mid-1970s, we were joined by Nabil G. Seidah, who had decided to apply his expertise in physical chemistry to biological problems. This was the start of an enduring research partnership which persists to this day and during which Nabil made many seminal discoveries.

The β-LPH/γ-LPH/β-MSH biosynthetic model greatly expanded when Hughes and Kosterliz serendipitously found that met-enkephalin was the pentapeptide 61–65 of β-LPH (6). Months later, β-END was isolated by us from human pituitaries (7) and by different groups for other species. The observation that this opioid peptide was a C-terminal fragment of β-LPH added to the importance of the latter molecule as a poly-hormone precursor giving rise to two biologically active molecules, i.e., β-MSH and β-END. At about the same time, different groups established, via elegant pulse-chase experiments as well as by cDNA cloning and sequencing, that β-LPH was part of a much larger precursor containing also ACTH, itself a precursor to α-MSH. Concomitantly, we also established by pulse-chase analysis the detailed biosynthetic pathway of their common precursor protein (8). This was a period of great effervescence to which participated, among others, Philippe Crine, Christina Gianoulakis, Guy Boileau, and Michael Dennis. They were succeeded by a core of collaborators, Claude Lazure, Majambu Mbikay, Ajoy Basak, Annik Prat, and Janice Mayne, who are still members of the group.

In the meantime, we had named the common precursor proopiomelanocortin, POMC in short (9). Years later, back in his native France, Xavier Bertagna amusingly interpreted the acronym to mean peptide of Michel Chrétien. The coincidence was unintended; but it is somewhat pleasing, considering the important part POMC-derived peptides have played in my scientific career.

The rest is an ongoing scientific story which revealed serendipity in action at many turns. It took us 23 years to identify the first two proprotein convertases, PC1 and PC2 (10). Today, the field of maturation endoproteolysis has expanded so much that it has become a fundamental process in many aspects of biology, from virus to human, in health and disease (11). From an enzymatic system that was initially proposed for a few peptide hormones, it embraces neuropeptides and a wide array of important proteins including growth factors, receptors, viral envelope proteins, and transcription factors. The convertases themselves came about with many surprises of their own, but the most unexpected results came from the last two members of the family, proprotein-convertase-serine-kexin 8 (PCSK8) (12) and PCSK9 (13). Both cleave at nonbasic residues and play crucial roles in cholesterol homeostasis. While PCSK8 activates the sterol-response element-binding proteins (SREBPs), PCSK9 downregulates the level of low-density lipoprotein receptor (LDL-R). Moreover, PCSK9 is highly polymorphic in human: some mutations are associated with autosomal dominant hypercholesterolemia, while others are accompanied by cardioprotective hypocholesterolemia. As a result, PCSK9 has more clinical relevance than do all the other PCs combined. Our recent finding that hypocholesterolemic mutations are also present in many Canadian families makes it very close to home.

In his 2002 publication in *Nature Reviews* (14), Dr Gary Thomas, of the Vollum Institute, in Portland, Oregon, has most beautifully captured in his own words the importance of the discovery of post-translational endoproteolytic modification of secretory proteins, calling it "as revolutionary as those of Krebs and Fischer, which showed that protein phosphorylation, is a universal modification in signal transduction."

Acknowledgments

Serendipity in my career would not have become rewarding opportunities without the guidance of my mentors Dr Jacques Genest, Dr Roger Boucher, Dr John S.L. Browne, and Dr Arthur Gagnon in Montréal; Dr George Thorn and Dr George Cahill in

Boston; Dr Choh Hao Li, Dr John Ramachandran, Dr Jon Dixon and David Chung at the University of California at San Francisco and Berkeley. I am deeply indebted to my long-time and closest collaborators Dr Nabil G. Seidah, Mrs Suzanne Benjannet, and Dr Majambu Mbikay. Over the years, we were joined by many others: Drs Claude Lazure, Ajoy Basak, Annik Prat, Mycieslaw Marcinkiewicz, Martin Lis, Robert Day, Abdel-Majid Khatib, and Janice Mayne. All along, I benefited from the administrative support of three skillful and devoted assistants: Mmes Diane Marcil, Sylvie Émond, and Denise Joanisse. Many post-doctoral fellows or then-graduate students have made invaluable contributions to our research: alphabetically by last name, Drs Younes Anini, Xavier Bertagna, Guy Boileau, Peter Burbach, John Chan, Gilles Croissandeau, Philippe Crine, Michael Dennis, Christina Gianoulakis, François Gilbert, Francis Gossard, Kuo Liang Hsi, Haruo Iguchi, François Jean, Guy Lambert, Normand Larivière, Richard Leduc, Chao Lin Lu, Rami Morcos, Joanne Paquin, Didier Pélaprat, Marie-Laure Raffin-Sanson Hélène Scherrer, Glen Smith, Philipp Stawowy, and Philippe Touraine. Some devoted technicians have worked with us for decades: Jim Rochemont, Andrew Chen, Francine Sirois, Odette Théberge, Marie-Claude Guérinot, and Josée Hamelin.

References

1. Biron, P., Chrétien, M., Koiw, E., and Genest, J. (1962) Effects of angiotensin infusions on aldosterone and electrolyte excretion in normal subjects and patients with hypertension and adrenocortical disorders *Br Med J* **1**, 1569–75.
2. Chrétien, M. (1962) *Angiotensin blood levels in humans.* M.Sc. Thesis, McGill University, Montreal
3. Genest, J., Boucher, R., De Champlain, J., Veyrat, R., Chrétien, M., Biron, P., Tremblay, G., Roy, P., and Cartier, P. (1964) Studies on the renin-angiotensin system in hypertensive patients *Can Med Assoc J* **90**, 263–8.
4. Li, C. H., Barnafi, L., Chrétien, M., and Chung, D. (1965) Isolation and amino-acid sequence of beta-LPH from sheep pituitary glands *Nature* **208**, 1093–4.
5. Chrétien, M., and Li, C. H. (1967) Isolation, purification, and characterization of gamma-lipotropic hormone from sheep pituitary glands *Can J Biochem* **45**, 1163–74.
6. Hughes, J., Smith, T. W., Kosterlitz, H. W., Fothergill, L. A., Morgan, B. A., and Morris, H. R. (1975) Identification of two related pentapeptides from the brain with potent opiate agonist activity *Nature* **258**, 577–80.
7. Chrétien, M., Benjannet, S., Dragon, N., Seidah, N. G., and Lis, M. (1976) Isolation of peptides with opiate activity from sheep and human pituitaries: Relationship to beta-lipotropin *Biochem Biophys Res Commun* **72**, 472–8.
8. Crine, P., Gianoulakis, C., Seidah, N. G., Gossard, F., Pezalla, P. D., Lis, M., and Chrétien, M. (1978) Biosynthesis of beta-endorphin from beta-lipotropin and a larger molecular weight precursor in rat pars intermedia *Proc Natl Acad Sci USA* **75**, 4719–23.
9. Chrétien, M., Benjannet, S., Gossard, F., Gianoulakis, C., Crine, P., Lis, M., and Seidah, N. G. (1979) From beta-lipotropin to beta-endorphin and 'pro-opio-melanocortin' *Can J Biochem* **57**, 1111–21.
10. Seidah, N. G., Gaspar, L., Mion, P., Marcinkiewicz, M., Mbikay, M., and Chrétien, M. (1990) cDNA sequence of two distinct pituitary proteins homologous to Kex2 and furin gene products: Tissue-specific mRNAs encoding candidates for pro-hormone processing proteinases *DNA Cell Biol* **9**, 415–24.
11. Chrétien, M., Seidah, N. G., Basak, A., and Mbikay, M. (2008) Proprotein convertases as

therapeutic targets *Expert Opin Ther Targets* **12**, 1289–300.

12. Seidah, N. G., Mowla, S. J., Hamelin, J., Mamarbachi, A. M., Benjannet, S., Touré, B. B., Basak, A., Munzer, J. S., Marcinkiewicz, J., Zhong, M., Barale, J. C., Lazure, C., Murphy, R. A., Chrétien, M., and Marcinkiewicz, M. (1999) Mammalian subtilisin/kexin isozyme SKI-1: A widely expressed proprotein convertase with a unique cleavage specificity and cellular localization *Proc Natl Acad Sci USA* **96**, 1321–6.
13. Seidah, N. G., Benjannet, S., Wickham, L., Marcinkiewicz, J., Jasmin, S. B., Stifani, S., Basak, A., Prat, A., and Chrétien, M. (2003) The secretory proprotein convertase neural apoptosis-regulated convertase 1 (NARC-1): Liver regeneration and neuronal differentiation *Proc Natl Acad Sci USA* **100**, 928–33.
14. Thomas, G. (2002) Furin at the cutting edge: From protein traffic to embryogenesis and disease *Nat Rev Mol Cell Biol* **3**, 753–66.

Section II

Biochemistry and Cell Biology

Chapter 3

The Proprotein Convertases, 20 Years Later

Nabil G. Seidah

Abstract

The proprotein convertases (PCs) are secretory mammalian serine proteinases related to bacterial subtilisin-like enzymes. The family of PCs comprises nine members, PC1/3, PC2, furin, PC4, PC5/6, PACE4, PC7, SKI-1/S1P, and PCSK9 (**Fig. 3.1**). While the first seven PCs cleave after single or paired basic residues, the last two cleave at non-basic residues and the last one PCSK9 only cleaves one substrate, itself, for its activation. The targets and substrates of these convertases are very varied covering many aspects of cellular biology and communication. While it took more than 22 years to begin to identify the first member in 1989–1990, in less than 14 years they were all characterized. So where are we 20 years later in 2011? We have now reached a level of maturity needed to begin to unravel the mechanisms behind the complex physiological functions of these PCs both in health and disease states. We are still far away from comprehensively understanding the various ramifications of their roles and to identify their physiological substrates unequivocally. How do these enzymes function in vivo? Are there other partners to be identified that would modulate their activity and/or cellular localization? Would non-toxic inhibitors/silencers of some PCs provide alternative therapies to control some pathologies and improve human health? Are there human SNPs or mutations in these PCs that correlate with disease, and can these help define the finesses of their functions and/or cellular sorting? The more we know about a given field, the more questions will arise, until we are convinced that we have cornered the important angles. And yet the future may well reserve for us many surprises that may allow new leaps in our understanding of the fascinating biology of these phylogenetically ancient eukaryotic proteases (**Fig. 3.2**) implicated in health and disease, which traffic through the cells via multiple sorting pathways (**Fig. 3.3**).

Key words: Proprotein convertases, limited proteolysis, secretory proteins, single and pairs of basic residues, cancer metastasis, viral infections, neural and endocrine disorders, gene knockout, cholesterol metabolism, dyslipidemia.

1. Introduction

Whenever an important breakthrough in a given scientific discipline has been achieved, it becomes critical to summarize the historical perspectives of the discovery and to put those into context

M. Mbikay, N.G. Seidah (eds.), *Proprotein Convertases*, Methods in Molecular Biology 768,
DOI 10.1007/978-1-61779-204-5_3, © Springer Science+Business Media, LLC 2011

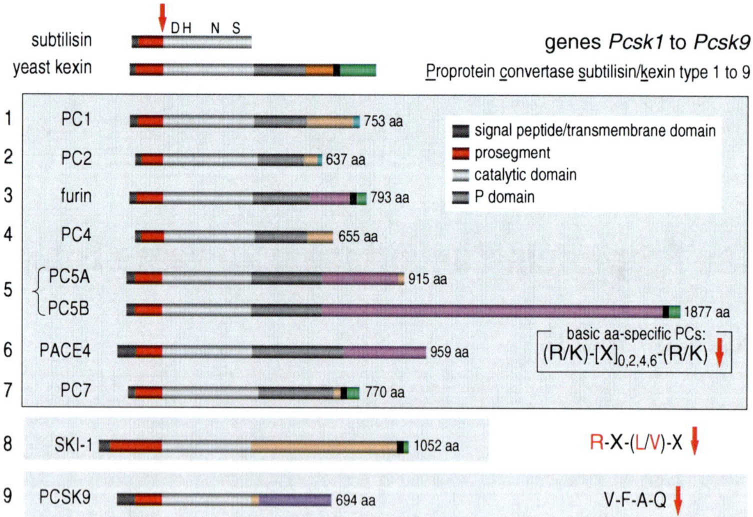

Fig. 3.1. The proprotein convertase family.

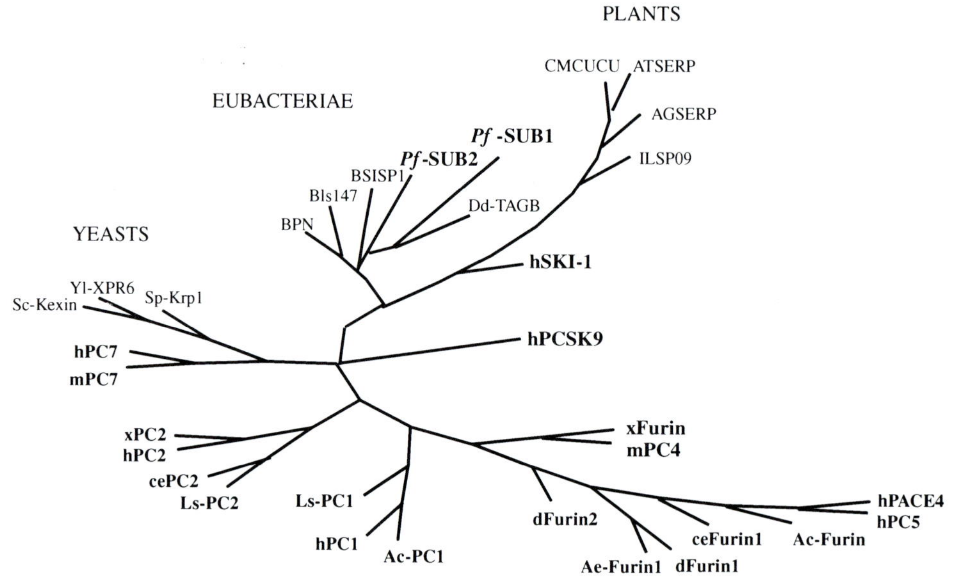

Fig. 3.2. Phylogenetic analysis of mammalian proprotein convertases.

with the present knowledge in this particular field. This is well exemplified by the discovery over the last 33 years of the proteases, their substrates, and post-translational modification (PTM) enzymes implicated in the shaping of the active form(s) of secretory polypeptides and proteins. The diversification of the genome information provided by such modifications is enormous and has played a major role in the evolution of the species.

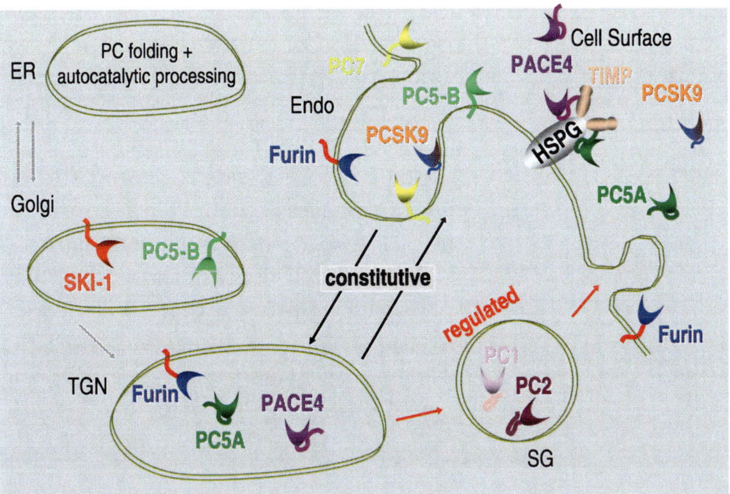

Fig. 3.3. Proprotein convertases in the secretory pathway.

Ever since the mid-1960s, a large effort was devoted toward the definition of the biosynthetic pathway and primary structural analysis of various precursors, including neural and endocrine peptides and their PTMs, coupled with the analysis of the cellular morphology and definition of the functions of the various organelles within the secretory pathway. All these monumental efforts by many talented international investigators in the field culminated with a general model in which polypeptide hormones and active proteins are often initially synthesized as relatively inactive precursors, which for maximal activation require one or more PTMs, including site-specific proteolytic cleavage, C- and N-terminal amino acid (aa) trimming, C-terminal amidation, and multiple residue modifications such as N- and O-glycosylation, Tyr and oligosaccharide sulfation, disulfide bridge formation, N-terminal acetylation, Ser/Thr phosphorylation, and Ser octanoylation. All these analyses led to the concept that within the secretory pathway there must exist a specific proteolytic machinery that results in the limited initial cleavage of proprotein and prohormone precursors, usually C-terminal to single or pairs of basic residues, of the type LysArg and ArgArg and less frequently LysLys and ArgLys. However, it was also realized that cleavage after hydrophobic or small amino acids also occurs in some cases, suggesting the presence of more than one type of proteases and/or similar enzymes with different specificities.

Where do such cleavages occur? Biosynthetic and immunocytochemical analyses of various precursor processing suggested that depending on the substrate this can occur in at least four different compartments, including the trans-Golgi network (TGN), cell surface, endosomes, and secretory granules. It turns out that the first three processes mostly occur in proteins trafficking through the so-called constitutive secretory pathway, whereas the last one concerns those proproteins that are processed in the regulated secretory pathway (**Fig. 3.3**), which involves most neural and endocrine polypeptide hormones (1). This spatial segregation of proteins before their processing turned out to be a very refined filtering process to allow for controlled limited proteolysis in a time- and space-dependent manner.

The isolation of the processing enzymes turned out to be a very arduous process that took over 22 years to begin to identify the cognate mammalian proteases implicated in the process of protein precursor activation. During these long and arduous years hunting for the correct processing enzymes, many false positives were reported, to unfortunately be relegated to the side once they were tested and scrutinized by biochemical and cellular criteria. The limitations of the methods used to purify the enzymes and the sensitivity and specificity of the substrates used to follow these activities were often behind the limited success achieved during the 1970s–1980s. Indeed, finding a needle in a haystack has always been a challenge, requiring a lot of effort, technological advances, and often serendipity. All of these ingredients were fundamental in the discovery of the nine-membered proprotein convertase (PC) family (**Fig. 3.1**), a saga that lasted almost a quarter of a century before the first light at the end of the tunnel appeared.

2. The First Glimpse of Light

In 1984, the identification of the yeast convertase kexin (also known as Kex2p) was indeed the culminating point that led to the molecular and genetic identification of the first prototype of the mammalian proprotein convertases. Kexin cleaves the precursor of pro-K1 killer toxin and pro-α-factor of the yeast *Saccharomyces cerevisiae* at the C-terminus of pairs of basic residues of the type ArgArg↓ and LysArg↓ (2, 3). In addition pro-K1 killer toxin is also cleaved after a single basic residue in the motif ProArg↓, by one or more enzymes (3). Kexin turned out to be a serine proteinase best related to bacterial subtilases rather than eukaryotic trypsin-like enzymes (4). This unexpected result explained many of the unfruitful attempts to identify the cognate mammalian PCs based on RT-PCR analyses using degenerate oligonucleotides derived from the consensus sequence of active sites of serine proteinases of the trypsin–chymotrypsin family (5).

How relevant is the type-I membrane-bound kexin to the mammalian proprotein convertases? Answers to this question quickly came from analyses of the processing of the mammalian pro-opiomelanocortin (POMC), the precursor of ACTH and β-endorphin, in mammalian cells overexpressing kexin (6). The data clearly showed that the yeast enzyme is capable of processing POMC into a set of products similar to those found in vivo, including β-LPH, γ-LPH, and β-endorphin, but not α-MSH. Furthermore, evidence was presented that kexin could not generate somatostatin-14 from mammalian pro-somatostatin, but rather kexin and another yeast enzyme, the aspartyl protease YAP3, cleaved in between the Arg↓Lys site to generate a Lys-extended somatostatin-14 (7). Interestingly, the monobasic cleavage generating somatostatin-28 was performed by YAP3. Finally, kexin was also shown to be able to act as a proalbumin converting enzyme (8). These data thus suggested that while kexin could be considered a prototype of the as yet undiscovered mammalian proprotein convertases, differences in specificity must exist that would require critical structural differences in mammalian PCs to produce the observed products in vivo.

3. Identification of Furin: The Beginning of an Active Era of Research

The suggestion in 1986 that the *fur* gene (*fes/fps* upstream region) cloned in the group of Wim Van de Ven by Anton Roebroek (9) was a mammalian homologue of kexin first appeared at the end of the discussion of a seminal manuscript by Robert Fuller et al. in 1988 (3). The striking similarity of furin to kexin (10, 11), especially in the catalytic serine subtilase domain, became apparent once the sequence of kexin became public (4). Quickly a number of groups began to analyze the specificity of the type-I membrane-bound furin and found that it can process intracellularly at the TGN, cell surface, and/or endosomes a large number of constitutively secreted substrates that include growth factors and their receptors, enzymes, surface glycoproteins of mammalian and viral and parasitic origin, blood coagulation factors, and even some polypeptide hormones (for comprehensive reviews, *see* (12–16)) The membrane-bound furin cycles from the cell surface back to the TGN through endosomes, a pathway regulated by various signals in its cytosolic tail (15). The furin gene (*PCSK3*) is localized to human chromosome 15 close to the *fes/feps* region (9).

Analysis of a large number of substrates processed and the various bonds cleaved suggest that furin best recognizes the sequence (R/K)-$[X]_{0,2,4,6}$-(R/K)↓P1′-P2′, with a large preference for a P1 Arg, and P1′ is usually a small amino acid with a preference for Asp and Glu, and an aliphatic aa (Ile, Val, Leu) in P2′ is best. Indeed, the crystal structure of furin first reported in 2003 confirmed this prediction (17). One of the best furin substrates contains the sequence RX(R/K)R↓EL, which has been inserted at strategic sites in mammalian proteins and viral glycoproteins for cleavage by endogenous furin (18, 19). Similar engineering strategies have been described for the production of active insulin from proinsulin in muscle (20).

Looking at the latest PubMed literature related to furin, it is amazing that in less than 20 years since its first discovery and that of the following members of the PC family, more than 1,630 reports have appeared that mention furin or directly relate to its activity and/or functions. What has come out is that furin is ubiquitously expressed at various levels in all tissues, thereby rationalizing its widespread role in the processing of various proteins, usually resulting in their activation (12–16), but sometimes in their inactivation as is the case of lipoprotein and endothelial lipases (21) and the proprotein convertase PCSK9 (18).

The varied physiological functions of furin suggested that animals lacking this enzyme may present serious developmental problems and/or anomalies. Indeed, mice lacking furin through inactivation of its gene, *Pcsk3* (*p*roprotein *c*onvertase *s*ubtilisin *k*exin type 3), die at about embryonic day 11 (E11). Embryos fail to undertake axial rotation and ventral closure needed to form a looping heart tube and a coherent primitive gut (22). Although only a few specific furin substrates have yet been identified in vivo, the overlap in the distribution of furin mRNA and that of some members of the transforming growth factor (TGF)β family (23, 24), such as TGFβ1 (25) and BMP4 (26) often best processed by furin (27, 28), is striking. A liver-specific furin KO and other tissues from adult mice, using the inducible Mx1-Cre transgene, resulted in mice with no phenotype, demonstrating redundancy with other PCs in these tissues (29). In contrast, in vivo studies showed that furin can uniquely process the Ac45 subunit of the vacuolar-type H^+-ATPase in pancreatic β-cells (30). Furthermore, conditional deletion of furin in T cells allowed for normal T-cell development but impaired the function of regulatory and effector T cells, which produced less TGFβ1 (31).

Whether inhibition of furin in the adult using protein (32) or small molecule (33, 34) inhibitors, RNAi or antibody approaches could be a useful therapy against certain pathologies such as cancers (35), and associated metastasis (36, 37) and/or viral/parasitic infections (38, 39) are under careful examination for possible short- and long-term toxicity effects. It must be mentioned here that heterozygote mice lacking one copy of furin are alive and show no obvious anomalies (22), and hence therapies aimed at reducing 50–70% of furin activity may still be viable. Furthermore, inhibition of cell-surface furin may not be as toxic as complete furin inhibition. Hence, the use of cell-impermeable approaches, such as inhibiting monoclonal antibodies or single-chain antibodies, may well turn out to be a feasible and therapeutically useful alternative to control some of the above deadly diseases. However, a word of caution should be taken into account. Namely, in some pathologies furin-like activity may be protective to the cell, as was recently demonstrated with the ability of furin-like enzymes to neutralize the HIV accessory protein Vpr and hence limit viral spread (40).

4. The Neural and Endocrine Convertases PC1/3 and PC2

Processing of most polypeptide hormone precursors occurs in immature secretory granules at acidic pHs (41–43). Therefore, it was expected that the cognate processing enzyme(s) implicated would be active in acidic conditions between pHs 5 and 6. While many attempts were made to isolate the cognate enzyme(s) using biochemical techniques, unfortunately these have all failed, mostly due to the low levels of the proteases and the lack of robust specific substrates sensitive to trace levels of enzymes. In my own laboratory, we had dissected 20,000 porcine pituitaries at a slaughterhouse in Saint-Hyacinthe close to Montreal, Quebec, with the hope of isolating enough enzyme for characterization by biochemical methods. This has led to the isolation and characterization of one of the many false positives (44), plasma kallikrein (45), which did cleave proenkephalin peptides correctly in vitro (46). One of the first reports closing in on the possible nature of the real proinsulin convertases used insulinoma granules as a source of enzymes, and the authors deduced that two proteases, possibly present in distinct subcellular compartments, may be involved in the generation of active insulin (47). However, the low levels of enzymes available precluded their characterization biochemically.

Technological advances are often behind new leaps in biology, and it is of no surprise that the introduction of the reverse transcriptase polymerase chain reaction (RT-PCR) played a major role in the identification of the two convertases implicated in the processing of most regulated polypeptide hormone precursors, as well as all the other PCs. This technique amplifies a single or few copies of a piece of DNA (generated for example by the action of reverse transcriptase on an mRNA pool) across several orders of magnitude, generating thousands to millions of copies of a particular DNA sequence. This method was first conceptualized by Kary B. Mullis in 1983 and later optimized and improved (48). For this discovery, he shared the 1993 Nobel Prize in Chemistry with Michael Smith who in 1978 had first introduced the use of oligonucleotides for site-directed mutagenesis (49). The automation of this technique only appeared in 1986–1987. This was just the time when I was spending a sabbatical year at the Pasteur Institute in Paris (1987, 1988) at the laboratory of François Rougeon who had first cloned the cDNA of the protease renin-2 from mouse submaxillary glands.

When I became familiar with the PCR technique and realized its potential, I decided to exploit it using both pituitary and heart atria extract as a source of mRNA. The choice of the pituitary was dictated by the production from pro-opiomelanocortin (POMC) of ACTH and β-LPH in the anterior lobe and α-MSH and β-endorphin in the intermediate lobe, likely by different convertases (50–52). In heart atria, we had shown that the precursor of atrial natriuretic factor (proANF) was processed at a single basic residue AGPR↓AL to produce the active atrial natriuretic factor (ANF; 53). The proANF convertase was reported much later to be corin, a type-II membrane-bound serine protease of the trypsin type (54).

The next leap was the use of sense and antisense degenerate oligonucleotides around the active site of proteases, a region found to be highly conserved between members of a single family. The thinking was that if the basic aa-specific processing enzyme(s) was of a trypsin type then use of degenerate oligonucleotides mimicking the active sites Ser, His, or Asp should by RT-PCR lead to the amplification of a DNA fragment of one or more processing enzymes. This technique was applied to DNA isolated by reverse transcriptase treatment of dissected rat pituitary (anterior and neurointermediate lobes) and heart atria. When we used degenerate oligonucleotides based on trypsin or serine proteases of blood coagulation factors, we invariably isolated DNAs similar to various trypsin-like enzymes, including plasma kallikrein and tryptase (unpublished results). It was thus, after frustrating 6 months of work, that the paper of Robert Fuller appeared, which alerted us on the possibility that a mammalian homologue of kexin could be furin (3). However, the genomic DNA sequence just published by Anton Roebroek suggested that the gene they had cloned could be an oncogene (9). Upon translating the 3′ end of the reported sequence, we quickly realized that the supposedly intronic sequence actually coded for a potential Asn of the oxyanion hole of subtilisin-like enzymes. We therefore designed degenerate oligonucleotides surrounding this Asn and possible Ser of active site subtilases and rapidly isolated from rat heart atria a furin sequence and from pituitary anterior and neurointermediate lobes two DNA fragments that could potentially code for part of the catalytic domain of two novel subtilisin-like enzymes. These *p*ituitary *c*onvertases (now known as *p*roprotein *c*onvertases, PCs) were given the names of PC1 and PC2, in the order in which their cDNAs were cloned (5). While the complete sequence of

PC2 was obtained from our cDNA library of AtT20 cells, that of PC1 was not possible to obtain, and we had to further clone the rest of the sequence from another AtT20 library and from two mouse insulinoma libraries (55). The chromosomal assignment of the genes coding for these convertases revealed their presence on mouse chromosomes 13 (*Pcsk1*, the PC1 gene) and 2 (*Pcsk2*, the PC2 gene), respectively (55), and the orthologous human chromosomes 5 (*PCSK1*) and 20 (*PCSK2*) (56).

The most inspiring result was obtained upon analyses of their mRNA expression by in situ hybridization. It became clear that PC1 is mostly expressed in the anterior lobe of the pituitary, including the corticotrophs where it co-localized with ACTH, while PC2 was highly concentrated in the intermediate lobe together with α-MSH (5, 55). This immediately suggested a model whereby PC1 would be critical for the formation of ACTH in the anterior lobe of the hypophysis, and PC2 must participate in the generation of α-MSH in the *pars intermedia*. This prediction based on cellular localization proved to be right upon biosynthetic analysis of cells co-expressing POMC and each of PC1 or PC2 (57).

Unknown to us and by a different approach, the group of Donald F. Steiner in Chicago independently cloned PC2 from a human insulinoma using degenerate oligonucleotides based on the kexin sequence (58) and later on also isolated PC1 (which they originally called PC3) (59). While it is an amazing coincidence that both groups named the second enzyme PC2, we now agreed to call the first enzyme PC1/3. Both enzymes later on were proven to be the processing enzymes of proinsulin (60) and implicated in POMC processing (61), validating the concept of distinct enzymes responsible for the cleavage at the B–C and A–C junctions of proinsulin (47) and agreeing with our own independent data (57).

These first exciting discoveries of PC1/3 and PC2 and the validation of their properties led to more than 500 publications to study their localizations, activation, functions in various organisms, cells, and their processing of a multitude of substrates. In essence, what has come out is that both PC1/3 and PC2 are sorted to mature secretory granules (62), likely via specific secondary structures in their C-terminal domains (63, 64), and that they are responsible for the processing of most pro-neural and endocrine peptides in a complex combinatorial fashion. It seems that PC1/3 gets activated first and hence acts on substrate before PC2. Furthermore, what also became apparent is that both PC1/3 and PC2 are under the control of endogenous inhibitors/chaperones. In the case of PC1/3 it seems that proSAAS could be processed by PC1/3 into a polypeptide product that acts as a potent inhibitor of the enzyme, thereby regulating its *in trans* enzymatic activity on other substrates (65, 66).

However, this may not be valid in all tissues (67). Interestingly, during our protein extractions of human pituitaries in search for the processing enzymes (68), we stumbled on a peptide we originally called 7B2 for its elution position on the HPLC (69, 70). While 7B2 was discovered in 1982, almost 8 years before PC2, we gained a lot of information on its *pan neuronal* and endocrine expression and its multifunctional role in the central nervous system (71–73). However, work in frogs by Gerard J.M. Martens re-discovered 7B2 and suggested that it may negatively regulate the activity of PC2 in the *pars intermedia* (74). This was confirmed by Iris Lindberg (75, 76) and by us (77, 78). It now seems that 7B2 is first needed as a chaperone to assist the folding of proPC2 in the ER. The complex proPC2–pro7B2 then exits the ER and the pro7B2 is first cleaved at ***RRKRR***$_{182}$↓S*V*N by furin in the TGN (79), generating a C-terminal 31 aa CT-peptide that is a potent inhibitor of PC2. The proPC2 is then autocatalytically processed into PC2 within the acidic environment of immature secretory granules and in turn cleaves the CT-peptide at VVA***KK***$_{189}$↓S*V*P, generating an inactive form to finally liberate the active enzyme PC2, which can then act *in trans* on other substrates (73, 80).

The physiological importance of PC1/3 (81) and PC2 (82) was deduced from studies of the phenotypes of their gene knockout in mice and the discovery of two human patients with defects in PC1/3 (83, 84). In all cases mice were viable, thus suggesting that individually the genes of these convertases are not essential for life. Nevertheless, even though PC2-null mice appear normal at birth, they exhibit retarded growth. Analysis of these mice reveals chronic fasting, hypoglycemia, and a deficiency in circulating glucagon. PC2 is known to process various neuroendocrine precursors, and many of these were not fully processed in PC2-null mice, including prosomatostatin, neuronal proCCK, neurotensin, neuromedin N, prodynorphin, proorphanin FQ/nociceptin, and POMC-derived peptides.

Contrary to PC2 (82), PC1/3 gene disruption results in severe developmental abnormalities (81). The PC1/3-null mice exhibit growth retardation. The adult mutant mice are about 60% of the normal size and phenotypically resemble those that have mutant growth hormone-releasing hormone (GHRH) receptor. Interestingly, insulin growth factor 1 (IGF-1) and GHRH levels were significantly reduced along with pituitary GH mRNA levels, suggesting that this reduction contributes to the growth retardation observed in these mice. Similarly, analysis of several protein precursors known to be processed by PC1/3 revealed that these mice, like PC2 mutant mice, exhibit multiple defects in multiple hormone precursor processing events. These include the hypothalamic GHRH, pituitary POMC, proinsulin, and intestinal proglucagon. In contrast to PC2-null mice, PC1/3-null mice process normally pituitary POMC to adrenocorticotropic hormone (ACTH) and have normal levels of blood corticosterone. Like PC2-null mice, they also developed hyperproinsulinemia.

Since PC2 is the major convertase that cleaves POMC and proenkephalin to generate the morphinomimetic peptides β-endorphin and Met- and Leu-enkephalins, respectively, it was important to investigate the role of PC2 in pain perception. Unexpectedly, after a short forced swim in warm water, PC2-null mice were significantly less (rather than more) responsive to the stimuli than wild-type mice, an indication of increased opioid-mediated stress-induced analgesia (85). The enhanced analgesia in PC2-null mice may be caused by an accumulation of opioid precursor processing intermediates with potent analgesic effects or by loss of anti-opioid peptides. Thus, the presence of abnormal cocktails of pain neuropeptides in the brain of PC2 KO mice is likely to disturb pain perception mechanisms in ways that remain to be fully elucidated.

PC1/3 deficiency in a female patient compound heterozygote for both splicing and non-synonymous mutations resulted in very low expression of the protein (83). This subject exhibited neonatal obesity and abnormal glucose homeostasis, as well as the presence of other endocrine defects, including the presence of very high circulating levels of proinsulin and multiple forms of partially processed POMC (intermediate ACTH precursors), low-serum estradiol, follicle-stimulating hormone (FSH), and luteinizing hormone (LH). Another PC1/3 deficiency female subject presented severe diarrhea, which started on the third postnatal day. Clinical investigations revealed a defect in the absorption of monosaccharides and fat, revealing the role of PC1/3 in the small intestinal absorptive function (84). Although the phenotypes of the PC1/3-null mice differ from those observed in these patients (PC1/3-null mice are not obese), the findings confirmed the importance of PC1/3 as a key neuroendocrine convertase. Interestingly, obesity, hyperphagia, and increased metabolic efficiency were recently identified in PC1/3 mutant mice exhibiting a homozygote mutation N222D/N222D that results in ~60% decrease in PC1/3 activity, suggesting that it is the dose of PC1/3 and possibly reduced hypothalamic α-MSH that may define the obesity phenotype (86). Finally, a single nucleotide polymorphic (SNP) variation in PC1/3 resulting in a N221D mutation and partial loss of function has been linked to monogenic obesity in children and adults (87).

Peptidomic analyses of PC1/3 (88) and PC2 (89) KO mice showed that loss of PC1/3 is often compensated for by PC2, but the reverse is not always true. Thus, although PC1/3 deficiency results in severe neonatal abnormalities and a reduction in litter size, many neuropeptides can still be processed in its absence. This suggests that few PC1/3-specific substrates play major roles in mouse development, and that some are critical in the control of human metabolic diseases such as obesity (87).

Interestingly, mice lacking proSAAS (90) provided evidence that it is involved in the *prenatal* regulation of neuropeptide processing in vivo. However, *adult* mice lacking proSAAS have normal levels of all peptides detected using a peptidomics approach, suggesting that PC1/3 activity is not affected by the absence of proSAAS in adult mice. The data also showed that in adults proSAAS has other functions, e.g., body weight regulation, and these are not related to PC1/3 inhibition. Mice lacking 7B2 suggested that 7B2 is indeed required for activation of PC2 in vivo but that it has additional important functions in regulating pituitary hormone secretion (91). However, steroidal control of pituitary function is mouse strain dependent (92) and is therefore not a general phenomenon. Whether 7B2 may exhibit other functions in other mouse strains is yet to be discovered.

5. The Germ Cell-Specific PC4

In 1991, during our search by RT-PCR for other members of the PC family in mRNA extracts of various tissues, we identified in rat testis three different ~600 bp cDNAs potentially coding for three different PCs, now known as PC4, PC5/6, and PACE4. Shortly thereafter, a paper appeared that reported the complete cDNA sequence of PACE4 from a human hepatoma HepG2 cell line and an osteosarcoma cDNA library (93). We therefore concentrated on the characterization of PC4 (94) and PC5/6 (95). PC4 is expressed exclusively in male testicular germline pachytene spermatocytes and round spermatids, suggesting that it may play a specific physiological function in reproduction. In agreement, PC4 was detected in the acrosomal granules of round spermatids, in the acrosomal ridges of elongated spermatids, and on the sperm plasma membrane overlying the acrosome (96). In female mice, PC4 was expressed in macrophage-like cells of the ovary, and its levels are downregulated in activated macrophages, such as in inflammation (97). Later on, PC4 was also shown to be expressed in human placenta (98).

It took a lot of effort from my colleague Majambu Mbikay during a sabbatical year at the Jackson laboratories to obtain mice that lacked PC4 expression (99). This manuscript that appeared on June 24, 1997, was the first report on any convertase KO mouse. The in vivo fertility of homozygous mutant males was severely impaired in these mice, without any evident spermatogenic abnormality. In vitro, the fertilizing ability of *Pcsk4*-null spermatozoa was also found to be significantly reduced. Moreover, eggs fertilized by these spermatozoa failed to grow to the blastocyst stage. Sperm physiologic anomalies likely contribute to the severe subfertility of PC4-deficient male mice (96). These results suggested that PC4 in the male may be important for achieving fertilization and for supporting early embryonic development in mice.

PC4 is a very special convertase whose C-terminus is species specific (94). This may be necessary to ensure no cross-species fertilization. So far, one of the identified specific substrates of PC4 in the testis is pituitary adenylate cyclase-activating polypeptide (PACAP) and PC4 is its sole processing enzyme in the testis and ovary of mice (100). In vitro studies with purified enzyme concluded that the most probable sequence motif for recognition by PC4 is ***K***X***K***XX***R***↓ or ***K***XX***R***↓, where X is any amino acid other than cysteine and that it prefers proline at P3, P5, and/or P2′ positions. It was also revealed that PC4 is a good candidate processing enzyme for the growth factors IGF-1 and -2 and several ADAM proteins such as ADAM-1, -2, -3, and -5 (101). Intrauterine fetal growth restriction is a leading cause of perinatal mortality. Recent work has unraveled an unusual property of PC4 in the processing of IGF-II, which has been shown to be an important regulator of fetoplacental growth. Thus, PC4 cleaves pro-IGF-II to generate the intermediate processed form, IGF-II (aa 1–102), and, subsequently, mature IGF-II (aa 1–67), thereby regulating fetoplacental growth (98). In the future, SNP variant PC4 that could affect its activity may explain some of the pathophysiology of fetoplacental growth restriction. Specific inhibitors of PC4, such as those recently reported in flavonoids (102), may one day serve as male contraceptives.

6. The Widely Expressed Convertases PC5/6 and PACE4

The identification of PC5A was done in my laboratory in 1993 (95) at a similar time to that made by Nakayama's group (103), who called the enzyme PC6. We now agreed to call it PC5/6. The convertases PC5/6 (95, 104) and PACE4 (93) seem to form a class of their own based on their primary structures and their ability to bind the cell surface via their C-terminal Cys-rich domains (CRD), which bind tissue inhibitors of metalloproteases (TIMPs) and heparin sulfate proteoglycans (HSPGs) (105, 106) and in many cases inactivate HSPG-bound proteins such as endothelial and lipoprotein lipases (21) and possibly adhesion molecules. In the CNS, it was shown that PC5/6 can process the neural adhesion molecule L1 assisting in neuronal repair and migration (107).

The specific physiological substrates of PC5/6 and PACE4 need to be unraveled in vivo, since in cellular experiments and in vitro many of the substrates processed by either enzyme can also be cleaved by furin and/or PC7.

PC5/6 is expressed as two mRNA transcripts, PC5/6A (soluble 915 aa) (95) and PC5/6B (type-I membrane bound, 1877 aa) (103), generated by differential splicing of its exons. Quantitative analysis of the tissue distribution of PC5/6 mRNA by qPCR revealed that the small intestine and kidney are the richest source of PC5/6B, whereas all other tissues express mostly PC5/6A. Both transcripts share the first 20 exons encoding the signal peptide, prosegment, catalytic domain, P-domain, and the cysteine-rich domain (CRD). The 21st exon of PC5/6A, coding for its last 38 residues, is replaced by 18 additional exons encoding the last 1,000 residues for PC5/6B (104, 108). Thus, while the CRD of PC5/6A contains 44 cysteine residues arranged in five tandem repeats of the consensus motif Cys-X_{2-3}-Cys-X_{3-4}-Cys-X_{2-7}-Cys-X_{5-10}-Cys-X_2-Cys-X_{9-13}-Cys-X_{3-5}-Cys-X_{7-16} (as it is also found in its closest homologue PACE4), the extended CRD of PC5/6B contains 22 repeats of this cysteine-rich motif. As for furin and PC7, PC5/6B also has a transmembrane domain and a cytosolic tail and cycles from the cell surface back to the TGN through endosomes (109). This regulated transit into multiple compartments is controlled by several sorting signals in their cytosolic tails and their interactions with specific sorting adaptors.

Evidence has been presented that, different from the other convertases, PC5/6A and PACE4 are activated at the cell surface while in contact with HSPGs (106). Here, the second cleavage of the prosegment, needed for zymogen activation, occurs at the cell surface, thereby limiting the functions of active PC5/6A and PACE4 to the cell surface and/or extracellular matrix, thereby favoring substrates that are also bound to HSPGs.

The present data strongly suggest unique tissue-specific functions of PC5/6 and PACE4. Thus, PC5/6 mRNA was detected only in neuronal cells, whereas PACE4 mRNA was expressed in both neuronal and glial cells. In areas that are rich in neuropeptides such as cortex, hippocampus, and hypothalamus, mRNA levels of PC5/6 were high but those of PACE4 were low or undetectable (110). In regions such as the amygdaloid body and thalamus, distinct but complementary distributions of PC5/6 and PACE4 mRNAs were observed. The medial habenular and cerebellar Purkinje cells expressed very high levels of PACE4 mRNA. Ontogeny and tissue distribution analysis showed that PC5/6 expression is detected early during embryonic development, appearing first in extra-embryonic tissues (111). By E9, it is also specifically expressed in cells of the maternal–embryonic junction, where no other convertase is expressed (112). What are the precursors that need such specific processing events and what is the role of TIMPs and HSPGs in this process are open questions that may be resolved by tissue-specific KOs and by careful analysis of the cell-surface proteome of various tissues in the presence or absence of these convertases.

The complete knockout (KO) of PACE4 and PC5/6 genes in mice resulted in different phenotypes. Thus, while the PACE4 KO results in a 75% viable phenotype with bone morphogenesis defects (113), that of PC5/6 causes embryonic death at birth, with mice exhibiting multiple morphogenic defects likely related to impaired homeotic transformations (114, 115). Thus, newborns exhibited major defects in the anteroposterior axis with extra-thoraxic and -lumbar vertebrates (18 and 8 instead of 13 and 6, respectively) and a lack of tail (114). This phenotype had been reported for mice lacking the TGFβ-like factor, Gdf11, also known as BMP11. Both Gdf11- and PC5/6-deficient mice lack kidneys, although the phenotype was more penetrant in PC5/6 KO mice (100% agenesis versus 86% lacking one or two kidneys). We showed that Gdf11 is a favorite substrate of PC5/6, in part due to an Asn residue at the first position after the cleavage site (P1′) ($RSRR_{296}\downarrow NL$).

While PC5/6 deficiency perfectly mimics the Gdf11 one, it also results in other phenotypes, suggesting the lack of processing of other substrates: *Pcsk5*$^{-/-}$ newborns died earlier, in the first 2 h, versus the first 2 days following birth for *Gdf11*$^{-/-}$ mice, likely by asphyxiation (lung alveoli were collapsed). They also exhibited smaller size than WT, retarded ossification, severe hindlimb hypoplasia, abdominal herniation, and superficial and tissue hemorrhages, suggesting vascular fragility (114).

In collaboration with S. Batthacharya (115), magnetic resonance imaging revealed severe phenotypes reminiscent of those observed in patients exhibiting VACTERL (vertebral, anorectal, cardiac, tracheoesophageal, renal, limb) malformations. Finally, exon sequencing of control and VACTERL patients linked mutations in the human PC5 gene (*PCSK5*) to this syndrome (115). We proposed that PC5/6, at least in part via Gdf11, coordinately regulates caudal Hox paralogs, to control anteroposterior patterning, nephrogenesis, and skeletal and anorectal development.

We showed a downregulation of PC5/6 mRNA in human colon tumors at various stages (116). Since PC5/6 is very rich in intestine, we evaluated its role in tumorigenesis by crosses with an $Apc^{Min/+}$ mouse model, which develops numerous adenocarcinomas along the small intestine and fewer in the colon. Since PC5/6-deficient mice die at birth, we generated mice lacking or not lacking PC5/6 specifically in enterocytes (Villin-Cre transgene) and analyzed the number and size of the tumors. The lack of PC5/6 in enterocytes results in a significantly higher tumor number in the duodenum and a premature mortality of $Apc^{Min/+}$ mice, suggesting that intestinal PC5/6 is protective toward tumorigenesis, especially in mouse duodenum, and possibly in human colon (116).

Recent studies revealed that in the adult PACE4 plays an important role in myogenic differentiation through its association with the IGF-II pathway (117). Thus, while PC4 processes proIGF-II in testis (98), PACE4 seems to be its cognate convertase in muscle. Finally, it was also reported that PACE4 could process the TGFβ-like substrate Nodal and that its intracellular traffic is dictated by the 18 kDa EGF-containing GPI-anchored proteoglycan Cripto that directs its traffic through an unconventional secretory pathway directly from the ER to the cell surface and sorting to detergent-resistant membrane microdomains (118). Cripto is the first receptor that binds both a PC and its substrate (Nodal), thereby enhancing the processing step.

7. The Ubiquitously Expressed PC7

In 1996, the last and still least studied member of the basic aa-specific PCs was identified in our lab and called PC7 (119). Its gene (*PCSK7*) was found to be on human chromosome 11 and mouse chromosome 9. Independently, PC7 was also cloned from a human lymphoma library and called LPC, for lymphoma PC (120). However, the name PC7 is now retained. Tissue distribution analyses revealed it to be ubiquitously expressed in most tissue and cell lines analyzed (119). It must be noted that PC7 is the most ancient of the basic aa-specific convertases (**Fig. 3.2**) and yet the most conserved phylogenetically.

Biosynthetic analyses of rat or human PC7 revealed that the enzyme is first synthesized as a zymogen which within the endoplasmic reticulum (ER) rapidly undergoes an autocatalytic cleavage at K***R***A***KR***$_{140}$↓ (rat) (119) or R***R***A***KR***$_{141}$↓ (human) (120), thereby releasing the active protease that exits the ER and is then competent to process substrates *in trans*. Further studies revealed that PC7 undergoes a number of post-translational modifications including N-glycosylation (119) and cytosolic tail Cys-palmitoylation (121). A number of investigations aimed at defining the sequence recognition of PC7 versus furin or other convertases suggested that PC7 can also cleave overexpressed substrates at Arg↓ residues both in vitro (122–125) and in cell lines (126–130). Although furin and PC7 have been proposed as the major gp160 processing convertases, rat liver microsomal gp160 processing activity was essentially resolved from furin and only partially overlapped with PC7, and density distribution studies revealed that PC7 resides in lighter subcellular fractions than those containing furin (131). Interestingly, while overexpression of the prosegments of furin, PC5, and PC7 resulted in potent inhibitors of substrate cellular processing (132, 133), only the prosegment of PC7 is secreted into the medium (123, 132). The C-terminal KRA***KR***$_{140}$ motif in the prosegment of PC7 was critical for its observed inhibitory activity (134). Finally, the in vivo substrates of PC7 will remain to be defined, since the KO of PC7 results in viable mice (N.G. Seidah and D. Constam, unpublished results).

The function of the peptide-loading complex (PLC) is to facilitate loading of MHC class I (MHC-I) molecules with antigenic peptides in the ER and to drive the selection of these ligands toward a set of high-affinity binders. When the PLC fails to perform properly, as frequently observed in virus-infected or tumor cells, structurally unstable MHC-I peptide complexes are generated, which are prone to disintegrate instead of presenting antigens to cytotoxic T cells. Recently, it was reported that PC7, which is highly expressed in the immune system (119), may be implicated in antigen presentation, as the knockdown of its mRNA leads to lysosomal degradation of MHC-I (135). It has also been reported that PC7 may play a role in tumorigenesis (129, 136). It has yet to be proven if in vivo these are physiological functions of PC7, and what the degree of redundancy is with other members of the PC family.

8. SKI-1/S1P Activates Membrane-Bound Transcription Factors

The ubiquitously expressed SKI-1 (137) (also known as S1P) activates membrane-bound transcription factors implicated in the endoplasmic reticulum (ER) stress response (ATF6) (138) or the regulation of cholesterol and fatty acid synthesis (sterol regulatory element-binding protein (SREBP)-1 and -2) (139, 140). ProSKI-1/S1P is autocatalytically cleaved into a mature ~106 kDa membrane-bound form (137) and a secreted ~98 kDa shed form (141). Its *PCSK8* gene, ubiquitously expressed (137), is located on human chromosome 16 and mouse chromosome 8 (142). In contrast to basic-aa-specific PCs, SKI-1/S1P cleaves substrates in the general motif ***R***X(***V,L***)(K,F,L)↓ (137, 141, 143). In the absence of sterols, SKI-1/S1P cleaves the membrane-bound transcription factors sterol regulatory element-binding proteins (SREBPs) in their luminal loop (144), leading to release of a cytosolic basic helix-loop-helix transcription factor. In the nucleus, this activates transcription of LDLR and all the genes involved in cholesterol and fatty acid synthesis (144). In the presence of sterols, SREBP cleavage is inhibited and hence transcription of its target genes is reduced, while the reverse is true in the absence of sterols (144). Other transmembrane transcription factors cleaved by SKI-1/S1P include the ER stress response factor ATF6 and CREB-like transcription factors Luman and CREB4 (143–151). We developed in vitro fluorogenic assays and inhibitors of cellular SKI-1 activity (143, 152–155). Aside from transcription factors, the other known SKI-1/S1P substrates are viral glycoproteins, brain-derived neurotrophic factor (BDNF), and somatostatin (143–151).

Recently, novel functions of SKI-1/S1P have been identified:

- Global μ-array analysis of HepG2 cells stably expressing the specific SKI-1/S1P inhibitor R134E prosegment (152) revealed that SKI-1/S1P inhibition causes widespread changes in key metabolic pathways other than those involving cholesterol and fatty acid synthesis (156).
- Small molecule inhibitors of SKI-1/S1P have been developed and shown to reduce cholesterol and fatty acid synthesis in vivo and, therefore, represent a potential new class of therapeutic agents for dyslipidemia and for a variety of cardiometabolic risk factors associated with diabetes, obesity, and the metabolic syndrome (157, 158).
- Using various protease inhibitors our data revealed that SKI-1/S1P plays a direct and/or indirect role in assembly of functional nucleation complexes in primary bone mineralization (159).
- We, and others, have shown that SKI-1/S1P is critically important in the activation of hemorrhagic fever viruses such as Lassa virus (143, 146), lymphocytic choriomeningitis virus (160), and Crimean-Congo hemorrhagic fever virus (143, 161) glycoproteins.
- Very little information is available on the in vivo physiological roles of SKI-1. Lethality occurs at the blastocyst stage in *Pcsk8*$^{-/-}$ (the SKI-1/S1P gene) mice with the absence of inner cell mass formation (162). However, liver and cartilage conditional knockouts are viable: loss of SKI-1/S1P in liver causes ~50% reduction in the levels of circulating LDL-cholesterol (LDL-C) and fatty acids (148); cartilage-specific *Pcsk8* KO mice exhibited chondrodysplasia, lack of endochondral ossification, disorganization of the collagen network, and the engorgement/fragmentation of the ER in chondrocytes in a manner characteristic of ER stress (163).

9. PCSK9 Regulates LDL-Cholesterol Levels: Implication in the Metabolic Syndrome

Complications resulting from cardiovascular disorders are the main cause of death worldwide, affecting ~13 million individuals/year, as compared to ~6 million/year due to various forms of cancer (http://www.poodwaddle.com/clocks/worldclock/). The incidence of cardiovascular pathologies is expected to increase dramatically in the next two decades. Elevated plasma cholesterol levels result in excess cholesterol deposition in arterial vessel walls and are a major risk factor for atherosclerosis and premature death by coronary artery disease. In the blood, cholesterol is transported in lipoprotein particles, ~70% of which in human are low-density lipoproteins (LDL). LDL is constantly cleared by internalization into cells by the LDL receptor (LDLR), which binds and internalizes LDL via its unique apolipoprotein B (apoB) protein. Mutations in *LDLR* or *APOB* genes are major causes for the frequent autosomal dominant genetic disorder known as familial hypercholesterolemia (164, 165). Among important cholesterol-lowering drugs are "statins," which inhibit cellular cholesterol synthesis (166). However, more efficient strategies to further decrease levels of circulating LDL-C are needed (167, 168).

Originally named NARC-1 for "neural apoptosis-regulated convertase," PCSK9 was first discovered and characterized in our laboratory (169). In collaboration with C. Boileau in Paris, we established the association between single-point mutations in the *PCSK9* gene and autosomal dominant hypercholesterolemia in two French families (170). Thus, *PCSK9* is the third gene associated with familial hypercholesterolemia (170, 171), with *LDLR* and *APOB* as the other two (164, 165). Later, Cohen et al. showed that nonsense PCSK9 mutations are associated with hypocholesterolemia in ~2% of black subjects (172, 173). Up to ~7% of black Africans living around the equator exhibit the loss of one allele of PCSK9 (174). In summary, point mutations in PCSK9 (171) are associated with either familial hypercholesterolemia (18, 170, 175–178) (gain of function of PCSK9; GOF) or hypocholesterolemia (172, 173, 179, 180) (loss of function of PCSK9; LOF). Two women lacking functional PCSK9 exhibited an ~85% reduction in circulating cholesterol associated with LDL (LDL-C) (180, 181). *Pcsk9*$^{-/-}$ mice are also viable and exhibit an ~80% drop in circulating LDL-C (182, 183), emphasizing the therapeutical potential of a PCSK9 inhibitor/silencer.

PCSK9 is mostly expressed in hepatocytes and small intestinal enterocytes (169). By an as-yet unknown mechanism(s), and independent of its enzymatic activity, PCSK9 enhances the degradation of cell-surface LDLR (180, 184–187) in endosomes/

lysosomes (188), resulting in increased circulating LDL. Statins, the best cholesterol-lowering drugs (166), reduce cholesterol synthesis by inhibiting the rate-limiting HMG-CoA reductase. The resulting cellular cholesterol depletion leads to transcription of genes involved in cholesterol metabolism, including those of PCSK9 and LDLR (189). While upregulation of LDLR reduces circulating LDL, that of PCSK9 counterbalances it through degradation of the LDLR (189, 190). PCSK9 inhibition is thus a promising complement to statin therapy to lower LDL-C (167, 180, 190).

PCSK9 (692 aa in human) comprises a signal peptide (aa 1–30) followed by prosegment (Pro; aa 31–152), catalytic (aa 153–407), hinge region (HR; aa 408–452), and C-terminal Cys-His-rich domain (CHRD; aa 453–692) segments. Following translocation into the endoplasmic reticulum (ER), the prosegment is autocatalytically cleaved at the VFAQ$_{152}$↓SIP site (185). In PCs, the prosegment is an essential intramolecular chaperone and inhibitor, which is usually removed intracellularly to yield a fully active protease. Different from other PCs, PCSK9 is secreted as a stable, enzymatically inactive, non-covalent complex [Pro≡PCSK9] (169, 171, 185). In accordance, enhanced degradation of the LDLR (184–186) induced by PCSK9 does not require its catalytic activity (191, 192). In human (18) and mouse (183) plasma, both full-length PCSK9 (aa 152–692) and a truncated form PCSK9-ΔN_{218} (aa 219–692) can be detected (18, 193). The latter, which has no activity on LDLR, is likely generated by furin, since it efficiently cleaves PCSK9 ex vivo at RFHR$_{218}$↓ (18). Interestingly, the human GOF R218S, F216L (177), and R215H (194) mutations associated with hypercholesterolemia prevent such a cleavage (18) and presumably result in increased levels of active PCSK9. To optimize PCSK9 inactivation by furin, we designed a PCSK9-RRRR$_{218}$EL mutant, which resulted in the secretion of only the inactive PCSK9-ΔN_{218} (18). We also contributed to the setup of two ELISA assays of circulating human PCSK9 (193, 195, 196), revealing a good correlation between levels of PCSK9 and LDL-C in human plasma (193). Sequencing of *PCSK9* exons from individuals at the extremes of the PCSK9 distribution provided a database of PCSK9 mutations, which are valuable tools in structure–function analyses. Indeed, we have recently identified a novel LOF variant, R434W, associated with low levels of circulating PCSK9 and LDL-C (193). This mutation, which occurs in an exposed loop of the hinge region, does not prevent LDLR binding, but drastically reduces the ability of PCSK9 to enhance the degradation of the LDLR (193).

9.1. Structure of PCSK9 and Deduced LDLR-PCSK9 Interacting Domain

The crystal structure of PCSK9 revealed three separate domains: the prosegment and catalytic domain in tight complex and the spatially separated CHRD (197–199). In all three crystal structures, aa 31–60 of the prosegment and portions of the CHRD were unresolved, indicating their unstructured nature. Biochemical (200) and co-crystal structure (201) studies revealed that aa 153–156 and 367–381 directly interact with the EGF-A domain of the LDLR. The most severe mutation associated with hypercholesterolemia, D374Y (176), is within aa 367–381 and results in an ~25-fold higher affinity of PCSK9 toward LDLR (197). The shallow binding surface on PCSK9 is distant from its catalytic site, and the EGF-A domain of LDLR makes no contact with either the prosegment or the CHRD.

9.1.1. The Prosegment of PCSK9

Even though the prosegment does not bind the EGF-A domain of LDLR (201), it negatively regulates this interaction. The removal of its N-terminal acidic stretch (aa 31–53), which exhibits Tyr_{38} sulfation (169) and Ser_{47} phosphorylation (202), enhances the binding of PCSK9 to LDLR by ~sevenfold (201). Whether this unstructured acidic stretch (197–199) binds another domain of PCSK9 and/or interacts with another protein is yet to be defined.

9.1.2. The HR-CHRD of PCSK9

The HR is an exposed loop structure connecting the catalytic domain and the CHRD (193). The latter is composed of a six β-strand structure repeated three times and hence forming three subdomain modules M1, M2, and M3 (197). In the reported crystal structures (197–199), disordered segments include aa 573–584 (in M2), 660–667 (in M3), and the C-terminal aa 683–692 exhibiting Ser_{688} phosphorylation (202). A number of GOF (R469W, E482G, R496W, F515L, and H553R) and LOF (Q554E and the new one R434W) (193) mutations within the HR-CHRD were identified, but their underlying mechanisms are unknown.

9.2. Cellular Biology of PCSK9

9.2.1. PCSK9 Targets

Over 20 years experience with PCs led us to predict that PCSK9 should have more than one target (108). We thus first tested PCSK9 ex vivo activity on other members of the LDLR-like protein family. While LRP was not affected (185), the closest members to LDLR, i.e., VLDLR and ApoER2, were degraded faster in the presence of PCSK9 in a cell-type dependent fashion (203). We discovered that PCSK9 also enhances the degradation of the major hepatitis C virus (HCV) receptor, the tetraspanin protein CD81 (204).

9.2.2. PCSK9-Enhanced Degradation of the LDLR

The cell-surface localization of PCSK9 is dependent on the presence of the LDLR (205). The two proteins also co-localize in early and late endosomes (185). We previously developed an approach in which the fusion of a secretory protein of interest with the transmembrane domain and cytosolic tail (TM-CT) of the lysosomal protein Lamp1 results in an efficient degradation of its partners (105, 206). This strategy was applied to PCSK9 to better target its partners, including the LDLR, to degradative compartments. Accordingly, fusion of PCSK9 to the TM-CT of Lamp1 (PCSK9-Lamp1) resulted in super-active forms of PCSK9 capable of depleting cells from its targets (203).

9.2.3. Extracellular Versus Intracellular Pathways

The extracellular pathway is defined by the ability of extracellular PCSK9 to target the LDLR. Indeed, incubation of cells with PCSK9, but not PCSK9-ΔC, enhances the degradation of cell-surface LDLR in endosomes/lysosomes (207). Since the PCSK9-ΔC still interacts with the LDLR and is internalized, the CHRD is likely essential for the trafficking of [PCSK9≡LDLR] to endosomes/lysosomes. Internalization of cell-surface LDLR requires the adaptor protein ARH that binds its cytosolic tail on the cytoplasmic side of clathrin heavy chain-coated vesicles (196, 208). In accordance, the pharmaceutical company Amgen developed a clinically relevant monoclonal antibody that inhibits PCSK9 interaction with LDLR and results in an ~80% reduction of LDL-C that lasted for 2 weeks in monkey (209). However, we recently demonstrated the existence of an intracellular pathway (187). First, PCSK9 can degrade the LDLR in vivo (186) and ex vivo (210) in the absence of ARH. Second, siRNA knockdown of both a and b chains of clathrin light chains, which block exclusively the intracellular pathway by preventing the trafficking from the TGN to lysosomes (211), resulted in a drastically decreased LDLR degradation (210).

9.2.4. HR-CHRD Binding Proteins

In view of the critical importance of the HR-CHRD for targeting [PCSK9≡LDLR] to lysosomes, we hypothesized that it binds directly or indirectly a membrane-associated protein that would sort the complex to lysosomes. Accordingly, a Far-Western screen of endogenous interactors of PCSK9 revealed that annexin A2 (AnxA2) binds the HR-CHRD and inhibits the ability of PCSK9 to enhance the degradation of the LDLR (212). AnxA2 lacks a signal peptide but is found at the cell surface of endothelia (213), keratinocytes (214), and epithelial (215, 216) and tumor cells (217).

9.3. In Vivo Studies

9.3.1. PCSK9 Mouse Models

Knockout (KO; *Pcsk9*$^{-/-}$) mice exhibit higher levels of LDLR protein in liver and 42% less circulating total cholesterol, with an ~80% drop in LDL-C (182, 183). In contrast, transgenic mice overexpressing PCSK9 exhibit 5–15-fold higher levels of LDL-C (183, 196, 218). We also developed mice carrying conditional floxed alleles, in which the proximal promoter and exon 1 of *Pcsk9* are *fl*anked by *loxP* sites (*Pcsk9*$^{f/f}$). In mice expressing the Cre recombinase under the control of the albumin promoter, *Pcsk9* was specifically inactivated in hepatocytes by *loxP* sites recombination (*Pcsk9*$^{f/f}$ *Alb-cre*). Total KO and liver-specific KO (LivKO) mice exhibited 42 and 27% less circulating total cholesterol, respectively, indicating that hepatic PCSK9 is responsible for ~two-thirds of the phenotype. This suggested that the role of PCSK9 in cholesterol homeostasis is primarily mediated by its activity on LDLR, since liver accounts for ~70% of the body LDL-C clearance (219). Analysis of [*Pcsk9*$^{f/f}$ *Alb-cre*] livers demonstrated that PCSK9 expression is restricted to hepatocytes, from where circulating PCSK9 mostly originates. We generated double KO mice lacking both PCSK9 and LDLR (dKO; *Pcsk9*$^{-/-}$ *Ldlr*$^{-/-}$) and showed that their plasma lipid profile was identical to that of *Ldlr*$^{-/-}$ mice, confirming that PCSK9 activity on LDLR mediates most of its role in cholesterol homeostasis.

9.3.2. Circulating PCSK9

Human plasma contains ~100–200 ng/ml of PCSK9 (193, 196, 220). Its physiological role remains undefined, as well as that of its truncated form PCSK9-ΔN_{218} that represents ~50% of the PCSK9 species in mouse plasma. Analysis of transgenic lines that overexpress low or high levels of mouse PCSK9 in the liver indicated that only supra-physiological levels of circulating PCSK9 (~30-fold higher) increased circulating cholesterol (+60%); a threefold increase had no significant impact on circulating cholesterol. Moreover, data from transgenic mice expressing very high levels of human PCSK9 in kidney (218) or liver (196) or continuous infusions of recipient wild-type (WT) mice with recombinant human PCSK9 (221) indicated that microgram per milliliter amounts of circulating PCSK9 are required to significantly affect liver LDLR protein levels. Transgenic expression in kidney (32-fold the endogenous liver levels) led to 100% loss of LDLR protein in liver, but to only 50% loss in kidney (218). Thus, even at high levels, circulating PCSK9 reduces primarily liver LDLR with little effect on extrahepatic tissues, e.g., adrenals (196, 218).

9.3.3. Partial Hepatectomy (PHx)

To better understand the role of PCSK9 in liver, its major site of expression, we challenged this tissue by performing sham or PHx operations in WT and KO mice. Hepatectomized KO, but not WT, mice developed lesions, still visible 10 days after the liver had recovered its original mass. In addition, the proliferation of KO hepatocytes was delayed (183). Critically low levels of cholesterol may impede efficient liver regeneration. Indeed, HMG-CoA reductase mRNA levels were increased in KO regenerating livers only (2.5-fold at 72 h post-PHx). Also, when fed a high-cholesterol diet 1 week prior to PHx, KO mice no longer exhibited necrotic lesions.

9.3.4. Total Absence of PCSK9 May Affect β-Cell Function and Predispose to Diabetes

It was originally observed that the pancreatic insulin-producing β-TC3 cells express high levels of PCSK9 (169). LDLR is also highly expressed in insulin-producing pancreatic islet β-cells, possibly affecting the function of these cells. We recently showed that, compared to control mice, PCSK9-null male mice over

4 months of age carried more LDLR and less insulin in their pancreas; they were hypoinsulinemic, hyperglycemic, and glucose intolerant; their islets exhibited signs of malformation, apoptosis, and inflammation. Collectively, these observations suggested that PCSK9 may be necessary for the normal function of pancreatic islets (222).

10. Conclusions and Future Perspectives

The nine-membered family of the proprotein convertases (PCs) comprises seven basic amino acid-specific subtilisin-like serine proteinases, related to yeast kexin, known as PC1/3, PC2, furin, PC4, PC5/6, PACE4, and PC7, and two other subtilases that cleave at non-basic residues called SKI-1/S1P and PCSK9 (**Fig. 3.1**). The long and arduous task of identification of these processing enzymes is now over, as analysis of the genomes available failed to identify other potential members. While most PCs exert their functions through cleavage of substrates at either basic or non-basic aa, it is amazing that the last member PCSK9 only needs its enzymatic activity to autocatalytically process its prosegment in ER, which remains tightly associated with the catalytic subunit, resulting in an inactive protease. The absence of enzymatic activity may well explain the dominant pathological consequences of the lack or excess of PCSK9, which may be due to modified stoichiometric levels of protein–receptor complexes, such as PCSK9-LDLR. It is now the time to define the physiological functions of each PC, their substrates, and partners and to devise specific therapies aimed at controlling their levels. The development of specific inhibitors/modulators of convertases may find future applications in the control of some pathologies, e.g., hypercholesterolemia, cancer/metastasis, and viral infections.

Acknowledgments

The author thanks all present and past members of the Seidah laboratory for all their help during the arduous but exciting years of PC discovery and characterization. This research was supported by CIHR grants MOP-36496 and # CTP-82946 and MOP 36496, a Strauss Foundation grant, and a Canada Chair # 201652.

References

1. Gumbiner, B., and Kelly, R. B. (1982) Two distinct intracellular pathways transport secretory and membrane glycoproteins to the surface of pituitary tumor cells *Cell* **28**, 51–9.
2. Julius, D., Brake, A., Blair, L., Kunisawa, R., and Thorner, J. (1984) Isolation of the putative structural gene for the lysine-arginine-cleaving endopeptidase required for processing of yeast prepro-alpha-factor *Cell* **37**, 1075–89.
3. Fuller, R. S., Sterne, R. E., and Thorner, J. (1988) Enzymes required for yeast prohormone processing *Annu Rev Physiol* **50**, 345–62.
4. Mizuno, K., Nakamura, T., Ohshima, T., Tanaka, S., and Matsuo, H. (1988) Yeast KEX2 genes encodes an endopeptidase homologous to subtilisin-like serine proteases *Biochem Biophys Res Commun* **156**, 246–54.
5. Seidah, N. G., Gaspar, L., Mion, P., Marcinkiewicz, M., Mbikay, M., and Chrétien, M. (1990) cDNA sequence of two distinct pituitary proteins homologous to Kex2 and furin gene products: Tissue-specific mRNAs encoding candidates for prohormone processing proteinases *DNA Cell Biol* **9**, 789.
6. Thomas, G., Thorne, B. A., Thomas, L., Allen, R. G., Hruby, D. E., Fuller, R. et al. (1988) Yeast KEX2 endopeptidase correctly cleaves a neuroendocrine prohormone in mammalian cells *Science* **241**, 226–30.
7. Bourbonnais, Y., Germain, D., Ash, J., and Thomas, D. Y. (1994) Cleavage of prosomatostatins by the yeast Yap3 and Kex2 endoprotease *Biochimie* **76**, 226–33.
8. Bathurst, I. C., Brennan, S. O., Carrell, R. W., Cousens, L. S., Brake, A. J., and Barr, P. J. (1987) Yeast KEX2 protease has the properties of a human proalbumin converting enzyme *Science* **235**, 348–50.
9. Roebroek, A. J., Schalken, J. A., Bussemakers, M. J., van Heerikhuizen, H., Onnekink, C., Debruyne, F. M. et al. (1986) Characterization of human c-fes/fps reveals a new transcription unit (fur) in the immediately upstream region of the proto-oncogene *Mol Biol Rep* **11**, 117–25.
10. van den Ouweland, A. M., Van Groningen, J. J., Roebroek, A. J., Onnekink, C., and Van de Ven, W. J. (1989) Nucleotide sequence analysis of the human fur gene *Nucleic Acids Res* **17**, 7101–2.
11. Van de Ven, W. J., Voorberg, J., Fontijn, R., Pannekoek, H., van den Ouweland, A. M., van Duijnhoven, H. L. et al. (1990) Furin is a subtilisin-like proprotein processing enzyme in higher eukaryotes *Mol Biol Rep* **14**, 265–75.
12. Nakayama, K. (1997) Furin: A mammalian subtilisin/Kex2p-like endoprotease involved in processing of a wide variety of precursor proteins *Biochem J* **327**(Pt 3), 625–35.
13. Steiner, D. F. (1998) The proprotein convertases *Curr Opin Chem Biol* **2**, 31–9.
14. Seidah, N. G., and Chrétien, M. (1999) Proprotein and prohormone convertases: A family of subtilases generating diverse bioactive polypeptides *Brain Res* **848**, 45–62.
15. Thomas, G. (2002) Furin at the cutting edge: From protein traffic to embryogenesis and disease *Nat Rev Mol Cell Biol* **3**, 753–66.
16. Seidah, N. G., Mayer, G., Zaid, A., Rousselet, E., Nassoury, N., Poirier, S. et al. (2008) The activation and physiological functions of the proprotein convertases *Int J Biochem Cell Biol* **40**, 1111–25.
17. Henrich, S., Cameron, A., Bourenkov, G. P., Kiefersauer, R., Huber, R., Lindberg, I. et al. (2003) The crystal structure of the proprotein processing proteinase furin explains its stringent specificity *Nat Struct Biol* **10**, 520–6.
18. Benjannet, S., Rhainds, D., Hamelin, J., Nassoury, N., and Seidah, N. G. (2006) The proprotein convertase PCSK9 is inactivated by furin and/or PC5/6A: Functional consequences of natural mutations and post-translational modifications *J Biol Chem* **281**, 30561–72.
19. Rawling, J., Garcia-Barreno, B., and Melero, J. A. (2008) Insertion of the two cleavage sites of the respiratory syncytial virus fusion protein in Sendai virus fusion protein leads to enhanced cell-cell fusion and a decreased dependency on the HN attachment protein for activity *J Virol* **82**, 5986–98.
20. Groskreutz, D. J., Sliwkowski, M. X., and Gorman, C. M. (1994) Genetically engineered proinsulin constitutively processed and secreted as mature, active insulin *J Biol Chem* **269**, 6241–5.
21. Jin, W., Fuki, I. V., Seidah, N. G., Benjannet, S., Glick, J. M., and Rader, D. J. (2005) Proprotein convertases are responsible for proteolysis and inactivation of endothelial lipase *J Biol Chem* **280**, 36551–9.
22. Roebroek, A. J., Umans, L., Pauli, I. G., Robertson, E. J., van Leuven, F., Van de Ven, W. J. et al. (1998) Failure of ventral closure and axial rotation in embryos lacking the proprotein convertase Furin *Development* **125**, 4863–76.

23. Ducy, P., and Karsenty, G. (2000) The family of bone morphogenetic proteins *Kidney Int* **57**, 2207–14.
24. Karsenty, G. (1999) The genetic transformation of bone biology *Genes Dev* **13**, 3037–51.
25. Dickson, M. C., Martin, J. S., Cousins, F. M., Kulkarni, A. B., Karlsson, S., and Akhurst, R. J. (1995) Defective haematopoiesis and vasculogenesis in transforming growth factor-beta 1 knock out mice *Development* **121**, 1845–54.
26. Degnin, C., Jean, F., Thomas, G., and Christian, J. L. (2004) Cleavages within the prodomain direct intracellular trafficking and degradation of mature bone morphogenetic protein-4 *Mol Biol Cell* **15**, 5012–20.
27. Dubois, C. M., Blanchette, F., Laprise, M. H., Leduc, R., Grondin, F., and Seidah, N. G. (2001) Evidence that furin is an authentic transforming growth factor-beta1-converting enzyme *Am J Pathol* **158**, 305–16.
28. Cui, Y., Jean, F., Thomas, G., and Christian, J. L. (1998) BMP-4 is proteolytically activated by furin and/or PC6 during vertebrate embryonic development *EMBO J* **17**, 4735–43.
29. Roebroek, A. J., Taylor, N. A., Louagie, E., Pauli, I., Smeijers, L., Snellinx, A. et al. (2004) Limited redundancy of the proprotein convertase furin in mouse liver *J Biol Chem* **279**, 53442–50.
30. Louagie, E., Taylor, N. A., Flamez, D., Roebroek, A. J., Bright, N. A., Meulemans, S. et al. (2008) Role of furin in granular acidification in the endocrine pancreas: Identification of the V-ATPase subunit Ac45 as a candidate substrate *Proc Natl Acad Sci USA* **105**, 12319–24.
31. Pesu, M., Watford, W. T., Wei, L., Xu, L., Fuss, I., Strober, W. et al. (2008) T-cell-expressed proprotein convertase furin is essential for maintenance of peripheral immune tolerance *Nature* **455**, 246–50.
32. Jean, F., Thomas, L., Molloy, S. S., Liu, G., Jarvis, M. A., Nelson, J. A. et al. (2000) A protein-based therapeutic for human cytomegalovirus infection *Proc Natl Acad Sci USA* **97**, 2864–9.
33. Komiyama, T., Coppola, J. M., Larsen, M. J., van Dort, M. E., Ross, B. D., Day, R. et al. (2009) Inhibition of furin/proprotein convertase-catalyzed surface and intracellular processing by small molecules *J Biol Chem* **284**, 15729–38.
34. Jiao, G. S., Cregar, L., Wang, J., Millis, S. Z., Tang, C., O'Malley, S. et al. (2006) Synthetic small molecule furin inhibitors derived from 2,5-dideoxystreptamine *Proc Natl Acad Sci USA* **103**, 19707–12.
35. Coppola, J. M., Bhojani, M. S., Ross, B. D., and Rehemtulla, A. (2008) A small-molecule furin inhibitor inhibits cancer cell motility and invasiveness *Neoplasia* **10**, 363–70.
36. Bassi, D. E., Lopez, D. C., Mahloogi, H., Zucker, S., Thomas, G., and Klein-Szanto, A. J. (2001) Furin inhibition results in absent or decreased invasiveness and tumorigenicity of human cancer cells *Proc Natl Acad Sci USA* **98**, 10326–31.
37. Khatib, A. M., Siegfried, G., Chrétien, M., Metrakos, P., and Seidah, N. G. (2002) Proprotein convertases in tumor progression and malignancy: Novel targets in cancer therapy *Am J Pathol* **160**, 1921–35.
38. Shiryaev, S. A., Remacle, A. G., Ratnikov, B. I., Nelson, N. A., Savinov, A. Y., Wei, G. et al. (2007) Targeting host cell furin proprotein convertases as a therapeutic strategy against bacterial toxins and viral pathogens *J Biol Chem* **282**, 20847–53.
39. Ozden, S., Lucas-Hourani, M., Ceccaldi, P. E., Basak, A., Valentine, M., Benjannet, S. et al. (2008) Inhibition of chikungunya virus infection in cultured human muscle cells by furin inhibitors: Impairment of the maturation of the E2 surface glycoprotein *J Biol Chem* **283**, 21899–908.
40. Xiao, Y., Chen, G., Richard, J., Rougeau, N., Li, H., Seidah, N. G. et al. (2008) Cell-surface processing of extracellular human immunodeficiency virus type 1 Vpr by proprotein convertases *Virology* **372**, 384–97.
41. Docherty, K., and Steiner, D. F. (1982) Post-translational proteolysis in polypeptide hormone biosynthesis *Annu Rev Physiol* **44**, 625–38.
42. Loh, Y. P. (1987) Peptide precursor processing enzymes within secretory vesicles *Ann N Y Acad Sci* **493**, 292–307.
43. Kuliawat, R., and Arvan, P. (1994) Distinct molecular mechanisms for protein sorting within immature secretory granules of pancreatic beta-cells *J Cell Biol* **126**, 77–86.
44. Cromlish, J. A., Seidah, N. G., and Chrétien, M. (1986) Selective cleavage of human ACTH, beta-lipotropin, and the N-terminal glycopeptide at pairs of basic residues by IRCM-serine protease 1. Subcellular localization in small and large vesicles *J Biol Chem* **261**, 10859–70.
45. Seidah, N. G., Paquin, J., Hamelin, J., Benjannet, S., and Chrétien, M. (1988) Structural and immunological homology of human and porcine pituitary and plasma IRCM-serine protease 1 to plasma kallikrein: Marked selectivity for pairs of basic residues suggests a widespread role in pro-hormone

and pro-enzyme processing *Biochimie* **70**, 33–46.
46. Metters, K. M., Rossier, J., Paquin, J., Chrétien, M., and Seidah, N. G. (1988) Selective cleavage of proenkephalin-derived peptides (less than 23,300 daltons) by plasma kallikrein *J Biol Chem* **263**, 12543–53.
47. Davidson, H. W., Rhodes, C. J., and Hutton, J. C. (1988) Intraorganellar calcium and pH control proinsulin cleavage in the pancreatic beta cell via two distinct site-specific endopeptidases *Nature* **333**, 93–6.
48. Mullis, K. B., and Faloona, F. A. (1987) Specific synthesis of DNA in vitro via a polymerase-catalyzed chain reaction *Meth Enzymol* **155**, 335–50.
49. Hutchison, C. A., III, Phillips, S., Edgell, M. H., Gillam, S., Jahnke, P., and Smith, M. (1978) Mutagenesis at a specific position in a DNA sequence *J Biol Chem* **253**, 6551–60.
50. Crine, P., Gianoulakis, C., Seidah, N. G., Gossard, F., Pezalla, P. D., Lis, M. et al. (1978) Biosynthesis of beta-endorphin from beta-lipotropin and a larger molecular weight precursor in rat pars intermedia *Proc Natl Acad Sci USA* **75**, 4719–23.
51. Crine, P., Seidah, N. G., Routhier, R., Gossard, F., and Chrétien, M. (1980) Processing of two forms of the common precursor to alpha-melanotropin and beta-endorphin in the rat pars intermedia. Evidence for and partial characterization of new pituitary peptides *Eur J Biochem* **110**, 387–96.
52. Crine, P., Gossard, F., Seidah, N. G., Blanchette, L., Lis, M., and Chrétien, M. (1979) Concomitant synthesis of beta-endorphin and alpha-melanotropin from two forms of pro-opiomelanocortin in the rat pars intermedia *Proc Natl Acad Sci USA* **76**, 5085–9.
53. Seidah, N. G., Lazure, C., Chrétien, M., Thibault, G., Garcia, R., Cantin, M. et al. (1984) Amino acid sequence of homologous rat atrial peptides: Natriuretic activity of native and synthetic forms *Proc Natl Acad Sci USA* **81**, 2640–4.
54. Yan, W., Sheng, N., Seto, M., Morser, J., and Wu, Q. (1999) Corin, a mosaic transmembrane serine protease encoded by a novel cDNA from human heart *J Biol Chem* **274**, 14926–35.
55. Seidah, N. G., Marcinkiewicz, M., Benjannet, S., Gaspar, L., Beaubien, G., Mattei, M. G. et al. (1991) Cloning and primary sequence of a mouse candidate prohormone convertase PC1 homologous to PC2, Furin, and Kex2: Distinct chromosomal localization and messenger RNA distribution in brain and pituitary compared to PC2 *Mol Endocrinol* **5**, 111–22.
56. Seidah, N. G., Mattei, M. G., Gaspar, L., Benjannet, S., Mbikay, M., and Chrétien, M. (1991) Chromosomal assignments of the genes for neuroendocrine convertase PC1 (NEC1) to human 5q15-21, neuroendocrine convertase PC2 (NEC2) to human 20p11.1-11.2, and furin (mouse 7[D1-E2] region) *Genomics* **11**, 103–7.
57. Benjannet, S., Rondeau, N., Day, R., Chrétien, M., and Seidah, N. G. (1991) PC1 and PC2 are proprotein convertases capable of cleaving proopiomelanocortin at distinct pairs of basic residues *Proc Natl Acad Sci USA* **88**, 3564–8.
58. Smeekens, S. P., and Steiner, D. F. (1990) Identification of a human insulinoma cDNA encoding a novel mammalian protein structurally related to the yeast dibasic processing protease Kex2 *J Biol Chem* **265**, 2997–3000.
59. Smeekens, S. P., Avruch, A. S., LaMendola, J., Chan, S. J., and Steiner, D. F. (1991) Identification of a cDNA encoding a second putative prohormone convertase related to PC2 in AtT20 cells and islets of Langerhans *Proc Natl Acad Sci USA* **88**, 340–4.
60. Smeekens, S. P., Montag, A. G., Thomas, G., Albiges-Rizo, C., Carroll, R., Benig, M. et al. (1992) Proinsulin processing by the subtilisin-related proprotein convertases furin, PC2, and PC3 *Proc Natl Acad Sci USA* **89**, 8822–6.
61. Thomas, L., Leduc, R., Thorne, B. A., Smeekens, S. P., Steiner, D. F., and Thomas, G. (1991) Kex2-like endoproteases PC2 and PC3 accurately cleave a model prohormone in mammalian cells: Evidence for a common core of neuroendocrine processing enzymes *Proc Natl Acad Sci USA* **88**, 5297–301.
62. Malide, D., Seidah, N. G., Chrétien, M., and Bendayan, M. (1995) Electron microscopic immunocytochemical evidence for the involvement of the convertases PC1 and PC2 in the processing of proinsulin in pancreatic beta-cells *J Histochem Cytochem* **43**, 11–19.
63. Dikeakos, J. D., Mercure, C., Lacombe, M. J., Seidah, N. G., and Reudelhuber, T. L. (2007) PC1/3, PC2 and PC5/6A are targeted to dense core secretory granules by a common mechanism *FEBS J* **274**, 4094–102.
64. Dikeakos, J. D., Di, L. P., Lacombe, M. J., Ghirlando, R., Legault, P., Reudelhuber, T. L. et al. (2009) Functional and structural characterization of a dense core secretory granule sorting domain from the PC1/3 protease *Proc Natl Acad Sci USA* **106**, 7408–13.
65. Fricker, L. D., McKinzie, A. A., Sun, J., Curran, E., Qian, Y., Yan, L. et al.

(2000) Identification and characterization of proSAAS, a granin-like neuroendocrine peptide precursor that inhibits prohormone processing *J Neurosci* **20**, 639–48.

66. Qian, Y., Devi, L. A., Mzhavia, N., Munzer, S., Seidah, N. G., and Fricker, L. D. (2000) The C-terminal region of proSAAS is a potent inhibitor of prohormone convertase 1 *J Biol Chem* **275**, 23596–601.
67. Feng, Y., Reznik, S. E., and Fricker, L. D. (2002) ProSAAS and prohormone convertase 1 are broadly expressed during mouse development *Brain Res Gene Expr Patterns* **1**, 135–40.
68. Hsi, K. L., Seidah, N. G., Lu, C. L., and Chrétien, M. (1981) Reinvestigation of the N-terminal amino acid sequence of beta-lipotropin from human pituitary glands *Biochem Biophys Res Commun* **103**, 1329–35.
69. Hsi, K. L., Seidah, N. G., De Serres, G., and Chrétien, M. (1982) Isolation and NH2-terminal sequence of a novel porcine anterior pituitary polypeptide. Homology to proinsulin, secretin and Rous sarcoma virus transforming protein TVFV60 *FEBS Lett* **147**, 261–6.
70. Seidah, N. G., Hsi, K. L., De Serres, G., Rochemont, J., Hamelin, J., Antakly, T. et al. (1983) Isolation and NH2-terminal sequence of a highly conserved human and porcine pituitary protein belonging to a new superfamily. Immunocytochemical localization in pars distalis and pars nervosa of the pituitary and in the supraoptic nucleus of the hypothalamus *Arch Biochem Biophys* **225**, 525–34.
71. Marcinkiewicz, M., Benjannet, S., Cantin, M., Seidah, N. G., and Chrétien, M. (1986) CNS distribution of a novel pituitary protein '7B2': Localization in secretory and synaptic vesicles *Brain Res* **380**, 349–56.
72. Marcinkiewicz, M., Benjannet, S., Seidah, N. G., Cantin, M., and Chrétien, M. (1985) Immunocytochemical localization of a novel pituitary protein (7B2) within the rat brain and hypophysis *J Histochem Cytochem* **33**, 1219–26.
73. Mbikay, M., Seidah, N. G., and Chrétien, M. (2001) Neuroendocrine secretory protein 7B2: Structure, expression and functions *Biochem J* **357**, 329–42.
74. Martens, G. J., Braks, J. A., Eib, D. W., Zhou, Y., and Lindberg, I. (1994) The neuroendocrine polypeptide 7B2 is an endogenous inhibitor of prohormone convertase PC2 *Proc Natl Acad Sci USA* **91**, 5784–7.
75. van Horssen, A. M., van den Hurk, W. H., Bailyes, E. M., Hutton, J. C., Martens, G. J., and Lindberg, I. (1995) Identification of the region within the neuroendocrine polypeptide 7B2 responsible for the inhibition of prohormone convertase PC2 *J Biol Chem* **270**, 14292–6.
76. Zhu, X., and Lindberg, I. (1995) 7B2 facilitates the maturation of proPC2 in neuroendocrine cells and is required for the expression of enzymatic activity *J Cell Biol* **129**, 1641–50.
77. Benjannet, S., Savaria, D., Chrétien, M., and Seidah, N. G. (1995) 7B2 is a specific intracellular binding protein of the prohormone convertase PC2 *J Neurochem* **64**, 2303–11.
78. Benjannet, S., Mamarbachi, A. M., Hamelin, J., Savaria, D., Munzer, J. S., Chrétien, M. et al. (1998) Residues unique to the pro-hormone convertase PC2 modulate its autoactivation, binding to 7B2 and enzymatic activity *FEBS Lett* **428**, 37–42.
79. Paquet, L., Bergeron, F., Boudreault, A., Seidah, N. G., Chrétien, M., Mbikay, M. et al. (1994) The neuroendocrine precursor 7B2 is a sulfated protein proteolytically processed by a ubiquitous furin-like convertase *J Biol Chem* **269**, 19279–85.
80. Fortenberry, Y., Liu, J., and Lindberg, I. (1999) The role of the 7B2 CT peptide in the inhibition of prohormone convertase 2 in endocrine cell lines *J Neurochem* **73**, 994–1003.
81. Zhu, X., Zhou, A., Dey, A., Norrbom, C., Carroll, R., Zhang, C. et al. (2002) Disruption of PC1/3 expression in mice causes dwarfism and multiple neuroendocrine peptide processing defects *Proc Natl Acad Sci USA* **99**, 10293–8.
82. Furuta, M., Yano, H., Zhou, A., Rouille, Y., Holst, J. J., Carroll, R. et al. (1997) Defective prohormone processing and altered pancreatic islet morphology in mice lacking active SPC2 *Proc Natl Acad Sci USA* **94**, 6646–51.
83. Jackson, R. S., Creemers, J. W., Ohagi, S., Raffin-Sanson, M. L., Sanders, L., Montague, C. T. et al. (1997) Obesity and impaired prohormone processing associated with mutations in the human prohormone convertase 1 gene *Nat Genet* **16**, 303–6.
84. Jackson, R. S., Creemers, J. W., Farooqi, I. S., Raffin-Sanson, M. L., Varro, A., Dockray, G. J. et al. (2003) Small-intestinal dysfunction accompanies the complex endocrinopathy of human proprotein convertase 1 deficiency *J Clin Invest* **112**, 1550–60.
85. Croissandeau, G., Wahnon, F., Yashpal, K., Seidah, N. G., Coderre, T. J., Chrétien, M. et al. (2006) Increased stress-induced analgesia in mice lacking the proneuropeptide convertase PC2 *Neurosci Lett* **406**, 71–5.

86. Lloyd, D. J., Bohan, S., and Gekakis, N. (2006) Obesity, hyperphagia and increased metabolic efficiency in Pc1 mutant mice *Hum Mol Genet* **15**, 1884–93.
87. Benzinou, M., Creemers, J. W., Choquet, H., Lobbens, S., Dina, C., Durand, E. et al. (2008) Common nonsynonymous variants in PCSK1 confer risk of obesity *Nat Genet* **40**, 943–5.
88. Wardman, J. H., Zhang, X., Gagnon, S., Castro, L. M., Zhu, X., Steiner, D. F. et al. (2010) Analysis of peptides in prohormone convertase 1/3 null mouse brain using quantitative peptidomics *J Neurochem* **114**, 215–25.
89. Zhang, X., Pan, H., Peng, B., Steiner, D. F., Pintar, J. E., and Fricker, L. D. (2010) Neuropeptidomic analysis establishes a major role for prohormone convertase-2 in neuropeptide biosynthesis *J Neurochem* **112**, 1168–79.
90. Morgan, D. J., Wei, S., Gomes, I., Czyzyk, T., Mzhavia, N., Pan, H. et al. (2010) The propeptide precursor proSAAS is involved in fetal neuropeptide processing and body weight regulation *J Neurochem* **113**, 1275–84.
91. Westphal, C. H., Muller, L., Zhou, A., Zhu, X., Bonner-Weir, S., Schambelan, M. et al. (1999) The neuroendocrine protein 7B2 is required for peptide hormone processing in vivo and provides a novel mechanism for pituitary Cushing's disease *Cell* **96**, 689–700.
92. Lee, S. N., Peng, B., Desjardins, R., Pintar, J. E., Day, R., and Lindberg, I. (2007) Strain-specific steroidal control of pituitary function *J Endocrinol* **192**, 515–25.
93. Kiefer, M. C., Tucker, J. E., Joh, R., Landsberg, K. E., Saltman, D., and Barr, P. J. (1991) Identification of a second human subtilisin-like protease gene in the fes/fps region of chromosome 15 *DNA Cell Biol* **10**, 757–69.
94. Seidah, N. G., Day, R., Hamelin, J., Gaspar, A., Collard, M. W., and Chrétien, M. (1992) Testicular expression of PC4 in the rat: Molecular diversity of a novel germ cell-specific Kex2/subtilisin-like proprotein convertase *Mol Endocrinol* **6**, 1559–70.
95. Lusson, J., Vieau, D., Hamelin, J., Day, R., Chrétien, M., and Seidah, N. G. (1993) cDNA structure of the mouse and rat subtilisin/kexin like PC5: A candidate proprotein convertase expressed in endocrine and nonendocrine cells *Proc Natl Acad Sci USA* **90**, 6691–5.
96. Gyamera-Acheampong, C., Tantibhedhyangkul, J., Weerachatyanukul, W., Tadros, H., Xu, H., Van de Loo, J. W. et al. (2006) Sperm from mice genetically deficient for the PCSK4 proteinase exhibit accelerated capacitation, precocious acrosome reaction, reduced binding to egg zona pellucida, and impaired fertilizing ability *Biol Reprod* **74**, 666–73.
97. Tadros, H., Chrétien, M., and Mbikay, M. (2001) The testicular germ-cell protease PC4 is also expressed in macrophage-like cells of the ovary *J Reprod Immunol* **49**, 133–52.
98. Qiu, Q., Basak, A., Mbikay, M., Tsang, B. K., and Gruslin, A. (2005) Role of pro-IGF-II processing by proprotein convertase 4 in human placental development *Proc Natl Acad Sci USA* **102**, 11047–52.
99. Mbikay, M., Tadros, H., Ishida, N., Lerner, C. P., De Lamirande, E., Chen, A. et al. (1997) Impaired fertility in mice deficient for the testicular germ-cell protease PC4 *Proc Natl Acad Sci USA* **94**, 6842–6.
100. Li, M., Mbikay, M., and Arimura, A. (2000) Pituitary adenylate cyclase-activating polypeptide precursor is processed solely by prohormone convertase 4 in the gonads *Endocrinology* **141**, 3723–30.
101. Basak, S., Chrétien, M., Mbikay, M., and Basak, A. (2004) In vitro elucidation of substrate specificity and bioassay of proprotein convertase 4 using intramolecularly quenched fluorogenic peptides *Biochem J* **380**, 505–14.
102. Majumdar, S., Mohanta, B. C., Chowdhury, D. R., Banik, R., Dinda, B., and Basak, A. (2010) Proprotein convertase inhibitory activities of flavonoids isolated from oroxylum indicum *Curr Med Chem* **17**, 2049–58.
103. Nakagawa, T., Hosaka, M., Torii, S., Watanabe, T., Murakami, K., and Nakayama, K. (1993) Identification and functional expression of a new member of the mammalian Kex2-like processing endoprotease family: Its striking structural similarity to PACE4 *J Biochem (Tokyo)* **113**, 132–5.
104. Nakagawa, T., Murakami, K., and Nakayama, K. (1993) Identification of an isoform with an extremely large Cys-rich region of PC6, a Kex2-like processing endoprotease *FEBS Lett* **327**, 165–71.
105. Nour, N., Mayer, G., Mort, J. S., Salvas, A., Mbikay, M., Morrison, C. J. et al. (2005) The cysteine-rich domain of the secreted proprotein convertases PC5A and PACE4 functions as a cell surface anchor and interacts with tissue inhibitors of metalloproteinases *Mol Biol Cell* **16**, 5215–26.
106. Mayer, G., Hamelin, J., Asselin, M. C., Pasquato, A., Marcinkiewicz, E., Tang, M. et al. (2008) The regulated cell surface

zymogen activation of the proprotein convertase PC5A directs the processing of its secretory substrates *J Biol Chem* **283**, 2373–84.

107. Kalus, I., Schnegelsberg, B., Seidah, N. G., Kleene, R., and Schachner, M. (2003) The proprotein convertase PC5A and a metalloprotease are involved in the proteolytic processing of the neural adhesion molecule L1 *J Biol Chem* **278**, 10381–8.
108. Seidah, N. G., and Prat, A. (2002) Precursor convertases in the secretory pathway, cytosol and extracellular milieu *Essays Biochem* **38**, 79–94.
109. Xiang, Y., Molloy, S. S., Thomas, L., and Thomas, G. (2000) The PC6B cytoplasmic domain contains two acidic clusters that direct sorting to distinct trans-Golgi network/endosomal compartments *Mol Biol Cell* **11**, 1257–73.
110. Dong, W., Marcinkiewicz, M., Vieau, D., Chrétien, M., Seidah, N. G., and Day, R. (1995) Distinct mRNA expression of the highly homologous convertases PC5 and PACE4 in the rat brain and pituitary *J Neurosci* **15**, 1778–96.
111. Zheng, M., Seidah, N. G., and Pintar, J. E. (1997) The developmental expression in the rat CNS and peripheral tissues of proteases PC5 and PACE4 mRNAs: Comparison with other proprotein processing enzymes *Dev Biol* **181**, 268–83.
112. Essalmani, R., Hamelin, J., Marcinkiewicz, J., Chamberland, A., Mbikay, M., Chrétien, M. et al. (2006) Deletion of the gene encoding proprotein convertase 5/6 causes early embryonic lethality in the mouse *Mol Cell Biol* **26**, 354–61.
113. Constam, D. B., and Robertson, E. J. (2000) SPC4/PACE4 regulates a TGFbeta signaling network during axis formation *Genes Dev* **14**, 1146–55.
114. Essalmani, R., Zaid, A., Marcinkiewicz, J., Chamberland, A., Pasquato, A., Seidah, N. G. et al. (2008) In vivo functions of the proprotein convertase PC5/6 during mouse development: Gdf11 is a likely substrate *Proc Natl Acad Sci USA* **105**, 5750–5.
115. Szumska, D., Pieles, G., Essalmani, R., Bilski, M., Mesnard, D., Kaur, K. et al. (2008) VACTERL/caudal regression/Currarino syndrome-like malformations in mice with mutation in the proprotein convertase Pcsk5 *Genes Dev* **22**, 1465–77.
116. Sun, X., Essalmani, R., Seidah, N. G., and Prat, A. (2009) The proprotein convertase PC5/6 is protective against intestinal tumorigenesis: In vivo mouse model *Mol Cancer* **8**, 73.
117. Yuasa, K., Masuda, T., Yoshikawa, C., Nagahama, M., Matsuda, Y., and Tsuji, A. (2009) Subtilisin-like proprotein convertase PACE4 is required for skeletal muscle differentiation *J Biochem (Tokyo)* **146**, 407–15.
118. Blanchet, M. H., Le Good, J. A., Mesnard, D., Oorschot, V., Baflast, S., Minchiotti, G. et al. (2008) Cripto recruits Furin and PACE4 and controls Nodal trafficking during proteolytic maturation *EMBO J* **27**, 2580–91.
119. Seidah, N. G., Hamelin, J., Mamarbachi, M., Dong, W., Tardos, H., Mbikay, M. et al. (1996) cDNA structure, tissue distribution, and chromosomal localization of rat PC7, a novel mammalian proprotein convertase closest to yeast kexin-like proteinases *Proc Natl Acad Sci USA* **93**, 3388–93.
120. Meerabux, J., Yaspo, M. L., Roebroek, A. J., Van de Ven, W. J., Lister, T. A., and Young, B. D. (1996) A new member of the proprotein convertase gene family (LPC) is located at a chromosome translocation breakpoint in lymphomas *Cancer Res* **56**, 448–51.
121. Van de Loo, J. W., Teuchert, M., Pauli, I., Plets, E., Van de Ven, W. J., and Creemers, J. W. (2000) Dynamic palmitoylation of lymphoma proprotein convertase prolongs its half-life, but is not essential for trans-Golgi network localization *Biochem J* **352**(Pt 3), 827–33.
122. Munzer, J. S., Basak, A., Zhong, M., Mamarbachi, A., Hamelin, J., Savaria, D. et al. (1997) In vitro characterization of the novel proprotein convertase PC7 *J Biol Chem* **272**, 19672–81.
123. Seidah, N. G. (2004) Proprotein Convertase 7. In: *Handbook of Proteolytic Enzymes,* 2nd Edition. Barrett, A. J., Rawlings, N. D. and Woessner, J. F., eds., Academic: San Diego, CA, pp. 1877–80.
124. Basak, A., Zhong, M., Munzer, J. S., Chrétien, M., and Seidah, N. G. (2001) Implication of the proprotein convertases furin, PC5 and PC7 in the cleavage of surface glycoproteins of Hong Kong, Ebola and respiratory syncytial viruses: A comparative analysis with fluorogenic peptides *Biochem J* **353**, 537–45.
125. Fugere, M., Appel, J., Houghten, R. A., Lindberg, I., and Day, R. (2007) Short polybasic peptide sequences are potent inhibitors of PC5/6 and PC7: Use of positional scanning-synthetic peptide combinatorial libraries as a tool for the optimization of inhibitory sequences *Mol Pharmacol* **71**, 323–32.
126. Decroly, E., Benjannet, S., Savaria, D., and Seidah, N. G. (1997) Comparative functional role of PC7 and furin in the processing of the

HIV envelope glycoprotein gp160 *FEBS Lett* **405**, 68–72.
127. Lopez-Perez, E., Seidah, N. G., and Checler, F. (1999) Proprotein convertase activity contributes to the processing of the Alzheimer's beta-amyloid precursor protein in human cells: Evidence for a role of the prohormone convertase PC7 in the constitutive alpha-secretase pathway *J Neurochem* **73**, 2056–62.
128. Scamuffa, N., Basak, A., Lalou, C., Wargnier, A., Marcinkiewicz, J., Siegfried, G. et al. (2008) Regulation of prohepcidin processing and activity by the subtilisin-like proprotein convertases Furin, PC5, PACE4 and PC7 *Gut* **57**, 1573–82.
129. Siegfried, G., Basak, A., Cromlish, J. A., Benjannet, S., Marcinkiewicz, J., Chrétien, M. et al. (2003) The secretory proprotein convertases furin, PC5, and PC7 activate VEGF-C to induce tumorigenesis *J Clin Invest* **111**, 1723–32.
130. Van de Loo, J. W., Creemers, J. W., Bright, N. A., Young, B. D., Roebroek, A. J., and Van de Ven, W. J. (1997) Biosynthesis, distinct post-translational modifications, and functional characterization of lymphoma proprotein convertase *J Biol Chem* **272**, 27116–23.
131. Wouters, S., Decroly, E., Vandenbranden, M., Shober, D., Fuchs, R., Morel, V. et al. (1999) Occurrence of an HIV-1 gp160 endoproteolytic activity in low-density vesicles and evidence for a distinct density distribution from endogenously expressed furin and PC7/LPC convertases *FEBS Lett* **456**, 97–102.
132. Zhong, M., Munzer, J. S., Basak, A., Benjannet, S., Mowla, S. J., Decroly, E. et al. (1999) The prosegments of furin and PC7 as potent inhibitors of proprotein convertases. In vitro and *ex vivo* assessment of their efficacy and selectivity *J Biol Chem* **274**, 33913–20.
133. Nour, N., Basak, A., Chrétien, M., and Seidah, N. G. (2003) Structure-function analysis of the prosegment of the proprotein convertase PC5A *J Biol Chem* **278**, 2886–95.
134. Bhattacharjya, S., Xu, P., Zhong, M., Chrétien, M., Seidah, N. G., and Ni, F. (2000) Inhibitory activity and structural characterization of a C-terminal peptide fragment derived from the prosegment of the proprotein convertase PC7 *Biochemistry* **39**, 2868–77.
135. Leonhardt, R. M., Fiegl, D., Rufer, E., Karger, A., Bettin, B., and Knittler, M. R. (2010) Post-endoplasmic reticulum rescue of unstable MHC class I requires proprotein convertase PC7 *J Immunol* **184**, 2985–98.
136. Bassi, D. E., Fu, J., Lopez, D. C., and Klein-Szanto, A. J. (2005) Proprotein convertases: "Master switches" in the regulation of tumor growth and progression *Mol Carcinog* **44**, 151–61.
137. Seidah, N. G., Mowla, S. J., Hamelin, J., Mamarbachi, A. M., Benjannet, S., Toure, B. B. et al. (1999) Mammalian subtilisin/kexin isozyme SKI-1: A widely expressed proprotein convertase with a unique cleavage specificity and cellular localization *Proc Natl Acad Sci USA* **96**, 1321–6.
138. Haze, K., Yoshida, H., Yanagi, H., Yura, T., and Mori, K. (1999) Mammalian transcription factor ATF6 is synthesized as a transmembrane protein and activated by proteolysis in response to endoplasmic reticulum stress *Mol Biol Cell* **10**, 3787–99.
139. Cheng, D., Espenshade, P. J., Slaughter, C. A., Jaen, J. C., Brown, M. S., and Goldstein, J. L. (1999) Secreted site-1 protease cleaves peptides corresponding to luminal loop of sterol regulatory element-binding proteins *J Biol Chem* **274**, 22805–12.
140. Horton, J. D., Goldstein, J. L., and Brown, M. S. (2002) SREBPs: Activators of the complete program of cholesterol and fatty acid synthesis in the liver *J Clin Invest* **109**, 1125–31.
141. Elagoz, A., Benjannet, S., Mammarbassi, A., Wickham, L., and Seidah, N. G. (2002) Biosynthesis and cellular trafficking of the convertase SKI-1/S1P: Ectodomain shedding requires SKI-1 activity *J Biol Chem* **277**, 11265–75.
142. Tadros, H., Seidah, N. G., Chrétien, M., and Mbikay, M. (2002) Genetic mapping of the gene for SKI-1/S1P protease (locus symbol Mbtps1) to mouse chromosome 8 *DNA Seq* **13**, 109–11.
143. Pasquato, A., Pullikotil, P., Asselin, M. C., Vacatello, M., Paolillo, L., Ghezzo, F. et al. (2006) The Proprotein Convertase SKI-1/S1P: In vitro analysis of lassa virus glycoprotein-derived substrates and *ex vivo* validation of irreversible peptide inhibitors *J Biol Chem* **281**, 23471–81.
144. Sakai, J., Rawson, R. B., Espenshade, P. J., Cheng, D., Seegmiller, A. C., Goldstein, J. L. et al. (1998) Molecular identification of the sterol-regulated luminal protease that cleaves SREBPs and controls lipid composition of animal cells *Mol Cell* **2**, 505–14.
145. Brown, M. S., and Goldstein, J. L. (1997) The SREBP pathway: Regulation of cholesterol metabolism by proteolysis of a membrane-bound transcription factor *Cell* **89**, 331–40.

146. Lenz, O., ter Meulen, J., Klenk, H. D., Seidah, N. G., and Garten, W. (2001) The Lassa virus glycoprotein precursor GP-C is proteolytically processed by subtilase SKI-1/S1P *Proc Natl Acad Sci USA* **98**, 12701–5.
147. Ye, J., Rawson, R. B., Komuro, R., Chen, X., Dave, U. P., Prywes, R. et al. (2000) ER stress induces cleavage of membrane-bound ATF6 by the same proteases that process SREBPs *Mol Biol Cell* **6**, 1355–64.
148. Yang, J., Goldstein, J. L., Hammer, R. E., Moon, Y. A., Brown, M. S., and Horton, J. D. (2001) Decreased lipid synthesis in livers of mice with disrupted Site-1 protease gene *Proc Natl Acad Sci USA* **98**, 13607–12.
149. Mouchantaf, R., Watt, H. L., Sulea, T., Seidah, N. G., Alturaihi, H., Patel, Y. C. et al. (2004) Prosomatostatin is proteolytically processed at the amino terminal segment by subtilase SKI-1 *Regul Pept* **120**, 133–40.
150. Lu, R., Yang, P., O'Hare, P., and Misra, V. (1997) Luman, a new member of the CREB/ATF family, binds to herpes simplex virus VP16-associated host cellular factor *Mol Cell Biol* **17**, 5117–26.
151. Stirling, J., and O'Hare, P. (2006) CREB4, a transmembrane bZip transcription factor and potential new substrate for regulation and cleavage by S1P *Mol Biol Cell* **17**, 413–26.
152. Pullikotil, P., Vincent, M., Nichol, S. T., and Seidah, N. G. (2004) Development of protein-based inhibitors of the proprotein of convertase SKI-1/S1P: Processing of SREBP-2, ATF6, and a viral glycoprotein *J Biol Chem* **279**, 17338–47.
153. Basak, A., Chrétien, M., and Seidah, N. G. (2002) A rapid fluorometric assay for the proteolytic activity of SKI-1/S1P based on the surface glycoprotein of the hemorrhagic fever Lassa virus *FEBS Lett* **514**, 333–9.
154. Toure, B. B., Munzer, J. S., Basak, A., Benjannet, S., Rochemont, J., Lazure, C. et al. (2000) Biosynthesis and enzymatic characterization of human SKI-1/S1P and the processing of its inhibitory prosegment *J Biol Chem* **275**, 2349–58.
155. Okada, T., Haze, K., Nadanaka, S., Yoshida, H., Seidah, N. G., Hirano, Y. et al. (2003) A serine protease inhibitor prevents endoplasmic reticulum stress-induced cleavage but not transport of the membrane-bound transcription factor ATF6 *J Biol Chem* **278**, 31024–32.
156. De Windt, A., Rai, M., Bernier, L., Thelen, K., Soini, J., Lefebvre, C. et al. (2007) Gene set enrichment analysis reveals several globally affected pathways due to SKI-1/S1P inhibition in HepG2 cells *DNA Cell Biol* **26**, 765–72.
157. Hawkins, J. L., Robbins, M. D., Warren, L. C., Xia, D., Petras, S. F., Valentine, J. J. et al. (2008) Pharmacologic inhibition of site 1 protease activity inhibits sterol regulatory element-binding protein processing and reduces lipogenic enzyme gene expression and lipid synthesis in cultured cells and experimental animals *J Pharmacol Exp Ther* **326**, 801–8.
158. Hay, B. A., Abrams, B., Zumbrunn, A. Y., Valentine, J. J., Warren, L. C., Petras, S. F. et al. (2007) Aminopyrrolidineamide inhibitors of site-1 protease *Bioorg Med Chem Lett* **17**, 4411–14.
159. Gorski, J. P., Huffman, N. T., Cui, C., Henderson, E. P., Midura, R. J., and Seidah, N. G. (2009) Potential role of proprotein convertase SKI-1 in the mineralization of primary bone *Cell Tissue Organ* **189**, 25–32.
160. Beyer, W. R., Popplau, D., Garten, W., Von Laer, D., and Lenz, O. (2003) Endoproteolytic processing of the lymphocytic choriomeningitis virus glycoprotein by the subtilase SKI-1/S1P *J Virol* **77**, 2866–72.
161. Vincent, M. J., Sanchez, A. J., Erickson, B. R., Basak, A., Chrétien, M., Seidah, N. G. et al. (2003) Crimean-Congo hemorrhagic fever virus glycoprotein proteolytic processing by subtilase SKI-1 *J Virol* **77**, 8640–9.
162. Mitchell, K. J., Pinson, K. I., Kelly, O. G., Brennan, J., Zupicich, J., Scherz, P. et al. (2001) Functional analysis of secreted and transmembrane proteins critical to mouse development *Nat Genet* **28**, 241–9.
163. Patra, D., Xing, X., Davies, S., Bryan, J., Franz, C., Hunziker, E. B. et al. (2007) Site-1 protease is essential for endochondral bone formation in mice *J Cell Biol* **179**, 687–700.
164. Brown, M. S., and Goldstein, J. L. (1974) Familial hypercholesterolemia: Defective binding of lipoproteins to cultured fibroblasts associated with impaired regulation of 3-hydroxy-3-methylglutaryl coenzyme A reductase activity *Proc Natl Acad Sci USA* **71**, 788–92.
165. Maxfield, F. R., and Tabas, I. (2005) Role of cholesterol and lipid organization in disease *Nature* **438**, 612–21.
166. Briel, M., Nordmann, A. J., and Bucher, H. C. (2005) Statin therapy for prevention and treatment of acute and chronic cardiovascular disease: Update on recent trials and meta-analyses *Curr Opin Lipidol* **16**, 601–5.
167. Brown, M. S., and Goldstein, J. L. (2006) Biomedicine. Lowering LDL – not only how low, but how long? *Science* **311**, 1721–3.

168. Tall, A. R. (2006) Protease variants, LDL, and coronary heart disease *New Engl J Med* **354**, 1310–12.
169. Seidah, N. G., Benjannet, S., Wickham, L., Marcinkiewicz, J., Jasmin, S. B., Stifani, S. et al. (2003) The secretory proprotein convertase neural apoptosis-regulated convertase 1 (NARC-1): Liver regeneration and neuronal differentiation *Proc Natl Acad Sci USA* **100**, 928–33.
170. Abifadel, M., Varret, M., Rabes, J. P., Allard, D., Ouguerram, K., Devillers, M. et al. (2003) Mutations in PCSK9 cause autosomal dominant hypercholesterolemia *Nat Genet* **34**, 154–6.
171. Seidah, N. G., and Prat, A. (2007) The proprotein convertases are potential targets in the treatment of dyslipidemia *J Mol Med* **85**, 685–96.
172. Cohen, J., Pertsemlidis, A., Kotowski, I. K., Graham, R., Garcia, C. K., and Hobbs, H. H. (2005) Low LDL cholesterol in individuals of African descent resulting from frequent nonsense mutations in PCSK9 *Nat Genet* **37**, 161–5.
173. Kotowski, I. K., Pertsemlidis, A., Luke, A., Cooper, R. S., Vega, G. L., Cohen, J. C. et al. (2006) A spectrum of PCSK9 alleles contributes to plasma levels of low-density lipoprotein cholesterol *Am J Hum Genet* **78**, 410–22.
174. Sirois, F., Gbeha, E., Sanni, A., Chrétien, M., Labuda, D., and Mbikay, M. (2008) Ethnic differences in the frequency of the cardioprotective C679X PCSK9 mutation in a West African population *Genet Test* **12**, 377–80.
175. Leren, T. P. (2004) Mutations in the PCSK9 gene in Norwegian subjects with autosomal dominant hypercholesterolemia *Clin Genet* **65**, 419–22.
176. Timms, K. M., Wagner, S., Samuels, M. E., Forbey, K., Goldfine, H., Jammulapati, S. et al. (2004) A mutation in PCSK9 causing autosomal-dominant hypercholesterolemia in a Utah pedigree *Hum Genet* **114**, 349–53.
177. Allard, D., Amsellem, S., Abifadel, M., Trillard, M., Devillers, M., Luc, G. et al. (2005) Novel mutations of the PCSK9 gene cause variable phenotype of autosomal dominant hypercholesterolemia *Hum Mutat* **26**, 497–506.
178. Naoumova, R. P., Tosi, I., Patel, D., Neuwirth, C., Horswell, S. D., Marais, A. D. et al. (2005) Severe hypercholesterolemia in four British families with the D374Y mutation in the PCSK9 gene: Long-term follow-up and treatment response *Arterioscler Thromb Vasc Biol* **25**, 2654–60.
179. Berge, K. E., Ose, L., and Leren, T. P. (2006) Missense mutations in the PCSK9 gene are associated with hypocholesterolemia and possibly increased response to statin therapy *Arterioscler Thromb Vasc Biol* **26**, 1094–100.
180. Zhao, Z., Tuakli-Wosornu, Y., Lagace, T. A., Kinch, L., Grishin, N. V., Horton, J. D. et al. (2006) Molecular characterization of loss-of-function mutations in PCSK9 and identification of a compound heterozygote *Am J Hum Genet* **79**, 514–23.
181. Hooper, A. J., Marais, A. D., Tanyanyiwa, D. M., and Burnett, J. R. (2007) The C679X mutation in PCSK9 is present and lowers blood cholesterol in a Southern African population *Atherosclerosis* **193**, 445–8.
182. Rashid, S., Curtis, D. E., Garuti, R., Anderson, N. N., Bashmakov, Y., Ho, Y. K. et al. (2005) Decreased plasma cholesterol and hypersensitivity to statins in mice lacking Pcsk9 *Proc Natl Acad Sci USA* **102**, 5374–9.
183. Zaid, A., Roubtsova, A., Essalmani, R., Marcinkiewicz, J., Chamberland, A., Hamelin, J. et al. (2008) Proprotein convertase subtilisin/kexin type 9 (PCSK9): Hepatocyte-specific low-density lipoprotein receptor degradation and critical role in mouse liver regeneration *Hepatology* **48**, 646–54.
184. Maxwell, K. N., and Breslow, J. L. (2004) Adenoviral-mediated expression of Pcsk9 in mice results in a low-density lipoprotein receptor knockout phenotype *Proc Natl Acad Sci USA* **101**, 7100–5.
185. Benjannet, S., Rhainds, D., Essalmani, R., Mayne, J., Wickham, L., Jin, W. et al. (2004) NARC-1/PCSK9 and its natural mutants: Zymogen cleavage and effects on the low density lipoprotein (LDL) receptor and LDL cholesterol *J Biol Chem* **279**, 48865–75.
186. Park, S. W., Moon, Y. A., and Horton, J. D. (2004) Post-transcriptional regulation of low density lipoprotein receptor protein by proprotein convertase subtilisin/kexin type 9a in mouse liver *J Biol Chem* **279**, 50630–8.
187. Maxwell, K. N., Fisher, E. A., and Breslow, J. L. (2005) Overexpression of PCSK9 accelerates the degradation of the LDLR in a post-endoplasmic reticulum compartment *Proc Natl Acad Sci USA* **102**, 2069–74.
188. Nassoury, N., Blasiole, D. A., Tebon, O. A., Benjannet, S., Hamelin, J., Poupon, V. et al. (2007) The cellular trafficking of the secretory proprotein convertase PCSK9 and its dependence on the LDLR *Traffic* **8**, 718–32.
189. Dubuc, G., Chamberland, A., Wassef, H., Davignon, J., Seidah, N. G., Bernier, L. et al.

(2004) Statins upregulate PCSK9, the gene encoding the proprotein convertase neural apoptosis-regulated convertase-1 implicated in familial hypercholesterolemia *Arterioscler Thromb Vasc Biol* **24**, 1454–9.

190. Attie, A. D., and Seidah, N. G. (2005) Dual regulation of the LDL receptor – some clarity and new questions *Cell Metab* **1**, 290–2.

191. McNutt, M. C., Lagace, T. A., and Horton, J. D. (2007) Catalytic activity is not required for secreted PCSK9 to reduce low density lipoprotein receptors in HepG2 cells *J Biol Chem* **282**, 20799–803.

192. Li, J., Tumanut, C., Gavigan, J. A., Huang, W. J., Hampton, E. N., Tumanut, R. et al. (2007) Secreted PCSK9 promotes LDL receptor degradation independently of proteolytic activity *Biochem J* **406**, 203–7.

193. Dubuc, G., Tremblay, M., Pare, G., Jacques, H., Hamelin, J., Benjannet, S. et al. (2010) A new method for measurement of total plasma PCSK9: Clinical applications *J Lipid Res* **51**, 140–9.

194. Cameron, J., Holla, O. L., Laerdahl, J. K., Kulseth, M. A., Ranheim, T., Rognes, T. et al. (2008) Characterization of novel mutations in the catalytic domain of the PCSK9 gene *J Intern Med* **263**, 420–31.

195. Mayne, J., Raymond, A., Chaplin, A., Cousins, M., Kaefer, N., Gyamera-Acheampong, C. et al. (2007) Plasma PCSK9 levels correlate with cholesterol in men but not in women *Biochem Biophys Res Commun* **361**, 451–6.

196. Lagace, T. A., Curtis, D. E., Garuti, R., McNutt, M. C., Park, S. W., Prather, H. B. et al. (2006) Secreted PCSK9 decreases the number of LDL receptors in hepatocytes and in livers of parabiotic mice *J Clin Invest* **116**, 2995–3005.

197. Cunningham, D., Danley, D. E., Geoghegan, K. F., Griffor, M. C., Hawkins, J. L., Subashi, T. A. et al. (2007) Structural and biophysical studies of PCSK9 and its mutants linked to familial hypercholesterolemia *Nat Struct Biol* **14**, 413–19.

198. Hampton, E. N., Knuth, M. W., Li, J., Harris, J. L., Lesley, S. A., and Spraggon, G. (2007) The self-inhibited structure of full-length PCSK9 at 1.9 A reveals structural homology with resistin within the C-terminal domain *Proc Natl Acad Sci USA* **104**, 14604–9.

199. Piper, D. E., Jackson, S., Liu, Q., Romanow, W. G., Shetterly, S., Thibault, S. T. et al. (2007) The crystal structure of PCSK9: A regulator of plasma LDL-cholesterol *Structure* **15**, 545–52.

200. Zhang, D. W., Lagace, T. A., Garuti, R., Zhao, Z., McDonald, M., Horton, J. D. et al. (2007) Binding of proprotein convertase subtilisin/kexin type 9 to epidermal growth factor-like repeat a of low density lipoprotein receptor decreases receptor recycling and increases degradation *J Biol Chem* **282**, 18602–12.

201. Kwon, H. J., Lagace, T. A., McNutt, M. C., Horton, J. D., and Deisenhofer, J. (2008) Molecular basis for LDL receptor recognition by PCSK9 *Proc Natl Acad Sci USA* **105**, 1820–5.

202. Dewpura, T., Raymond, A., Hamelin, J., Seidah, N. G., Mbikay, M., Chrétien, M. et al. (2008) PCSK9 is phosphorylated by a Golgi casein kinase-like kinase *ex vivo* and circulates as a phosphoprotein in humans *FEBS J* **275**, 3480–93.

203. Poirier, S., Mayer, G., Benjannet, S., Bergeron, E., Marcinkiewicz, J., Nassoury, N. et al. (2008) The proprotein convertase PCSK9 induces the degradation of low density lipoprotein receptor (LDLR) and its closest family members VLDLR and ApoER2 *J Biol Chem* **283**, 2363–72.

204. Labonte, P., Begley, S., Guevin, C., Asselin, M. C., Nassoury, N., Mayer, G. et al. (2009) PCSK9 impedes hepatitis C virus infection in vitro and modulates liver CD81 expression *Hepatology* **50**, 17–24.

205. Cameron, J., Holla, O. L., Ranheim, T., Kulseth, M. A., Berge, K. E., and Leren, T. P. (2006) Effect of mutations in the PCSK9 gene on the cell surface LDL receptors *Hum Mol Genet* **15**, 1551–8.

206. Conesa, M., Prat, A., Mort, J. S., Marvaldi, J., Lissitzky, J. C., and Seidah, N. G. (2003) Down-regulation of alphav/beta3 integrin via misrouting to lysosomes by overexpression of a beta3Lamp1 fusion protein *Biochem J* **370**, 703–11.

207. Zhang, D. W., Garuti, R., Tang, W. J., Cohen, J. C., and Hobbs, H. H. (2008) Structural requirements for PCSK9-mediated degradation of the low-density lipoprotein receptor *Proc Natl Acad Sci USA* **105**, 13045–50.

208. Qian, Y. W., Schmidt, R. J., Zhang, Y., Chu, S., Lin, A., Wang, H. et al. (2007) Secreted PCSK9 downregulates low density lipoprotein receptor through receptor-mediated endocytosis *J Lipid Res* **48**, 1488–98.

209. Chan, J. C., Piper, D. E., Cao, Q., Liu, D., King, C., Wang, W. et al. (2009) A proprotein convertase subtilisin/kexin type 9 neutralizing antibody reduces serum cholesterol

in mice and nonhuman primates *Proc Natl Acad Sci USA* **106**, 9820–5.

210. Poirier, S., Mayer, G., Poupon, V., McPherson, P. S., Desjardins, R., Ly, K. et al. (2009) Dissection of the endogenous cellular pathways of PCSK9-induced LDLR degradation: Evidence for an intracellular route *J Biol Chem* **284**, 28856–64.
211. Poupon, V., Girard, M., Legendre-Guillemin, V., Thomas, S., Bourbonniere, L., Philie, J. et al. (2008) Clathrin light chains function in mannose phosphate receptor trafficking via regulation of actin assembly *Proc Natl Acad Sci USA* **105**, 168–73.
212. Mayer, G., Poirier, S., and Seidah, N. G. (2008) Annexin A2 is a C-terminal PCSK9-binding protein that regulates endogenous low density lipoprotein receptor levels *J Biol Chem* **283**, 31791–801.
213. Cesarman, G. M., Guevara, C. A., and Hajjar, K. A. (1994) An endothelial cell receptor for plasminogen/tissue plasminogen activator (t-PA) II. Annexin II-mediated enhancement of t-PA-dependent plasminogen activation *J Biol Chem* **269**, 21198–203.
214. Ma, A. S., Bell, D. J., Mittal, A. A., and Harrison, H. H. (1994) Immunocytochemical detection of extracellular annexin II in cultured human skin keratinocytes and isolation of annexin II isoforms enriched in the extracellular pool *J Cell Sci* **107**(Pt 7), 1973–84.
215. Chung, C. Y., and Erickson, H. P. (1994) Cell surface annexin II is a high affinity receptor for the alternatively spliced segment of tenascin-C *J Cell Biol* **126**, 539–48.
216. Patchell, B. J., Wojcik, K. R., Yang, T. L., White, S. R., and Dorscheid, D. R. (2007) Glycosylation and annexin II cell surface translocation mediate airway epithelial wound repair *Am J Physiol Lung Cell Mol Physiol* **293**, L354–63.
217. Yeatman, T. J., Updyke, T. V., Kaetzel, M. A., Dedman, J. R., and Nicolson, G. L. (1993) Expression of annexins on the surfaces of non-metastatic and metastatic human and rodent tumor cells *Clin Exp Metastasis* **11**, 37–44.
218. Luo, Y., Warren, L., Xia, D., Jensen, H., Sand, T., Petras, S. et al. (2008) Function and distribution of circulating human PCSK9 expressed extrahepatically in transgenic mice *J Lipid Res* **50**, 1581–8.
219. Dietschy, J. M., Turley, S. D., and Spady, D. K. (1993) Role of liver in the maintenance of cholesterol and low density lipoprotein homeostasis in different animal species, including humans *J Lipid Res* **34**, 1637–59.
220. Alborn, W. E., Cao, G., Careskey, H. E., Qian, Y. W., Subramaniam, D. R., Davies, J. et al. (2007) Serum proprotein convertase subtilisin kexin type 9 is correlated directly with serum LDL cholesterol *Clin Chem* **53**, 1814–19.
221. Grefhorst, A., McNutt, M. C., Lagace, T. A., and Horton, J. D. (2008) Plasma PCSK9 preferentially reduces liver LDL receptors in mice *J Lipid Res* **49**, 1303–11.
222. Mbikay, M., Sirois, F., Mayne, J., Wang, G. S., Chen, A., Dewpura, T. et al. (2010) PCSK9-deficient mice exhibit impaired glucose tolerance and pancreatic islet abnormalities *FEBS Lett* **584**, 701–6.

Chapter 4

Insights from Bacterial Subtilases into the Mechanisms of Intramolecular Chaperone-Mediated Activation of Furin

Ujwal Shinde and Gary Thomas

Abstract

Prokaryotic subtilisins and eukaryotic proprotein convertases (PCs) are two homologous protease subfamilies that belong to the larger ubiquitous super-family called *subtilases*. Members of the subtilase super-family are produced as zymogens wherein their propeptide domains function as dedicated intramolecular chaperones (IMCs) that facilitate correct folding and regulate precise activation of their cognate catalytic domains. The molecular and cellular determinants that modulate IMC-dependent folding and activation of PCs are poorly understood. In this chapter we review what we have learned from the folding and activation of prokaryotic subtilisin, discuss how this has molded our understanding of furin maturation, and foray into the concept of pH sensors, which may represent a paradigm that PCs (and possibly other IMC-dependent eukaryotic proteins) follow for regulating their biological functions using the pH gradient in the secretory pathway.

Key words: Intramolecular chaperones, pH sensors, subtilases, proprotein convertases, protease activation and regulation, secretory pathway, histidine protonation.

1. Introduction

The limited proteolysis of an inactive precursor is a regulatory mechanism responsible for the generation of biologically active proteins and peptides (1). Proprotein convertases (PCs), which include seven mammalian Ca^{2+}-dependent endoproteases, furin, PC1/PC3, PC2, PC4, PACE4, PC5/PC6, and PC7/LPC/PC8, represent one such family that has been identified in all eukaryotes (2–6). Although they potentially share overlapping cleavage specificity and function, each PC has its own specific set of protein substrates that are generally cleaved at a pair of basic residues,

M. Mbikay, N.G. Seidah (eds.), *Proprotein Convertases*, Methods in Molecular Biology 768, DOI 10.1007/978-1-61779-204-5_4, © Springer Science+Business Media, LLC 2011

such as a Lys-Arg (7, 8). More recently, SKI/S1P (9, 10) and NARC-1/PCSK9 (8), which share sequence similarity with PCs, have also been identified as enzymes using this regulatory mechanism (11). While SKI/S1P is a protease, NARC-1/PCSK9 does not appear to display proteolytic activity toward other substrates other that its own IMC domain and functions as a chaperone that binds to LDL receptor and targets it for lysosomal degradation (12). Together, these nine proteases belong to the subtilisin-like super-family called *subtilases* (13).

Subtilases constitute the largest family of serine proteases after chymotrypsin (14) and contain several divergent proteases found in prokaryotes, archaea, eukaryotes, and viruses. Until the determination of the sequence of bacterial subtilisin (15), it was believed that all serine-type peptidases would be homologous to chymotrypsin. The subsequent X-ray structure established that subtilisin is clearly different and unrelated to chymotrypsin (16) and now corresponds to the family S8 according to the MEROPS database (14). The subtilase family is divided into two subfamilies, with prokaryotic subtilisins the archetype for subfamily S8A and eukaryotic kexin (3–6) the archetype for subfamily S8B. Because of the prokaryotic subtilisins' broad specificity, their ability to hydrolyze both native and denatured proteins, their catalytic activity under alkaline conditions, and their remarkable stability, they are widely use in detergents, cosmetics, food processing, skin care ointments, and contact lens cleaners and for research purposes in synthetic organic chemistry (17). Such commercial importance provided the momentum to gather extensive biophysical, biochemical, and structural information and has made prokaryotic subtilisins the prototype model for the subtilase super-family. Until 2003, the only structural information on PCs was gleaned through homology models derived using high-resolution crystallographic data of prokaryotic subtilisins as templates (18, 19). The recent high-resolution X-ray structures of furin (20) and kexin (21) have transformed our understanding of the basis of remarkable specificity displayed by eukaryotic PCs (22) when compared with their promiscuous prokaryotic counterparts. Simultaneously, they may potentially provide us with the means to better understand the structural and functional evolution of subtilases within a cellular context.

Furin, which is a constitutively expressed protease and the most intensively studied member of the PC family, can catalyze proteolytic maturation of a diverse repertoire of proprotein substrates within the cellular secretory pathway (2). Since most enzymes are exquisitely pH sensitive, the pH of each secretory and endocytic pathway compartment critically determines and regulates coordinated biochemical reactions (23). These compartments within eukaryotic cells therefore serve to segregate specific biosynthetic and catalytic functions within membrane-limited organelles. Such compartmentalization likely evolved from the necessity to optimize performance of individual metabolic pathways by providing unique environmental conditions and to enable energy storage in the form of electrochemical gradients across the dielectric membrane (24). PCs and their substrates are synthesized in the lumen of endoplasmic reticulum (ER), wherein they undergo correct folding and often have to traverse the changing pH of the secretory pathway compartments together en route to their final destination (1, 11, 25, 26). Since premature protease activity can lead to inappropriate protein activation, sorting, or degradation, PCs and many of their substrates are synthesized as inactive zymogens (27). Upon reaching their correct cellular compartments these zymogens undergo activation usually through proteolysis. The synthesis of proteases as zymogens enables cells with the means to spatially and temporally regulate the catalytic activities of PCs. However, the molecular and cellular determinants that modulate activation of PCs are poorly understood. In this chapter we review what we have learned from the folding and activation of prokaryotic subtilisin, discuss how this has molded our understanding of furin maturation (25, 28), and foray into the concept of pH sensors (26), which may represent a paradigm that PCs (and possibly other propeptide-dependent eukaryotic proteins) follow for regulating their biological functions using the pH gradient in the secretory pathway.

2. Propeptide-Mediated Folding of Bacterial Subtilisin

Bacterial subtilisins constitute a large class of microbial serine proteases, among which subtilisin E (*Bacillus subtilis*), subtilisin BPN (*Bacillus amyloliquefaciens*), and subtilisin Carlsberg (*Bacillus licheniformis*) are the most extensively studied (29). The timely cloning of the genes and their ease of expression, purification, and crystallization and subsequent high-resolution X-ray studies have made subtilisin a model system for protein engineering studies (17). An analysis of the cDNA for the subtilisin E gene suggests that subtilisins are synthesized as zymogens, with an approximately 77-residue propeptide that is located between the signal sequence and the protease domain (30). Propeptide deletion results in robust expression of inactive subtilisin E. Furthermore, if the catalytic domains of any of these bacterial subtilisins are denatured using chaotropes, the unfolded proteins fail to refold into their catalytically active native states even when these chaotropes are removed (31). Further studies established that the addition of the 77-residue propeptide to the folding reaction results in a robust recovery of catalytic activity (32). This establishes that the inability of the catalytic domain to spontaneously refold is because the propeptide is essential for folding the catalytic domain to its native state. Subsequently, the propeptide is removed by two distinct autoproteolytic cleavages, each with a different pH optimum (33, 34), which results in the maturation of the zymogen into enzymatically active subtilisin (35, 36). Additionally, subtilisin propeptides are also effective inhibitors of the cognate catalytic domains and may hence additionally function downstream of the folding process as regulators of enzymatic activity (35).

3. The Concept of Intramolecular Chaperones

Propeptide-mediated folding mechanisms have since been demonstrated to exist in various unrelated proteases, suggesting that such folding pathways may have evolved through convergent evolution (37, 38). Consequently, propeptides are termed as intramolecular chaperones (IMCs) (38) to differentiate them from molecular chaperones (MCs) (39–41). IMCs differ from MCs in a number of ways. Unlike MCs, which fold diverse substrates into thermodynamically stable states in an energy-dependent manner, IMCs are highly substrate specific and can mediate folding in an energy-independent manner. Upon completion of folding, the IMCs are proteolytically degraded, which effectively destroys part of the folding information. This makes the process irreversible by forcing the IMC to function as a as single-turnover catalyst (42). Hence IMCs appear to facilitate folding in an inefficient manner when compared with MCs, which are true multi-turnover catalysts (29).

3.1. Examples of IMC-Mediated Folding

Although initially discovered in bacterial proteases (43), subsequent work establishes the rather ubiquitous existence of IMC-dependent folding in a variety of proteins that include both proteases and non-proteases from prokaryotes, eukaryotes, archaea, and viruses. A few of the classic and emerging examples include the following:

(1) α-Lytic protease is a chymotrypsin-like serine protease that is secreted by the gram-negative soil bacterium *Lysobacter enzymogenes* and serves to lyse and degrade microorganisms. α-Lytic protease is secreted with a 166-residue propeptide and a 33-residue signal sequence. The 198-residue protease belongs to the same family as the mammalian digestive serine proteases, trypsin and chymotrypsin (44). Several studies clearly establish that the propeptide functions as both a chaperone and an

inhibitor of the protease and ensures its folding to an active, secretion-competent, stable conformation. Interestingly, the eukaryotic homologues trypsin and chymotrypsin that display low sequence identity, but adopt similar three-dimensional scaffolds, can fold independent of an IMC domain (45, 46) and have provided valuable insights into understanding the overall mechanism of IMC-dependent folding.

(2) Carboxypeptidase Y (CPY) from *Saccharomyces cerevisiae* is a serine carboxypeptidase that is used extensively as a marker for protein transport and vacuolar sorting in yeasts. This protease, which is synthesized as a pre-pro-protease with a 91-residue propeptide, folds and cleaves its IMC in the endoplasmic reticulum (ER), resulting in an inhibited complex (47). Upon its translocation to the yeast vacuole the enzyme is activated *in trans* by another serine protease, Proteinase A (48, 49). Guanidine hydrochloride denatured pro-CPY can be rapidly and efficiently refolded by dilution into a suitable buffer. Under identical conditions, mature CPY fails to refold to an enzymatically active form and suggests that the propeptide is required for correct folding of the mature protein (50). Folding of mature CPY in the absence of the propeptide results in the formation of a molten globule-like intermediate state, similar to that observed in the case of α-lytic protease and SbtE (51).

(3) Proteinase A (PrA) from *S. cerevisiae* is a vacuolar aspartic endo-proteinase (329 residues) that is vital for sporulation and viability during nitrogen starvation. The protease is secreted with a 54-residue propeptide that is proteolytically removed in the vacuole. Although the pro-PrA has been difficult to purify, studies suggest that the propeptide directly assists protease folding (52).

(4) Procathepsin-L, a member of the large family of cysteine proteinases, was the first example of propeptide-assisted folding in this family of enzymes (53). Studies demonstrate that loss of protease activity is directly proportional to truncations within the propeptide domain, and that the complete cognate propeptide is required for correctly folded cathepsin-L (54). Subsequently, it was established that cathepsin-S and cathepsin-B also required the presence of their cognate propeptides for productive folding (55). Folding of cathepsin-S under varying conditions of pH, time, redox state, and ionic strength did not compensate for the loss of propeptide (54). Similar to the subtilisin family (42), studies on cathepsins demonstrate that mutations in the propeptide directly affect protease function. This is highlighted by the hereditary disease pycnodysostosis,

caused by "loss-of-function" mutations in the cathepsin-K gene, one of which is directly localized in the propeptide domain (56).

(5) Endosialidases (endoNs) are tailspike proteins of bacteriophages that bind and specifically degrade the α-2,8-linked polysialic acid (polySia) capsules of their hosts. Recently, an IMC was identified in tailspike proteins of evolutionarily distant viruses, which require a C-terminal chaperone for correct folding. The structure of catalytic domain of coliphage K1F endoN reveals a functional trimer whose folding is mediated by a C-terminal IMC domain. Release of the IMC confers kinetic stability to the folded catalytic domain of endoNF (57). The recent crystal structures of the IMC domain in its pre-cleaved and cleaved isolated forms reveal that tentacle-like protrusions enfold the polypeptide chains of the precursor protein during the folding process. Upon completion of the assembly, correctly folded β-helices trigger a serine–lysine catalytic dyad to autoproteolytically release the mature protein. Interestingly, sequence analysis shows a conservation of the intramolecular chaperones in functionally unrelated proteins sharing β-helices as a common structural motif. Such conserved chaperone domains are interchangeable between pre-proteins and release themselves after protein folding (58).

(6) Elastase, an important virulence factor in the opportunistic pathogen *Pseudomonas aeruginosa*, is a thermolysin-like neutral zinc metalloprotease (TNP) that is synthesized with an amino-terminal propeptide (174 residues). Elastase was the first TNP family member that was demonstrated to require its propeptide for both folding and secretion (59). Subsequent studies on other TNPs such as thermolysin, the prototype of this family of proteases, and pro-aminopeptidase processing protease (PA protease), demonstrated that their N-terminal propeptides function as IMCs as well (60). Analysis of propeptides of TNPs demonstrates the presence of two conserved regions within the propeptide domains that may be critical for function. Mutations within the two conserved regions directly affect the chaperone function (59). Interestingly, the propeptide of vibriolysin, another TNP, has been shown to chaperone the folding of PA protease even though they share only 36% sequence identity.

(7) Several growth factors and neuropeptides such as transforming growth factor-β1, activin A (61), nerve growth factor, and amphiregulin (62, 63); hormones such as insulin (64); certain glycoproteins like von Willebrand factor (65);

and bacterial pancreatic trypsin inhibitor (BPTI) (66) also depend on propeptides for folding assistance. The above examples effectively highlight the wide scaffolds that fold using the assistance of IMC domains. Furthermore, apart from IMCs that directly catalyze the folding process, there exist other propeptides that can indirectly assist in folding. For example, the propeptide of barnase interacts with the molecular chaperone GroEL and thus ensures productive folding of barnase (61). Further, the propeptides in matrix metalloproteases (MMPs) contain a conserved cysteine residue where the sulfhydryl group is coordinated by the catalytic Zn^{2+} ion, thus maintaining these proteases in a catalytically inactive state. Proteolytic cleavage within the propeptide triggers a conformational change and releases shielding of the catalytic cleft in MMPs by interrupting the coordination between Zn^{2+} ions and cysteine residue (67). Thus, based on their roles in protein folding, propeptides have been grouped into two major classes, Class I and II (29, 43). Class I propeptides directly catalyze the folding, while Class II propeptides function in oligomerization, protein transport, localization, etc., and are indirectly involved in folding. The IMC of SbtE is a stereotypical Class I propeptide as it directly functions to catalyze the folding process (29).

To date, IMCs have been identified in all four major classes of proteases: serine, cysteine, aspartyl, and metalloproteases (37, 43). It is important to note that in the above proteases the IMC domains also function as potent protease inhibitors of their cognate catalytic domains and have to undergo activation to produce a catalytically active enzyme. While the bacterial proteases are mostly secreted extracellularly, the eukaryotic proteases undergo activation mostly in subcellular compartments of extreme pH (38, 68). *Why do specific protease sequences require IMCs to fold to their native states? What are the structural and functional determinants that may have driven specific protein families to choose between IMC- and MC-dependent folding pathways? Do IMCs employ a common mechanism to assist the folding of these varied scaffolds? What are the functional implications of the inhibitory functions of IMCs within proteases?* Answering these questions will have enormous implications for the fundamental understanding of protein folding in general and regulation of cellular proteases in particular and will also facilitate the rational design of protein-specific chaperones.

3.2. Similarities and Differences in IMCs

Concurrent studies on subtilases, α-lytic protease, and carboxypeptidase established that productive folding mediated by the IMC is a kinetically driven process (69). This conservation of function across unrelated protease families suggests that IMCs have evolved through convergent multiple parallel pathways and may share a common mechanism of action (37). Since IMC mechanisms appear to have evolved through convergent evolution, significant sequence similarity between IMC domains of non-homologous protein families would not be expected. Interestingly, even among homologous families, sequence similarity between IMC domains is significantly lower than that observed between the cognate catalytic domains (38). Nevertheless, sequence analyses of the subtilisin IMCs highlight some unique characteristics that may be critical for function. Alignment of known IMC sequences from subtilases helped identify two small hydrophobic motifs, N1 and N2, that appear to be conserved within such propeptides (70). Interestingly, when one compares the IMC sequence from subtilisin E with aqualysin, POIA1 (71), or with a designed peptide chaperone ProD which was computationally forced to diverge from the IMC of subtilisin, a high degree of sequence conservation appears isolated within motifs N1 and N2 (72). Random mutagenesis using error-prone PCR along with an activity-based genetic screening technique demonstrated that substitutions within motifs N1 and N2 were often deleterious (73, 74). NMR spectroscopy showed that while the subtilisin E IMC domain is largely unstructured, motifs N1 and N2 can display conformational rigidity (75). Together, these experiments suggest that the individual motifs may be critical for nucleating folding, while the non-conserved segments between these motifs may be responsible for functional specificity toward their cognate catalytic domains.

Even though the percentages are restricted to a few homologues from subtilisin, α-lytic protease, and cathepsin family of proteases, the trend of lower sequence conservation among IMC domains compared to their cognate catalytic domains generally holds true for other IMC-dependent systems as well (29). When one compares the percent of charges among the two domains within individual proteins, IMCs also contain more charged amino acids when compared to their respective catalytic domains (38). For example, while 12% of residues in the mature domain of subtilisin E are charged, the IMC sequence has 36% charged residues. Also, the charge on the IMC of SbtE directly complements a pocket around the substrate-binding site (76). Similar trends in asymmetric charge distribution are observed in most IMC-dependent proteins, including α-lytic protease (mature 10%; IMC contains 22%), carboxypeptidase Y (mature 20%; IMC contains 30%), and proteinase A (mature 20%; IMC contains 31%). Establishing why IMCs have evolved to be charged and if the charge on IMCs was selected with their chaperone function would give further insights into the nature of kinetic barriers on folding pathways.

Since structures evolve slower than sequences, proteins with highly divergent sequences may adopt similar structural folds. To explore this possibility, the sequence–structure relation within subtilases was analyzed (72). It is easier to reconcile the differences in the extent of sequence conservation among the IMCs and their cognate protease domains in different protein families. Since protease domains are catalytic units that facilitate similar chemical reactions, they require similar structural organization and precise spatial orientation of key residues. In this regard, the catalytic region of an enzyme represents a small portion of the entire protein. How variable IMC domains can mediate folding of structurally conserved catalytic domains is, however, more difficult to reconcile – especially since the IMCs are folding catalysts that help to attain structurally similar native states. It is possible that certain structural folds are necessary to bestow a specific function to a domain in spite of sequence variation. Since propeptides may impart structural information to their catalytic domains, propeptides within one family could adopt similar structural folds, despite digressions in their primary sequences (42).

3.3. IMCs as Potent Catalytic Inhibitors

Since the catalysis of folding requires the IMC (the catalyst) to interact with its cognate catalytic domain (the reactant), the nature of these interactions was investigated in subtilisin using isolated wild-type and mutant IMC domains (33, 73, 77). Studies indicate that the *entire* IMC functions as a *slow-binding competitive inhibitor* of subtilisin (42). In general, slow-binding inhibition is evident when initially weakly associated enzyme–substrate complexes undergo conformational changes that enhance affinity within these complexes. In case of the subtilisin, this involves the transition of the isolated IMC domain from an intrinsically unstructured state to a well-defined α–β conformation upon forming a stoichiometric complex with the catalytic domain (78). Since incorrect spatial and temporal proteolysis within a cell may be lethal, these inhibitory properties were hypothesized to be a mechanism that prevents premature protease activation. In general, cells appear to have evolved two distinct mechanisms to control activity of proteases. The first involves co-evolution of specific endogenous inhibitors, typically within compartments spatially distinct from those containing active enzymes. The second involves proteases being synthesized as inactive or less active precursors, which become activated by limited intra- or intermolecular proteolysis cleaving off a small peptide. Interestingly, these two different protease regulatory mechanisms appear combined in the chaperoning capabilities of IMC-dependent systems. Not surprisingly, the chaperoning ability and inhibitory functions of IMC domains correlate well, and IMC variants with diminished binding affinity are often weaker chaperones (73, 79, 80). However, this correlation is not always true, as the propeptide of aqualysin (a closely related thermostable homologue of subtilisin)

(81) and a computationally designed synthetic peptide are potent slow-binding inhibitors (72), yet much weaker chaperones of subtilisin. An examination of other convergently evolved IMC-dependent protease systems establishes that the slow-binding inhibition function of IMC domains appears to be a common theme within proteinase A, α-lytic protease, carboxypeptidase Y, elastase, and the cathepsin family. The important role of this slow-binding inhibition and the convergence of these two distinct protease control mechanisms will both be addressed later.

4. Understanding IMC-Dependent Folding and Maturation of Bacterial Subtilisin

Over the past three decades, numerous genetic, biochemical, and structural analyses of bacterial subtilisins along with complementary data from other proteases have provided insights into the mechanism of IMC-mediated protein folding. Structural data on the bacterial subtilisins offer interesting snapshots into the gradual transition of the polypeptide from an unfolded state, through an inhibited complex, to an active protease. Complementary biophysical and biochemical studies have helped to elaborate reasons for non-productive folding of the isolated protease domains and to elucidate a general mechanism for how IMCs may function in this pathway.

4.1. The Structure of SbtE and Pro-SbtE Complex

Several high-resolution crystal structures of both SbtE and SbtE in complex with its IMC domain have been solved (76, 82, 83). The structure of SbtE (**Fig. 4.1a**) is comprised of three β-sheets and nine α-helices, with Asp_{32}-His_{64}-Ser_{221} forming the catalytic triad. The largest β-sheet is comprised of seven parallel β-strands and is flanked on one side by three helices and on the other by two. The substrate-binding site (**Fig. 4.1b**) is a surface channel that accommodates six residues (P4–P2′). In SbtE, the substrate-binding pocket is large, hydrophobic, and made of main-chain residues from Ser_{125}-Leu_{126}-Gly_{127} and main- and side-chain residues of Ala_{152}-Ala_{153}-Gly_{154} and Gly_{166}. The Gly_{166} is at the bottom of the pocket for P1 and is critical for specificity. The P1–P4 substrate backbone forms the central β-strand in an antiparallel β-sheet with the protease residues 100–102 and 125–127. Further, SbtE displays two calcium-binding sites (**Fig. 4.1a**), a high-affinity site that is well conserved (A-site) in most subtilases and a weak affinity site that is less conserved (B-site) (13). Calcium at A-site is coordinated in pentagonal-bipyramidal geometry by the loop comprised of residues 75–83, Gln_2 at the N-terminus, and an Asp at position 41. The seven coordination distances range from 2.3 to 2.6 Å with the Asp being the closest (**Fig. 4.1c**). The

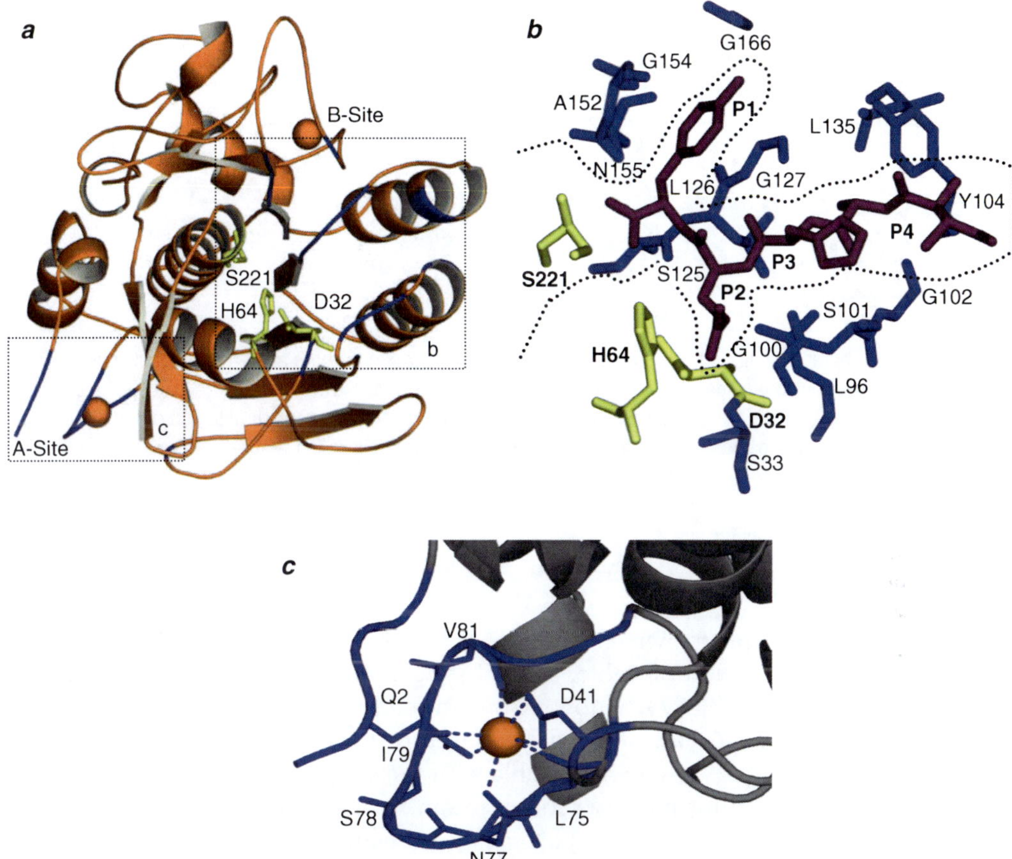

Fig. 4.1. Structural organization of SbtE. (**a**) Structure of SbtE (ISCJ) is depicted together with the substrate-binding site (**b**) and the calcium (*orange*)-binding A-site (**c**). SbtE also has a second calcium-binding site (B-site) of medium affinity. The catalytic residues are highlighted in *yellow*. (**b**) The substrate-binding site is highlighted with an inhibitor (*magenta*) bound in the S1–S4 pocket. Residues lining the substrate-binding pocket are highlighted in *blue*. **(c)** Calcium binding at the A-site is coordinated by residues from a loop comprised of residues 75–83, an N-terminal Gln, and an Asp.

second calcium makes contact with the main-chain carbonyl oxygen atoms of residues 169, 171, and 174 in a shallow crevice near the surface of the molecule and is coordinated in a distorted pentagonal bipyramid (84). These two calcium-binding sites together make SbtE an extremely stable protease in the absence of any cysteines or stabilizing disulfides in its structure (85). The lack of cysteine residues in SbtE is advantageous because it allows the probing of specific interactions during the folding process. This approach has been used to identify a precise non-native interaction by engineering two cysteine residues, which are distal in the native protein but are proximal during folding, and form a specific intramolecular molecular disulfide bond under oxidative folding conditions (86).

Engineering a $S_{221}C$ substitution at the active site of Pro-SbtE blocks the maturation process subsequent to the first

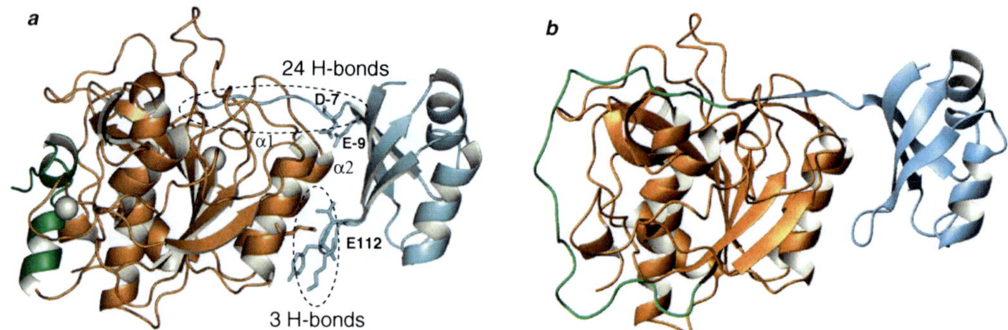

Fig. 4.2. Structural characterization of Pro-SbtE complex. (**a**) Crystal structure of the autoprocessed, inhibited complex (1SCJ). The IMC (*blue*) docks against two helices (α1 and α2) in the protease (*orange*) and occludes the substrate-binding site. The binding interface is stabilized by 27 hydrogen bonds in a asymmetric distribution, of which 3 are contributed by E_{112} in protease. D_{-7} and E_{-9} form helix caps for the two helices (α1 and α2). The calcium ions (*white*) and the N-terminal helix (*green*) are also highlighted. (**b**) Modeled structure of the propeptide–protease complex prior to cleavage. The N-terminus (*green*) of the protease (*orange*) is bound to the IMC (*blue*) at the active site and hence the calcium-binding A-site is not fully formed.

proteolytic cleavage of the IMC. This variant facilitates the isolation of stable, cleaved stoichiometric Pro:S_{221}C–SbtE complex (77, 87), whose X-ray structure (**Fig. 4.2a**) has been solved (76, 82, 88). The structure of the mature domain in the complex is superimposable with the structure of the isolated protease domain with a root mean square deviation of 0.46 Å, when the Cα atoms were compared. While the isolated IMC is largely unstructured, the IMC in complex with the protease folds into a single domain with a four-stranded anti-parallel β-sheet and two three-turn helices, forming an α+β plait. The structured inhibitory IMC packs against two surface helices (α1 and α2; **Fig. 4.2a**) of the protease domain formed by residues Tyr_{104}-Asn_{117} and Ser_{132}-Ser_{145}. Further, residues –1 to –7 from the IMC domain directly interact with the substrate-binding region to complete a three-stranded β-sheet with β-strands from the protease domain. In all, there are 27 hydrogen bonds (**Fig. 4.2a**) at the IMC–subtilisin interface. Interestingly, 24 hydrogen bonds stabilize the interaction of residues –1 to –9 (*Note*: the nine C-terminal residues of the subtilisin IMC; the cleavage site locates between residues –1 and 1) with the active site and with the substrate-binding regions. These include the three hydrogen bonds that stabilize the backbone amide groups of $Glu_{(-9)}$ and $Asp_{(-7)}$ that form helix caps for the two SbtE interaction helices (α1 and α2). In contrast, the remaining 68 residues of the propeptide are stabilized by only three hydrogen bonds, between the backbone amides of residues –34 to –36 and the carboxylate group of the Glu_{112} from the protease domain. The significance of this asymmetric distribution of hydrogen bonds at the IMC–protease interface is unknown. Another noteworthy point is that while the average B-factor of the side-chain and the main-chain atoms is 15.0 and 13.6 Å^2 in the protease, it is 35.5 and 33.8 Å^2, respectively, in the IMC (76). This approximately twofold increase in the B-factors within the IMC domain may be biologically significant and necessary for function.

An interesting insight from the structure of the Pro-S_{221} C-SbtE is that the N-terminus of the protease is more than 20 Å away from the active site. Earlier biochemical studies had clearly established that the processing of IMC was intramolecular (89). Based on this, the structure of Pro-SbtE (**Fig. 4.2b**) just prior to cleavage was modeled with the N-terminus of the protease bound to the C-terminus of the IMC at the active site. The lack of a preferred conformation in N-terminal residues (residues 1–6) of SbtE and the recovery of activity in a H_{64}A-SbtE active-site mutant with a $E_{(-2)}$H substitution in IMC, through substrate-assisted catalysis, substantiated the model (87). A recent crystal structure of an active-site mutant of a subtilisin homologue, pro-kumamolysin, shows the propeptide bound to the protease prior to cleavage and confirms this model proposed for the uncleaved precursor (90).

4.2. Mechanism of IMC-Mediated Structural Acquisition

4.2.1. How Do Polypeptides Enhance the Rates of Folding of Their Cognate Catalytic Domains?

4.2.1.1. Kinetic and Thermodynamic Characterization

To establish how IMCs function in the folding pathway, folding of α-lytic protease and SbtBPN was compared in the presence and absence of their IMC (45, 91). Folding of SbtBPN in the absence of its IMC resulted in formation of a non-functional, structured state that is stable for weeks. This intermediate displayed a hydrodynamic volume intermediate between that of the fully folded and fully unfolded protease. Circular dichroism spectra of the intermediate in the far-UV region (190–250 nm) corresponded to a well-defined secondary structure with a minimum at 208 and 222 nm. However, in the near-UV region (250–320 nm) the intermediate displayed no amplitude, suggesting a lack of well-formed tertiary packing. This was also confirmed through an NMR study that established a strongly reduced dispersion in the amide and methyl regions of the ^{1}H NMR spectrum compared with the fully folded protease. A noteworthy point, however, was that the intermediate appeared to bind calcium with a stoichiometry of 1, but with an affinity intermediate to affinities of A-site and B-site (91). Similar behavior was also observed with the intermediate state of α-lytic protease folded in the absence of its IMC (68). Although the intermediate state was extremely stable for weeks, addition of cognate IMCs yielded active native protease.

Hence, these studies suggested that in the absence of IMC, the protease folds to a kinetically trapped state with properties of a classical "molten globule" intermediate (92). A high-energy barrier between the molten globule intermediate and transition state limits the folding to a native state (**Fig. 4.3**). Addition of

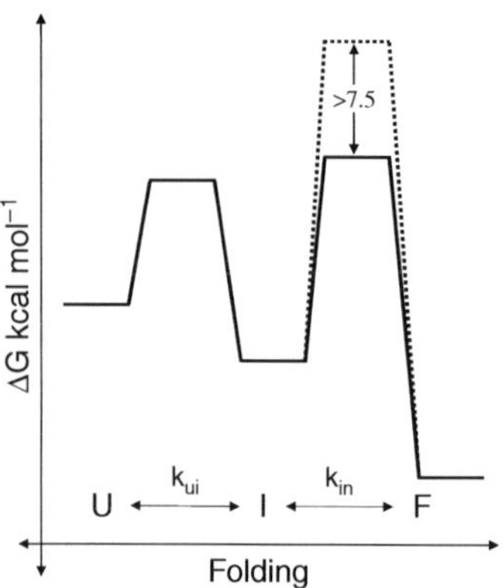

Fig. 4.3. IMCs lower kinetic barriers on the folding pathway. (**a**) Unfolded Pro-S_{221} C-SbtBPN (U) folds to the native state (F) through a stable molten globule intermediate (I). Kinetic studies establish that the high activation energy barrier to protease folding (*broken line*) is lowered by >7.5 kcal/mol in the presence of the IMC (*solid black line*).

the IMC lowers this barrier to enable productive folding (45, 91, 93, 94). Based on these observations, it was established that *the IMC functions to overcome a kinetic barrier on the folding pathway, and that the observed intermediate is either on-pathway or in equilibrium with a conformation on the folding pathway.*

To establish the relevance of the observed intermediate to the biological *in-cis* folding, refolding of full-length Pro-SbtE and Pro-SbtBPN was analyzed (93, 95). To avoid complications of proteolysis and to inhibit IMC processing, both studies were done using a $Ser_{221}Ala$ active-site variant that has lower proteolytic activity by six orders of magnitude. This variant represents the propeptide:protease complex just prior to cleavage as discussed earlier (**Fig. 4.2b**). The folded, but uncleaved, Pro-S_{221}A-SbtBPN binds calcium ions and adopts a compact conformation with an apparent molecular weight of 36 kDa. Equilibrium unfolding of fully folded Pro-S_{221}A-SbtBPN monitored through changes in circular dichroism, and fluorescence spectroscopy, followed a three-state unfolding curve. Most of the tertiary and a part of the secondary structure unfolded through an initial cooperative process while separated secondary structures followed a less cooperative second transition. Interestingly, the second unfolding transition was similar to the unfolding transition of the molten globule intermediate formed in the absence of IMC. Additionally, at higher denaturant concentrations Pro-S_{221}A-SbtBPN displayed properties similar to the trapped molten globule intermediate. These results, taken together, established that the equilibrium unfolding of the Pro-S_{221}A-SbtBPN occurs via an intermediate that is similar to the kinetically trapped molten globule intermediate. Further, three-state equilibrium-unfolding transition suggested that the polypeptide folds to a molten globule-like state and then transitions to the native state, through the assistance of the IMC. Thus it was argued that *the IMC functions only during the late stages of the folding pathway.* Another interesting finding was that the relative thermodynamic stability of the fully folded Pro-SbtBPN complex was only marginally higher than that of the SbtBPN intermediate (93, 95). The thermodynamic stability was also strongly dependent on salt concentrations, which was probably an effect of the highly charged propeptide and its effect on the electrostatic interactions of the solvent environment (37).

Most of the initial folding studies highlighted above employed slow dialysis for refolding and hence were not amenable to kinetic analysis (31). Subsequent optimization of a fast refolding that involves the rapid dilution of unfolded protein into denaturant free buffer paved the way for kinetic studies of IMC-mediated folding. Using the technique of refolding by rapid dilution, Eder and Fersht (93) established the kinetics of refolding of Pro-S_{221}A-SbtBPN by monitoring the increase in intrinsic fluorescence. Pro-S_{221}A-SbtBPN follows two-state kinetics with a rapid phase (65% amplitude as native) and slow phase (35% amplitude as native). While the rapid phase reflects the acquisition of structure in the intermediate, the slow phase established the kinetics of folding of the intermediate in the presence of the IMC. This slow phase followed a rate constant of 0.0047 s^{-1}. Initial studies had established a folding rate of $<1.4\times10^{-8}$ s^{-1} for SbtBPN refolded in the absence of IMC. Hence, taken together, this demonstrated that the IMC accelerates the kinetics of protease folding by at least five orders of magnitude. Based on equation [**1**],

$$\Delta\Delta G^{*} = -\mathrm{RT}\ln\left[\frac{k_{\mathrm{Pro-Sbt}}}{k_{\mathrm{Sbt}}}\right] \qquad [1]$$

the difference in transition state energies for folding with and without the IMC was estimated to be >7.5 kcal/mol (**Fig. 4.3**) (37). Similar studies with α-lytic protease established that the propeptide lowers the kinetic barrier to folding by 18 kcal/mol. With α-lytic protease, the relative thermodynamic stability of the native state relative to the unfolded and intermediate states was also established. Interestingly, these studies demonstrate that the native state is ~1 kcal/mol less stable than the unfolded state (45, 95). Hydrogen-exchange experiments establish that this state has extremely low conformational dynamics with >50% of the residues having a protection factor (P_f) $>10^4$ (46). Thus the protease appears to be in a kinetically trapped, thermodynamically unstable native state.

4.2.1.2. Calcium Deletion Variant

Both SbtE and SbtBPN have two well-defined calcium-binding sites, a high-affinity A-site, and a medium-affinity B-site (**Fig. 4.1a**). Studies establish that calcium binds to the A-site (**Fig. 4.1c**) with an affinity of $\sim 10^7$ M^{-1} and contributes significantly to the thermodynamic stability of the protease (84). What is particularly interesting is that the kinetic barrier to calcium dissociation is extremely high (~23 kcal/mol) and is higher than the binding free energy (96). This suggests that the binding of calcium and the accompanied lowering of conformational entropy might limit IMC-independent folding. To establish this, Bryan and coworkers created a variant ($SbtBPN_{\Delta cal}$) that had the calcium-binding loop (A-site: 75–83) removed. Upon refolding of $SbtBPN_{\Delta cal}$ under low ionic conditions, no activity was observed. However, folding of $SbtBPN_{\Delta cal}$ under high ionic conditions resulted in independent folding of $SbtBPN_{\Delta cal}$ to an active, native state (94). Folded $SbtBPN_{\Delta cal}$ is unstable and further mutations, including a disulfide bond, have to be introduced to enhance stability. However, the new variant has a structure similar to the wild-type protease except in the region of the A-site (97). This suggests that the calcium-binding A-site may be a critical factor that dictates the requirement for an IMC. However, it is interesting to note that productive folding is seen only under conditions of high ionic strength and that the calcium-independent protease variant is extremely unstable.

Thus, a high-energy barrier that separates the unfolded and native states limits the spontaneous folding of specific proteases. IMCs assist in the folding pathway by lowering these barriers to enable productive folding. Selection of such kinetic barriers on the folding pathway may provide a mechanism for evolution of optimal functional properties. Since most IMC-dependent proteases appear to function in harsh protease-rich conditions the presence of the IMC may enhance longevity through high unfolding energy barriers.

4.2.2. How Do IMCs Assist in Folding?

4.2.2.1. Stabilization of Folding Nucleus – "Side-On Model"

The 77-residue IMC domain is an intrinsically unstructured polypeptide that folds to an α–β conformation in the presence of the protease (75). Crystal structure of the inhibited Pro-S_{221} C-SbtE (**Fig. 4.2a**) demonstrated that the folded IMC interacts directly with two surface helices of the protease (76). Based on this, it was proposed that the α–β–α binding interface of the IMC may represent the folding nucleation motif of the protease, and that stabilization of this sub-structure upon binding of the IMC may help to induce folding (82). To establish the nature of the interaction between the IMC and protease, Bryan et al. analyzed the bimolecular folding of $SbtBPN_{\Delta cal}$ as given by equation [**2**],

$$\begin{aligned} \mathrm{Pro}^{\mathrm{U}} + \mathrm{SbtBPN}^{\mathrm{U}}_{\Delta\mathrm{cal}} \leftrightarrow \mathrm{Pro}^{\mathrm{F}} - \mathrm{SbtBPN}^{\mathrm{I}}_{\Delta\mathrm{cal}} \leftrightarrow \mathrm{Pro}^{\mathrm{F}} \\ -\mathrm{SbtBPN}^{\mathrm{F}}_{\Delta\mathrm{cal}} \rightarrow \mathrm{Pro}^{\mathrm{D}} + \mathrm{SbtBPN}^{\mathrm{F}}_{\Delta\mathrm{cal}} \end{aligned} \qquad [2]$$

where Pro^U and $SbtBPN^U_{\Delta cal}$ represent the unfolded propeptide and protease, respectively, $Pro^F - SbtBPN^I_{\Delta cal}$ represents the partially structured intermediate, $Pro^F - SbtBPN^F_{\Delta cal}$ represents the folded complex, Pro^D represents the degraded propeptide, and $SbtBPN^F_{\Delta cal}$ represents the free active protease. Reaction rates were determined using the differences in tryptophan content between the IMC and protease domains. $SbtBPN_{\Delta cal}$, with three tryptophan residues in its primary sequence, shows a 1.7-fold increase in intrinsic tryptophan fluorescence upon refolding. As the IMC has no tryptophan residues, any change in fluorescence reflects structural changes in the protease domain. Further, the binding of IMC to the protease increases the intrinsic fluorescence of the protease due to shielding of one of the tryptophan residues. By monitoring the rate of change in the intrinsic tryptophan fluorescence, the k_{on} and k_{off} for binding of IMC to $SbtBPN_{\Delta cal}$ and the rates of $SbtBPN_{\Delta cal}$ folding under increasing concentrations of the IMC were determined. These studies demonstrated that the formation of the initial complex ($Pro^F - SbtBPN^I_{\Delta cal}$) between IMC and $SbtBPN_{\Delta cal}$ was the rate-limiting step to folding. However, upon increasing residual structure in the isolated IMC domain the folding of $Pro^F - SbtBPN^I_{\Delta cal}$ to $Pro^F - SbtBPN^F_{\Delta cal}$ becomes rate limiting (79, 98). This established that structural content within the IMC might have a direct effect on its chaperoning function. Simultaneous studies that showed a direct correlation between inhibition constants of isolated IMC mutants and their chaperoning efficiency strengthened this conclusion (73, 99). Hence, it was proposed that the binding energy of the IMC contributes to stabilizing the α–β–α structure either by surmounting an entropic barrier through stabilization of native interactions or by overcoming an enthalpic barrier by breaking non-native interactions. While in the presence of an unstructured IMC, the binding energy may be diluted by its folding; the presence of a structured IMC ensures faster binding and folding.

Although this nucleation propagation mechanism of folding through a "side-on interaction" of the IMC with the protease seems possible, other studies, highlighted below, question this hypothesis. Moreover, the naturally occurring biological reaction is clearly unimolecular, and evolution of covalently linked protease domains may be to enhance efficiency and economy of IMC-mediated folding. Hence, while the bimolecular folding studies offer initial insights into propeptide-mediated folding, establishing similar principles in unimolecular folding is fundamental to the above hypothesis.

4.2.2.2. Stabilization of Folding Nucleus – "Top-On Model"

Studies based on unimolecular folding highlight a similar, but slightly varied, mechanism. Random mutagenesis helped identify a number of mutations in the IMC that affected the secretion of active protease. Second-site suppressor analysis for an $M_{(-60)}T$ mutation in the IMC identified a $S_{188}L$ substitution in SbtE that restored activity. Since Ser_{188} and Met_{-60} do not interact with each other and are ~47 Å apart in the folded Pro-SbtE complex, it was suggested that these residues may interact during the folding process. One possible way in which these two residues could interact is via a "top-on interaction" (**Fig. 4.4**). To test this hypothesis, cysteine residues were introduced at positions –60 and 188 and the folding of this double-cysteine variant ($M_{(-60)}C$-Pro-$S_{188}C$-SbtE) was analyzed under oxidizing and reducing conditions. Interestingly, folding under oxidizing conditions results in the formation of a cross-linked intermediate with stable secondary structure. A noteworthy point is that upon prolonged incubation with small peptide substrates such as AAPL-pNA and AAPF-pNA, the trapped intermediate displays catalytic activity. Addition of a reducing agent to the cross-linked intermediate triggers proteolysis of the IMC and results in a wild type-like native state. Hence, isolation of the stable cross-linked intermediate suggests that the IMC interacts with the protease in a "top-on orientation" during folding and, further, that this interaction results in productive folding (86).

Furthermore, the "side-on" interaction of the IMC with the protease is stabilized largely by three hydrogen bonds between E_{112} in the protease and IMC backbone amides (**Fig. 4.2a**). Disruption of these hydrogen bonds in the E_{112}A-SbtE variant does not affect IMC-mediated folding of the protease. However, the K_i of the IMC to E_{112}A-SbtE is lowered ~35-fold relative to that of SbtE. This suggests that the "side-on" interaction of the IMC is critical for inhibition and, further, that the inhibitory and

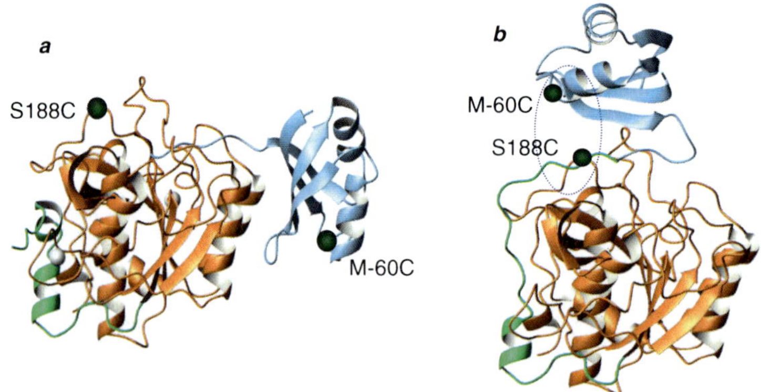

Fig. 4.4. IMC interacts in a "top-on" orientation. (**a**) M_{-60} in the IMC (*blue*) and S_{188} in the protease (*orange*), identified through second-site suppressor analysis, are 47 Å apart in the inhibited complex (1SCJ). (**b**) Cross-linking studies with a double-cysteine mutant helped isolate a stable cross-linked intermediate, based on which the "top-on" interaction between the propeptide (*blue*) and protease (*orange*) was modeled.

chaperone functions of the IMC may not be obligatorily linked (100). Additional evidence for the above comes from the fact that IMC of aqualysin is a 10-fold better inhibitor of SbtE, relative to its own IMC, but is unable to efficiently chaperone the folding of SbtE (81).

The above studies demonstrate that in unimolecular folding the "top-on" interaction may initiate folding while the "side-on" interaction is critical for inhibition. Thus the IMC may interact with the protease closer to the active-site region before it transits to the "site-on interaction" seen in the crystal structure. This movement of the IMC may be coincident with its cleavage by the protease.

4.2.2.3. Changes Coincident with IMC Cleavage

Subsequent to folding of the polypeptide to a structured state, the peptide bond between the IMC and protease domains is autoproteolyzed (89). However, the IMC remains bound to the protease as an inhibited complex. The solved crystal structure of the Pro-S_{221}C-SbtE (**Fig. 4.2a**) inhibited complex offers a structural snapshot of the polypeptide folding at the completion of autoprocessing (76).

To establish changes coincident with autoprocessing of the IMC, structural properties of the complex before and after autoprocessing were characterized (77) using ANS, a fluorescent dye that binds to exposed hydrophobic surfaces on structured proteins (101). ANS displayed a higher intensity and a shift toward a lower wavelength in the presence of the unautoprocessed complex (Pro-S_{221}A-SbtE) relative to processed complex (Pro-S_{221} C-SbtE). This suggested that there is a large decrease in exposed hydrophobic surface coincident with autoprocessing. Further, in the crystal structure of Pro-S_{221}C-SbtE, the N-terminus of the protease is at least 20 Å away from the active site. Based on this it was proposed that upon autoprocessing, the N-terminus of the protease folds back to form the N-terminal α-helix that contributes Gln_2 to calcium binding at the A-site (**Fig. 4.2**) (87). Recent studies establish that while the rates of folding and autoprocessing are independent of calcium, the stability of the autoprocessed complex and mature subtilisin shows a strong dependence on calcium. Further, the autoprocessed complex has much higher thermal, thermodynamic, and proteolytic stability relative to the unautoprocessed complex (85). This demonstrated that the A-site is indeed formed subsequent to cleavage, and that the IMC regulates its formation. The processing of the IMC triggers calcium binding and induces structural changes in the protease that serve to lock the protease in a more stable conformation. The release and degradation of the IMC from the stable inhibited complex are required for release of active protease (34). While this appears to be mediated through an autocatalytic activation, the precise mechanism of activation is unknown.

4.2.2.4. Alternative Approaches to SbtE Folding

While the above offer direct analyses of IMC-mediated folding, an alternative approach is to identify specific conditions that may allow folding of proteases in the absence of their IMCs. Studies demonstrated that folding low concentrations of the protease in 2M potassium acetate at pH 6.5, or refolding the protease from an acid-denatured state, allowed recovery of IMC-independent activity (102, 103). While the yield of active protease was initially low with potassium acetate, immobilization of the protease on a resin increased the folding efficiency (102). These experiments elucidate an interesting effect of electrostatics on the IMC-mediated folding process. In the case of the acid-denatured proteins, CD studies of the denatured protein established residual structure in the protease, even after denaturation. This further substantiates the hypothesis that stabilization of a folding nucleus may enable productive folding. Hence, the IMC appears to initiate folding by stabilizing a sub-structure within the protease. This may involve the interaction of the IMC with the protease in a "top-on" orientation (**Fig. 4.4**) (86). Upon acquisition of structure within the protease and formation of the catalytic triad, the IMC is cleaved in an autoprocessing reaction. This is coincident with the movement of the IMC to the "side-on" inhibitory orientation and the formation of a calcium-binding site (**Fig. 4.2**) (104). Release and degradation of the IMC from this inhibited complex releases an active native protease (**Fig. 4.5**).

4.2.3. Why Have Specific Proteins Evolved Dedicated IMCs to Mediate Their Folding?

Since IMCs function as single turnover catalysts, IMC-mediated folding pathways appear to be less efficient that MC-dependent multi-turnover pathways. *What is the functional advantage offered by IMC-dependent folding mechanisms?* One possibility is that the specific fold of the catalytic domain of subtilases mandates the requirement of IMC-dependent pathways. IMCs may hence serve as "bridges" to reach a specific conformational state. Since the principle of microscopic reversibility requires the mechanisms in the forward direction and reverse direction to be identical, and because protein folding is a reversible reaction, the presence of the folding catalyst – the covalently attached IMC – can also mediate unfolding under specific conditions. However, the proteolytic degradation of the IMC domain prevents microscopic reversibility once the native state is reached because the "bridge" between the unfolded and native state is destroyed, rendering the reverse pathway inaccessible. Such locked-in conformations may represent "high-energy," kinetically trapped native states that acquire structural information through their cognate IMCs. These are unlike the conventional thermodynamically stable, low-energy conformations found ubiquitously distributed in nature's conformational space. Since MCs do not

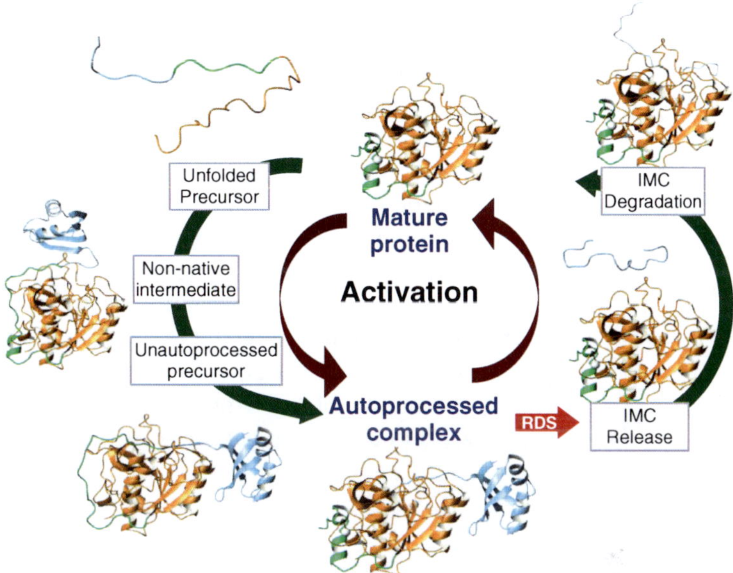

Fig. 4.5. IMC-mediated SbtE maturation – N-terminal helix of protease (*green*); calcium ion (*white*). Maturation of Pro-SbtE occurs in three stages: (1) A non-native top-on interaction between the protease (*orange*) and IMC (*blue*) punctuates the transition of the precursor from the unfolded state to a structured state (unautoprocessed precursor). (2) Once the active site is formed, the precursor autoproteolyzes to an inhibited, autoprocessed Pro:SbtE complex. (3) The release and degradation of the IMC from the complex release active protease that can subsequently trans-activate other proteases. Activation is the rate-limiting step to maturation.

impart structural information and only provide an appropriate environment for folding thermodynamically stable proteins, the kinetically trapped conformations may be inaccessible using MCs, hence mandating IMC-dependent folding pathways. Since the subtilase super-family contains several divergent IMC-dependent proteins, a detailed analysis of the sequence, structure, and function has the potential to provide insights into determinants that dictate various functions of the IMC domains.

The Nobel Laureate Christian B. Anfinsen established that all the information ribonuclease requires to fold into its native state resides within its amino acid sequence and that the native state of ribonuclease is independent of its folding pathway (105). Subsequently, experiments from several different laboratories directly supported the thermodynamic hypothesis by demonstrating that the folding/unfolding reactions of various small proteins are generally reversible and hold that the native conformations of proteins are at global free energy minima relative to all other states having identical bond chemistry. However, it is noteworthy that these evidences only argue that the native state is the lowest energy state within a conformational neighborhood which includes all kinetically accessible states (106). Hence the thermodynamic hypothesis is not falsifiable by experiment because conformations outside of this neighborhood cannot be accessed experimentally under normal conditions (69). However, if the final state of a system depends on the initial conditions, then the process is kinetically determined (**Fig. 4.6**). Evidence for the kinetic hypothesis emerged from conformational studies on plasminogen activator inhibitor 1 (PAI-1), a protease suicide inhibitor that is a member of the serpin family. Upon synthesis in vivo or after refolding following in vitro denaturation, PAI-1 initially adopts a kinetically stable state that functions as a potent protease inhibitor. Remarkably, this active form slowly converts to an inactive but thermodynamically more stable, latent form over a period of several hours (107). The latent form of PAI-1 can,

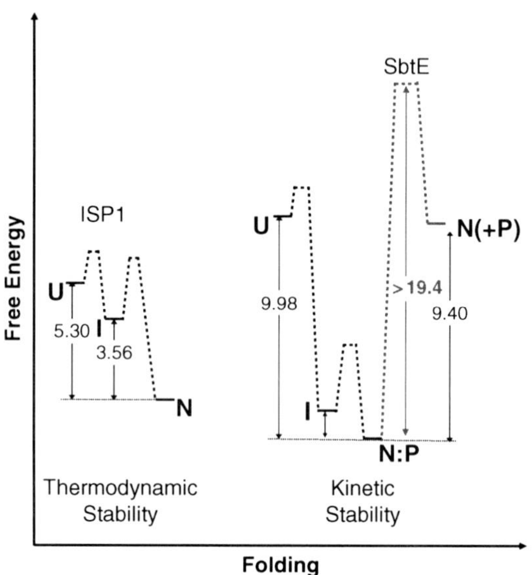

Fig. 4.6. The free energy diagram depicting thermodynamic stability of ISPs and kinetic stability of ESPs – Unfolded ISP1 (U) spontaneously adopts its thermodynamically stable native state (N) through a partially folded intermediate (I). The free energy difference between N and U is approximately 5.3 kcal/mol in the case of ISP1 and is lowest compared to all experimentally observed states (111). In case of ESPs, the unfolded IMC-subtilisin (U) undergoes rapid folding and autoproteolysis to give a thermodynamically stable IMC:SbtE complex (N:P) through an intermediate state (I) (77). The structure of the IMC:SbtE complex has been solved using X-ray crystallography (76). The activation energy for the spontaneous release of the IMC from this complex is energetically unfavorable (approximately 21 kcal/mol and is shown in a *broken light line*). The release of the first free protease molecule is a stochastic process (110) and the subsequent steps occur by trans-proteolysis. Once folded, the high activation energy barrier kinetically traps folded SbtE in its native state. The free energy difference between the N:P and N is approximately 9.4 kcal/mol of energy. The net free energy difference between the unfolded and folded SbtE is ~0.4 kcal/mol and is likely to preclude spontaneous folding in the absence of the IMC.

however, be converted back to the active inhibitory form by denaturation and renaturation. The crystal structures of the various conformers explain the structural basis of kinetic stability of the active form and the thermodynamic stability of the latent inactive form (108, 109). Hence the inhibitory form is the lowest energy state *accessible* during folding, but since it slowly converts to the latent form, it is clearly not the lowest energy state. Another compelling example of kinetic stability is evident from the studies of IMC-mediated and hence pathway-dependent folding of subtilisin (33, 42) and α-lytic protease (46), two evolutionarily unrelated prokaryotic serine proteases.

4.2.4. Kinetics of IMC-Mediated Maturation of Subtilisin

Based on structural and biochemical studies, the overall IMC-mediated maturation pathway (**Fig. 4.5**) can be described in terms of three distinct stages:

(1) Folding of the polypeptide to a structured state (77).

(2) Autoprocessing of the peptide bond between the IMC and protease domains resulting in a non-covalently associated IMC protease-inhibited complex (87).

(3) Release and degradation of IMC from the inhibition complex that results in an active protease (34, 100, 110).

As discussed above, most of the early studies focused on the kinetics of folding. Utilizing the optimization of folding by rapid dilution, and the isolation of folding mutants, Yabuta et al. carried out a thorough characterization of the kinetics of all the substages of the maturation pathway (34). These studies established that while in vitro folding and autoprocessing are rapid and reach completion in 30 min, the activation of the protease is not seen until ~240 min. Hence, activation and not protein folding is rate limiting to IMC-mediated maturation (34). As evident from the crystal structure, the protease is fully folded as an inhibited complex (**Fig. 4.2a**). However it appears that the release of the IMC from this complex is extremely slow. Establishing the reasons for this slow release and the mechanism of release would give further insights into IMC-mediated maturation. Further, during the maturation process, the IMC switches from a chaperone to an inhibitor and eventually to a proteolytic substrate (110). Establishing the energetics of each of these stages will give further insights into the high-energy kinetic barrier and how the IMC functions to modulate this barrier.

4.2.5. IMC-Regulated Activation of Subtilisin

Based on the current literature, a mechanistic model for precursor activation of bacterial subtilisin has been proposed (**Fig. 4.5**). The precursor of pro-subtilisin, which is secreted extracellularly into the media, produces mature subtilisin through a process that involves folding, followed by autoprocessing, and degradation, which represent two distinct steps of proteolysis (34, 100). As discussed earlier, folding of pro-subtilisin requires the presence of the IMC domain and is a rapid process that occurs through a partially structured non-native folding intermediate (86). The non-native intermediate then undergoes structural changes to give a native-like folded, but unautoprocessed precursor. The orientation of the IMC with subtilisin is critical for the autoprocessing reaction, which is also a fairly rapid process (100). Once autoprocessed, the IMC–subtilisin complex appears to be in a remarkably stable state due to the inhibitory function of the IMC (110). Release of the IMC is the rate-determining step (RDS) of the maturation reaction. Once a free protease molecule is formed, it can bind to the IMC domain in the autoprocessed complex and facilitate *trans* degradation. It is important for the IMC–subtilisin complex to interact with mature subtilisin, because this allows a rapid exponential activation. Moreover, the rate constants of the various steps of the maturation appear to be optimized for maximum yield of the protease domain (110). Consistent with this hypothesis, IMC variants that decrease the IMC affinity for their cognate catalytic domains are inadvertently less efficient chaperones. However, high affinity is alone not sufficient for efficient folding because despite displaying tighter affinity for subtilisin, the aqualysin IMC is a weaker chaperone compared to SubtE IMC (81). Nonetheless, the data suggest that folding and autoprocessing should be completed before the release of the first active subtilisin molecule. Hence, the IMC is essential for folding of the protease domain and the inhibitory function is required for the maximum efficiency of this process and serves to regulate protease activation.

The RDS of subtilisin activation was demonstrated to be stochastic in vitro (110). The energetics of activation establishes that the release of the tightly associated IMC domain from the catalytic domain is energetically unfavorable (**Fig. 4.6**) and is the primary cause of the associated stochastic behavior. Detailed analysis of stochastic activation shows that modulating the structure of the IMC through external solvent conditions can vary both the time and randomness of protease activation. This behavior of the protease correlates with the release-rebinding equilibrium of IMC and suggests that a delicate balance underlies IMC structure, release, and protease activation. Proteases are ubiquitous enzymes crucial for fundamental cellular processes and require deterministic activation mechanisms. The activation of subtilisin establishes that through selection of an intrinsically unstructured IMC domain, nature appears to have selected for a viable deterministic handle that controls a fundamentally random event and outlines an important mechanism for regulation of protease activation.

5. Understanding IMC-Independent Folding and Maturation of Subtilisin

A comprehensive search of the SWISSPROT database has identified two distinct subfamilies, intracellular serine proteases (ISPs) and extracellular serine proteases (ESPs), within the subtilase super-family (111). While ESPs and ISPs display a high level of sequence, structure, and functional conservation, the latter lack the classical subtilisin propeptide signature, which is essential for correct folding of ESPs (**Fig. 4.7**). Subsequent studies established that although determinants such as topology, contact order, and hydrophobicity that drive protein folding reactions are conserved, ESPs absolutely require the propeptide to fold into a kinetically trapped conformer. However, ISPs fold to a thermodynamically stable state more than 1 million times faster independent of an IMC (110). Moreover, the spectroscopic studies established that ISPs and ESPs fold into their native states through different intermediate states (**Fig. 4.6**). An evolutionary analysis of folding constraints in subtilases suggests that observed differences in folding pathways appear to be mediated through positive selection of specific residues that map mostly onto the protein surface. Together, these results suggest that closely related subtilases can fold through distinct pathways and mechanisms, and that fine sequence details can dictate the choice between IMC-dependent kinetic stability and IMC-independent thermodynamic stability (111).

Another important difference between ESPs and ISPs is their distribution of charged amino acids. The IMC domains of ESPs contain several basic amino acid residues that make the net isoelectric point of their precursors extremely alkaline (>9.0 pH). ISPs on the other hand have very small transient N-terminal extensions (**Fig. 4.7a**) and lack the motifs N1 and N2, which are signatures of IMC-dependent subtilases (**Fig. 4.7b**). In addition, the net isoelectric point of ISP precursors is highly acidic (<5.0 pH). This is because, despite similar net hydrophobicity, the protease domains of ISPs contain a significant percent of acid residues when compared with the relatively neutral ESPs. These residues appear to have evolved through positive selection and are distributed over the enzymes' molecular surface (**Fig. 4.7c**). As a consequence, ESPs and ISPs differ significantly in the properties of their solvent-accessible surfaces (111). This change in surface properties likely alters the folding pathways of the two homologues with ISPs becoming IMC independent while ESPs remain IMC dependent. **Figure 4.7d** depicts a model for how surface charges may contribute to rapid folding.

It is noteworthy that IMC-independent folding of subtilisin has also been achieved through extensive mutagenesis around

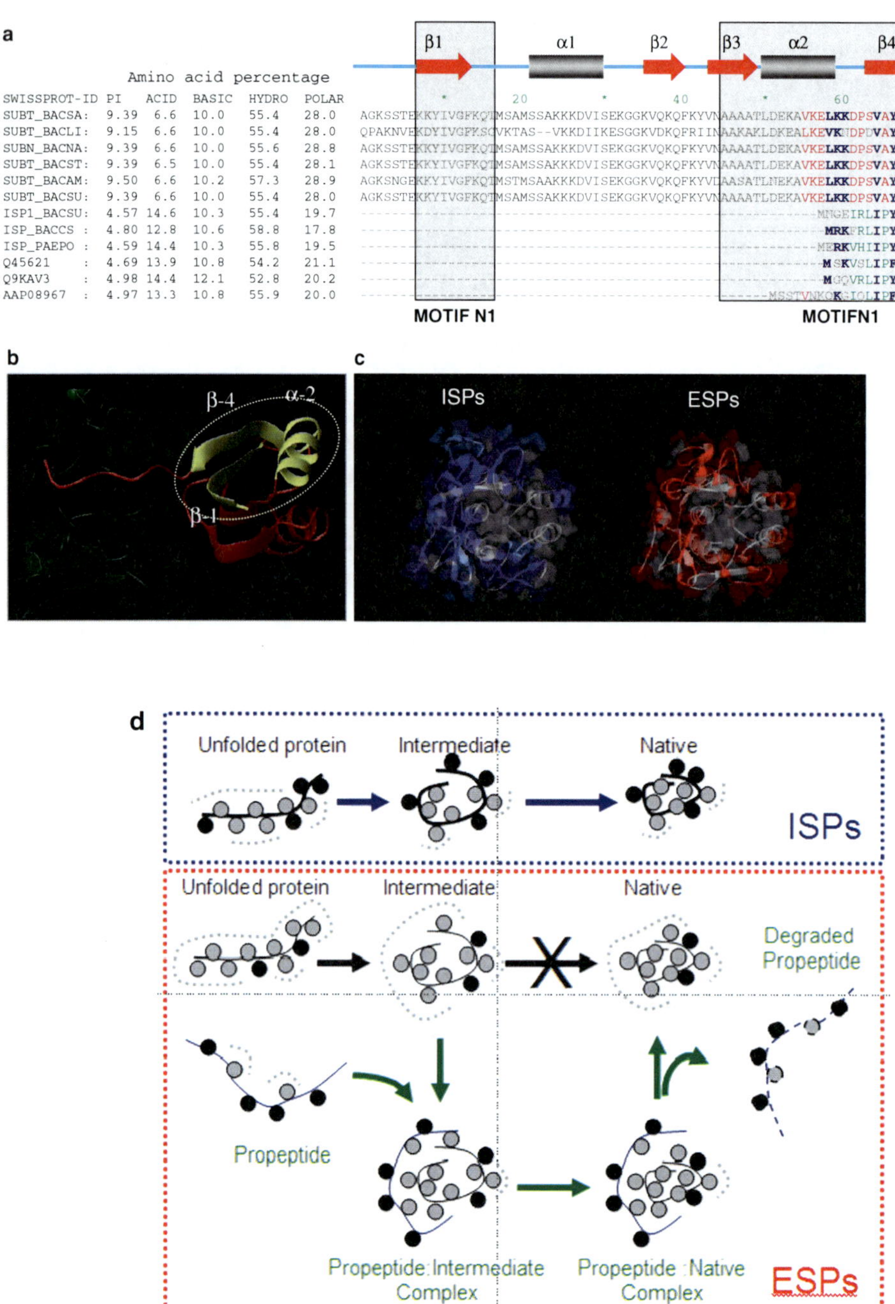

Fig. 4.7. (Continued)

the Ca^{2+}-binding site in the catalytic domain or through the use of high concentrations of organic acids. While such studies have provided valuable insights into the folding mechanisms of subtilisin (79, 88, 94, 99, 112), they provide little information regarding the evolutionary pressures that dictate the choice of specific folding pathways. It is striking that although ESPs and ISPs display roughly 50% sequence identity, the catalytic domains of PCs are more related with the corresponding domains of ESPs. Similar to ESPs, members of the PC family are all synthesized as multi-domain proteins and require their propeptides to function as IMCs, though it remains to be seen whether the catalytic domain of PCs actually displays kinetic stability. *Why do eukaryotic PCs require their IMCs to fold, rather than an IMC-independent pathway?* To better understand the role of IMCs in mediating folding and regulating activation of the catalytic domains of PCs, we will analyze their domain organization and compare that with bacterial subtilisins (**Fig. 4.8**).

6. Proprotein Convertases Are Homologues of IMC-Dependent Subtilases

PCs constitute a family of calcium-dependent serine endoproteases that catalyze cleavages of precursor proteins at sites containing doublets or clusters of basic amino acids to generate active proteins and peptides (11). Analogous to prokaryotic sub-

Fig. 4.7. (Continued) Comparison between ESPs and ISPs. (**a**) The first six Swiss-Prot accession numbers shown on the left side of the alignment are ESPs while the rest are ISPs. The amino acid composition represents the total percent across the entire protein and suggests that the total number of basic and hydrophobic residues is similar within ESPs and ISPs. However, ISPs have more acid residues when compared with ESPs. The ESPs have most of the basic residues located within the IMC domain while ISPs have the acid residues distributed within the catalytic domain. Motifs N1 and N2 that are conserved within ESPs are boxed and represent putative nucleation sites that initiate folding (38). These regions display conformational rigidity as seen using NMR spectroscopy (75) and constitute a sub-domain within the 3D structure. The secondary structure elements displayed above the alignment are based on the X-ray structure of SubtE (76). Residues that are conserved within ESPs (*red* typeface) and ISPs (*green* typeface) are colored separately while those that are conserved in both are indicated in *blue* typeface. (**b**) The folding nucleation sites (*yellow*) constituted by motifs N1 and N2 are superimposed on the structure of the IMC domain of SbtE (*red*). The C-terminus of the IMC domain interacts with the catalytic domain of SbtE (*green*) to inhibit catalytic activity. (**c**) Sequence conservation mapped onto structures. Residues conserved within both ESPs and ISPs are colored *gray*, while *red* or *blue* denotes residues that are conserved within only ESPs or ISPs, respectively. Residues that are conserved within ESPs, but without selection constraints in ISPs, are colored *cyan*. (**d**) Model for how changes in surface residues compensate for the loss of IMCs within ISPs. Hydrophobic residues (*gray circles*) initiate folding through a hydrophobic collapse. When hydrophobic surfaces are solvent exposed, the water tends to form a clathrate cage (*broken gray lines*). Both ESPs and ISPs are highly charged proteins but with extremely different isoelectric points. This charge (*black circles*) is localized within IMCs of ESPs, while it is distributed over the surface of the protease domain of ISPs. As a consequence, protease domains of ISPs are more polar than ESPs. This difference may help to enhance the conformational entropy of water around the folding polypeptide and assist the hydrophobic core in driving spontaneous folding of ISPs. When folding of ESPs is carried out in the absence of the propeptide, the protease domain adopts a partially structured molten globule intermediate with solvent-exposed hydrophobic surfaces (77). The partially formed hydrophobic core within the protease domain of ESPs may be insufficient to drive folding without the solvation assistance bestowed by the surface charge. This leads to stabilization of the molten globule intermediate that is prone to aggregation. This concept is consistent with findings that charge per se, and not its polarity, is critical for folding of subtilases (72, 81).

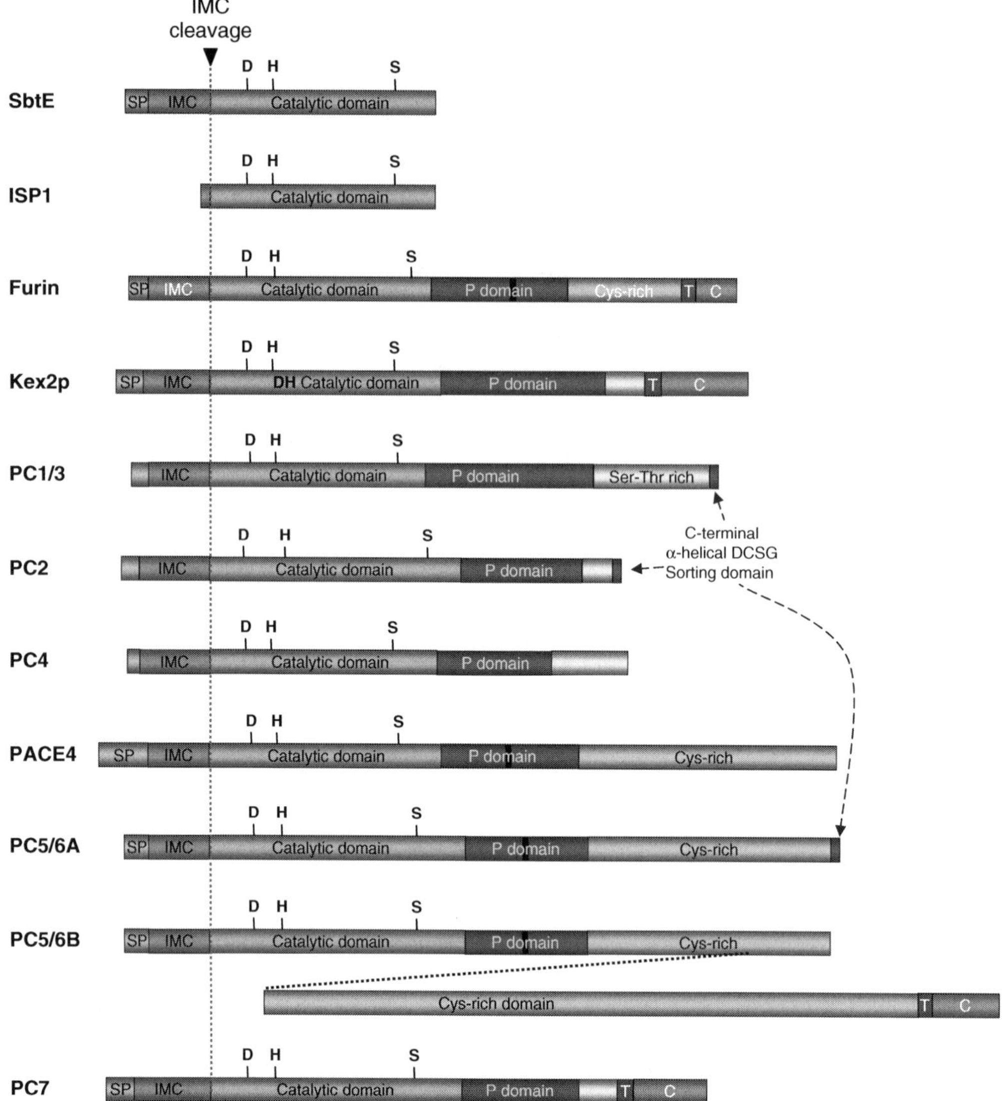

Fig. 4.8. The structural organization of PCs compared with their bacterial counterparts ESPs and ISPs. The letter T represents the trans-membrane region, SP for signal peptide, while C indicates the cytoplasmic domain. The letters D, H, and S represent the catalytic triad and other regions are marked as indicated. Proteolysis at the IMC cleavage site (*dotted grey line*) initiates protease activation.

tilisins, PCs require their propeptides to chaperone folding of their cognate catalytic domains (2, 28). In addition, these IMCs also help to transport the catalytic domains of PCs into specific organelles of the eukaryotic secretory pathway, a complexity that is absent within prokaryotic subtilisins. The eukaryotic secretory compartments maintain a pH gradient inside the cell, wherein the pH drops from ~pH 7.0 in the endoplasmic reticulum (ER) to pH 6.0 in the trans-Golgi network (TGN). The post-TGN compartments are further acidified, with the dense core secretory granules at pH ~5.5 and lysosomes at ~pH 5 (23). Because most enzymes are exquisitely pH sensitive, the pH of each organelle critically determines the coordinated biochemical reactions occurring along the endocytic and secretory pathways. Studies on furin, the most intensively analyzed member of the PC family, demonstrate that the 83-residue propeptide helps to fold the catalytic domain in the ER but requires two pH-dependent and compartment-specific proteolytic cleavages to produce active furin (25, 28). While this two-step activation is similar to bacterial subtilisins (42, 77, 78, 87), the discovery of the differential pH requirements of the two proteolytic cleavages in the IMC domain of furin suggests that the IMC of furin (25, 28) and possibly other PCs may have evolved to sense and respond to a specific local pH.

7. The Domain Organization of PCs

Due to their significant levels of sequence similarity between the catalytic domains of eukaryotic PCs and prokaryotic ESPs and ISPs, they are all members of the subtilase super-family (13, 113). **Figure 4.8** compares the domain organization between prokaryotic ESPs and ISPs and eukaryotic PCs. In addition to signal peptides, which direct translocation of pro-enzymes into the ER, PCs also contain IMC domains that are flanked by the signal peptidase cleavage sites at the amino-terminus and a pair of dibasic residues that constitute the autoproteolytic cleavage site at the carboxy-terminus.

7.1. The IMC Domain

The IMC domains in PCs, which are approximately 80–90 amino acid residues in length and are marginally longer than their prokaryotic counterparts, play a vital role in folding, transporting, activating, and regulating catalytic activities of PCs. In addition, the IMC domains of PCs also exhibit potent slow-binding inhibition of their cognate catalytic domains similar to bacterial subtilisins. In vitro studies demonstrate that the IMC:furin affinity is extremely tight ($K_{0.5}$ = 14 nM) (25). The only residue-specific information on the structure of the IMC domains comes from NMR studies of isolated mouse PC1 IMC domain (114). Although the sequence conservation between IMCs of eukaryotic PCs and prokaryotic subtilases is low, the IMC domains adopt similar folds when they form complexes with their cognate catalytic domains (**Fig. 4.9**). The relative orientation of the active sites and the IMC domains of bacterial and eukaryotic inhibition complexes are depicted in **Fig. 4.9b–d**. However, unlike bacterial subtilisins (78), the IMC domains of PCs appear to be partially to fully structured domains in their isolated forms (115).

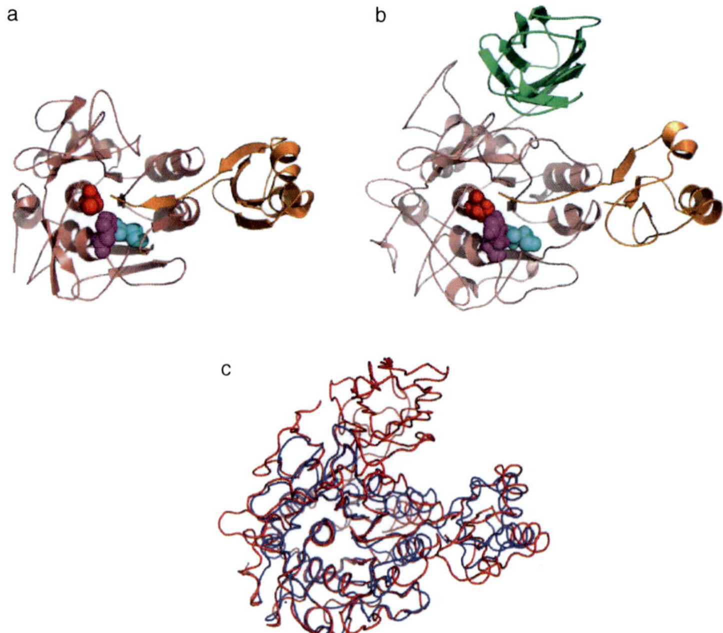

Fig. 4.9. Structural conservation among IMCs of SbtE, furin, and PC1. All subtilases have a well-conserved subtilisin-like catalytic domain and are stabilized by bound metal ions. *Yellow* spheres depict calcium ion, while the *orange* sphere represents sodium ion. (**a**) Crystal structure of subtilisin E (1SCJ), a bacterial alkaline serine protease. (**b**) Crystal structure of furin (1P8J), a eukaryotic subtilisin-like protease, with the P-domain and an inhibitor (*magenta*). (**c**) Structure of a Ak.1 protease (1DBI), a thermostable subtilase from *Bacillus* showing the location of several calcium ions.

The structure reveals that the IMC adopts an open sandwich antiparallel α–β, comprised of two α-helices packed against one side of a four-stranded twisted β-sheet (*see* **Fig. 4.9a**). The β-sheets of the IMC domain pack tightly against the two parallel surface α-helices. This interface is stabilized by a complementary surface created through a series of conserved hydrophobic residues. Additionally the carboxylates of Glu_{-7} and Asp_{-9} in the IMC domain form helix caps for the two parallel helices in the case of subtilisin complex (76, 82, 112). Initial structural models of IMC:PC complexes suggest that such capping may also exist within PCs and may play a role in their autocatalytic activation (22).

The observed slow binding in PCs may be a consequence of the structuring and binding of its floppy C-terminus of the propeptide to the catalytic domain and the subsequent reorientation of the propeptide:protease interface formed as a consequence of the bimolecular interaction. This was observed for PC5A which displays reduced catalytic activity when complexed with a peptide containing C-terminal residues from its own IMC domain (116). Since incorrect spatial and temporal proteolysis within a cell can be lethal, the inhibitory properties of IMCs were hypothesized to be a mechanism that prevents premature protease activation. In general, cells appear to have evolved two distinct mechanisms to control the activity of proteases. The first involves co-evolution of specific endogenous inhibitors, typically within compartments spatially distinct from those containing active enzymes. The second involves proteases being synthesized as inactive or less active zymogens, which become activated by limited intra- or intermolecular proteolysis cleaving off a small peptide. Interestingly, these two different regulatory mechanisms in proteases appear combined in the chaperoning capabilities of IMC-dependent systems and it is tempting to speculate that this combination enables facile spatiotemporal regulation of PC activation by exploiting the pH gradient of the secretory pathway.

To better understand the sequence–structure relationship in subtilases, a multiple sequence alignment of various bacterial subtilisin and eukaryotic PCs was conducted. The alignment also displays the earlier identified conserved motifs, N1 and N2, that are located at the amino- and carboxy-termini of the IMC domains (**Fig. 4.10a**). These motifs are more hydrophobic and are generally flanked by charged residues (38). As mentioned earlier, random mutagenesis studies using error-prone PCR demonstrate that substitutions within motifs N1 and N2 in bacterial subtilisin were often deleterious to folding (74). Circular dichroism studies suggest that the isolated subtilisin E IMC domain is largely unstructured (78). While NMR studies confirm this finding, they additionally suggest that motifs N1 and N2 display conformational rigidity (75). It is noteworthy that the unstructured IMC adopts a well-defined conformation when it binds to subtilisin wherein motifs N1 and N2 come together to form a α-β-α subdomain in this inhibition complex (**Fig. 4.10a, b**) and may be critical for initiating the folding process (76).

While the structures and functions of individual proteins may impose different constraints on their evolution, an analysis of patterns of divergence suggests that individual responses of most proteins are variations on a common set of selective constraints. In proteins with low divergence, mutations within the interior are under strong purifying selection that removes all but a few conservative changes. With increasing divergence, mutations in the interior become more widespread and closer in number to what is found in the intermediate and exposed regions. Since catalytic domains exhibit a higher degree of conservation within subtilases when compared to their IMCs, this suggests that the catalytic and IMC domains may have encountered different mutational frequencies and differential selective constraints. *However, why nature may have imposed differential selective constraints that alter both sequence and the asymmetrical distribution of charge in two functional domains within the same protein remains unclear.*

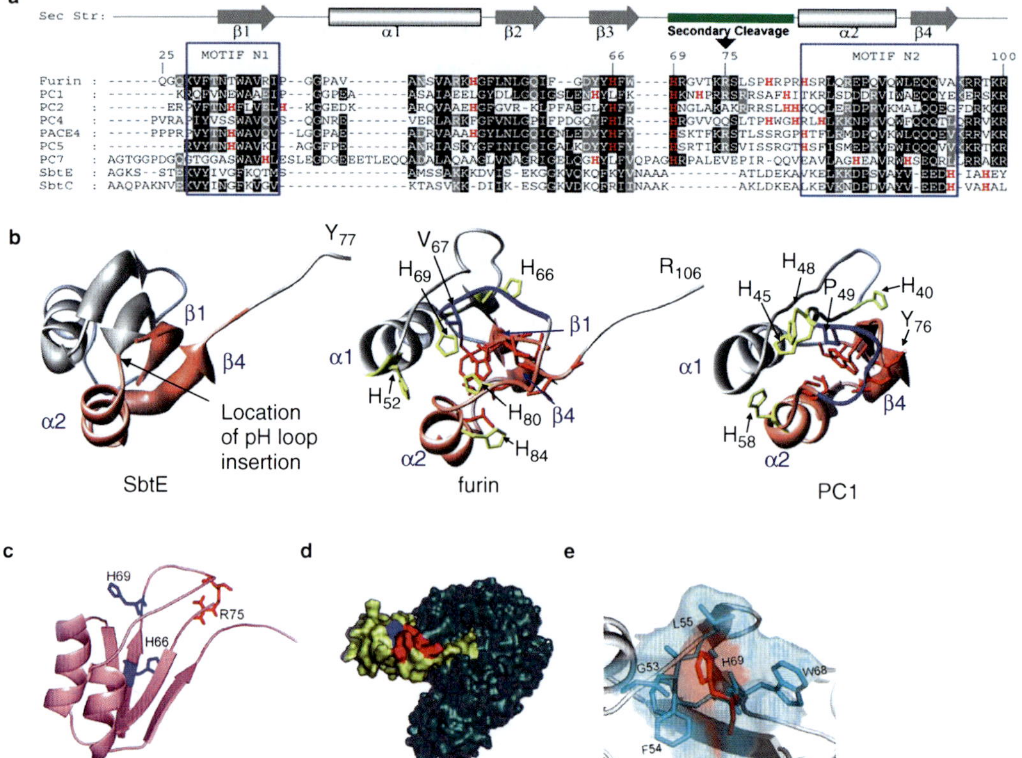

c d e

Fig. 4.10. A comparison between the IMC domains of various PCs with their bacterial counterparts. (**a**) Sequence alignment of the IMC domains of various PCs with bacterial SbtE and SbtBPN. Residues are numbered in reference to the furin IMC, which begins at Gln25. The secondary structure cartoon depicted above the alignment is based on the NMR structure of the IMC domain of PC1/3 (1kn6) (114) and the IMC domain of SbtE (1scj) (76). The pH-sensitive cleavage site loop contains secondary cleavage sites for furin and PC1. Motifs N1 and N2 are well conserved when compared with the rest of the protein (38) and represent the folding nucleation sites in SbtE (75). (**b**) *Pink* ribbons in the structures of SbtE, furin, and the IMC domain of PC1/3 depict the relative locations of motifs N1 and N2 mapped on the 3D structure of the IMC domain. The secondary cleavage site is also depicted by the *blue* tube structure. Histidine residues in the alignment are marked in *red* typeface and are depicted in the individual structures as well. (**c**) Ribbon representation of the IMC domain of furin obtained by homology modeling. His66 and His69 (*blue*) and Arg75 (*red*) are highlighted. (**d**) Surface representation of the propeptide–furin complex. The modeled furin IMC (*yellow*) is docked onto the active site of furin (*blue*). The internal IMC cleavage site (*red*) and His69 (*blue*) are highlighted. (**e**) Surface representation of the secondary cleavage site illustrates His69 (*red*) buried in the solvent-accessible pocket formed by Gly53, Phe54, Leu55, Phe67, and Trp68 hydrophobic residues (*blue*).

7.2. The Catalytic Domain

The greatest sequence similarity resides in the subtilisin-like catalytic domain, with the aspartate (Asp), histidine (His), and serine (Ser) residues which comprise the active site being invariant. Overall, the catalytic domains of the other PCs are 54–70% identical in sequence to furin, which is the most extensively studied PC. Although the sequence identity is lower (20–30%) when PCs and bacterial subtilases are compared, the recently solved X-ray structures establish that the catalytic domains of furin (20, 22) and kexin (21) adopt secondary and tertiary structures that are similar to bacterial ESPs and ISPs (**Fig. 4.9b–d**).

In addition, these structures also provide insights into their remarkable substrate specificity displayed by PCs when compared with their promiscuous bacterial counterparts. Not surprisingly, homology modeling suggests that the core of the catalytic domains of PCs, which consist of a highly twisted β-sheet comprised of seven parallel and one anti-parallel β-strands (**Fig. 4.9**), is highly conserved based on the sequence similarity within PCs (22). The β-sheet is surrounded by five adjacent and two peripheral α-helices and by two short β-hairpin loops. These core structural elements are connected through surface loops that are conformationally less well defined. The solvent-exposed surface of the two peripheral helices in subtilisin constitutes the interface that interacts with the IMC domain. The modeling studies further suggest that PCs exhibit a deep active-site cleft with a shape remarkably similar to furin and that the geometry and the charge distribution of their substrate-binding regions appear to be of utmost importance for the distinct specificities of the individual PCs (22).

7.3. The P-Domain

While eukaryotic PCs contain IMCs and catalytic domains that are conserved with corresponding domains within bacterial subtilisins, they also contain an approximately 130 residue P-domain (**Fig. 4.9c, d**) that is important for catalytic activity of PCs which is missing in their bacterial counterparts (4). The structure of the P-domain of PCs is organized as a separate eight-stranded β-barrel, whose strand connectivity is similar to that for jelly-roll β-barrels (**Fig. 4.9b**). The eight structurally conserved strands are arranged in two opposing four-stranded β-sheets, which pack in a small hydrophobic core. The catalytic domain and P-domain are folded into two separate but abutting domains that together form the characteristic two-domain structure of furin and kexin (20, 21). Both domains are covalently linked by the "inter-domain linker" and interact with each other through an interface that is approximately 1100 $Å^2$ to form the ectodomain of PCs. The few hydrophobic interface residues contributed by the catalytic domain are of special note, because most have polar/charged counterparts in the topologically related but P-domain-free bacterial subtilisins (22). These hydrophobic residues are mostly conserved in the PCs, obviously constrained by the common function to mediate interaction with the P-domain. Of particular note are a few charged interface residues, which are engaged in buried salt bridges that contribute to a part of the calcium-binding site. As a consequence the P-domain serves to stabilize calcium binding through this domain–domain interface.

7.4. The Variable C-Terminus of PCs

In addition to the signal peptide, propeptide, catalytic, and P-domains, furin, PACE4, PC4, PC5, and PC7 also contain a highly variable Cys-rich C-terminus (**Fig. 4.8**) (117). Moreover, the full-length furin, the PC5/6B isoform, and PC7 display additional trans-membrane and C-terminal cytoplasmic domains that are absent in PC1/3, PC2, and PC4 which terminate in long Ser/Thr-rich tails instead. PC1/3 and PC2 are predominantly expressed in neuroendocrine cells, PC4 is only found in testicular and ovarian germ cells, while the other mammalian PCs are more broadly distributed. PCs are often simultaneously expressed in various tissues, where they are mainly localized in the TGN, but are also found in other compartments of the secretory pathways and on the cell surface (11). Furin, a constitutively expressed member of the PC family, is localized primarily in the TGN but can traffic between two cycles: one at the TGN and the other at the cell surface where it interacts with its diverse repertoire of proprotein substrates. Moreover, furin is required to process these substrates precisely, even in the presence of other similar cleavage sites. Three of the PCs, PC1/3, PC2, and PC5/6A, are targeted to dense core secretory granules via α-helical domains located in their C-terminal tails (118).

8. Activation of Eukaryotic Subtilases

The activation pathway of PCs shows evolutionary conservation with the activation pathway of bacterial subtilisins, which use IMC cleavage to guide the folding of the catalytic domain and results in a structural reorganization of the folded proprotein. As seen from **Fig. 4.8**, furin, PC7, and the isoform PC5/6B are type-I membrane-bound proteases; PC1/3, PC2, and PC5/6A are packaged into dense core granules; PACE4 is localized in the extracellular matrix in osteoarthritic cartilages (119), while PC4 is constitutively secreted into the extracellular milieu. Furin can be shed from the membrane and along with PC5/6A can be found in the extracellular milieu. The sequences of IMC domains of PCs are 50–60% identical to furin and are flanked by the signal peptidase cleavage site on the amino-terminal side and by a conserved set of basic amino acids that comprise the autoproteolytic cleavage site on the carboxy-terminal side. Signal peptidase cleavage occurs after the signal peptide directs translocation of the precursor into the ER, where the IMC domain guides the folding of the catalytic triad into its correct conformation. With the exception of PC2, which is processed in the immature secretory granules, all PCs undergo the initial autocatalytic cleavage in the ER to generate a non-covalently associated heterodimeric

IMC:convertase complex. This complex exits the ER and sorts to specific secretory pathway compartments where it undergoes a second autocatalytic cleavage within the IMC domain to disinhibit the enzymatic domain. Furin and PC7 undergo this second cleavage in the TGN/endocytic system; PC5/6 and PACE4 appear to get activated in the TGN and/or at the cell surface while PC1/3 and PC2 are maximally active only in secretory granules. PC1/3's maximal activity in secretory granules is due to a second autoproteolytic cleavage of its C-terminal domain at a pair of arginine residues (R_{617}, R_{618}) (120).

This exquisitely well-orchestrated, multistep, compartment-specific pair of cleavages within the IMC domain is the most intensively studied in furin (2) and is accomplished by exploiting its own rules of substrate specificity to ensure correct spatial and temporal activation in the TGN, where the conditions of pH and calcium are optimal for the *in trans* cleavage of specific substrates. The first cleavage is rapid ($t_{1/2}$ =10 min) and takes place in the neutral pH of the ER within furin's consensus cleavage site –Arg–Thr–Lys–$Arg_{107}{\downarrow}$– after residue Arg_{107} (P1/P4 Arg canonical cleavage site) at the C-terminus of the IMC domain (121). This cleavage of the IMC can be blocked through an active-site substitution and results in the buildup of an intermediate in the ER–Golgi intermediate compartment (ERGIC)/*cis*-Golgi network. These data suggest that (i) the cleavage occurs through autoproteolysis, (ii) both the ER and ERGIC compartments participate in the initial steps of furin activation, and (iii) specific components of the cellular trafficking machinery detect formation of the furin: propeptide complex before directing their transit to late secretory pathway compartments. Recent studies suggest that PACS-1, a cytosolic sorting protein that connects the CK2-phosphorylated furin cytosolic domain to AP-1, is responsible for localizing the endoprotease to the TGN (122). After IMC cleavage occurs in the ER, the IMC domain remains associated with furin and functions as a potent autoinhibitor (IC_{50} = 14 nM) *in trans* during the transport of this IMC–furin complex to the late secretory compartments – TGN/endosomes – where the mildly acidic pH promotes a second, slower intramolecular autoproteolytic cleavage ($t_{1/2} \sim 90$ min) at His-Arg-Gly-Val-Thr-Lys-$Arg_{75}{\downarrow}$, an internal site (P1/P6 Arg cleavage site) within the IMC domain. Surprisingly, mutation of the P1/P6 Arg cleavage site to a P1/P4 Arg canonical furin site fails to yield mature, active furin and instead causes the accumulation of inactive profurin in the ER (25, 28). This second-site cleavage is rate limiting for activation of bacterial and human subtilisin-like proteases (34, 110). Cleavage at Arg_{75} is followed by the rapid dissociation of the cleaved IMC fragments that releases the inhibition of furin's catalytic triad allowing it to process its diverse repertoire of endogenous substrates (25). Together, these studies suggest that the ordered, compartment-specific cleavages of the furin propeptide are necessary to guide the folding and activation of the endoprotease, and that this activation process could be controlled by a pH sensor.

As discussed earlier, despite significant sequence and structural conservation with both PCs and ESPs, members of the prokaryotic ISP subfamily lack propeptide domains. *How do ISPs prevent premature proteolysis inside the cell?* An analysis of the ISPs reveals that they too are produced as zymogens with very small propeptide domains (**Fig. 4.7a**). Unlike ESPs, the propeptide domains of ISPs are roughly 20–25 residues long and do not function as IMCs. Nonetheless, these propeptides are potent inhibitors of the ISP activity and are removed through autoproteolysis (Subbian et al. unpublished data). Hence the expression of ISPs as zymogens can serve to prevent premature activation. These IMC-independent ISPs are involved in a variety of regulatory cellular functions such as DNA packing, genetic competence, and protein secretion and differ from ESPs which are scavenging proteases (123–126). Although PCs more closely align with ISPs in terms of regulating biological function, the amino acid sequences of PCs more closely resemble ESPs. Clearly both prokaryotic prototype ESPs and ISPs are produced as zymogens. From an evolutionary standpoint it is interesting to note that eukaryotic PCs require their IMC to chaperone correct folding. *Why do eukaryotic PCs, like ESPs, require their IMCs to fold, rather than an IMC-independent pathway that exists in ISPs?* We hypothesize that the choice of an IMC-dependent pathway was driven by a necessity to modulate folding, to allow interactions with the transport machinery, and to regulate the compartment-specific activation.

9. IMC-Encoded pH Sensors in Eukaryotic Subtilases

The required exposure of the IMC–furin complex to pH 6 in order to complete enzyme activation suggests a role for one or more histidine residues on the IMC domain to serve as a pH sensor controlling furin activation. Histidine is a unique amino acid that contains an imidazole ring for its side chain. The unprotonated imidazole is nucleophilic and can serve as a general base, while the protonated form can serve as a general acid. Hence protonation of the histidine imidazole ring, which has a p*K*a of 6.0, has profound effects on histidine chemistry under physiological conditions. Such pH-sensing roles for histidines are well established for generating allosteric changes, including control of O_2/CO_2 exchange by hemoglobin, gating of electrogenic molecules, and the pH-dependent conformational changes within class II MHC molecules that promote ligand exchange.

Analysis of the furin IMC predicts two His residues adjacent to the secondary cleavage site, which are either strictly (His_{69}) or partially (His_{66}) conserved within all PC family members (**Fig. 4.10a**). This cleavage site maps onto a loop between strands β3 and α2 and is part of a distinct hydrophobic pocket on the surface of the IMC abutting the catalytic domain (**Fig. 4.10b, d**). Although the cleavage site lies in a surface loop, H_{69} is buried at the center of a well-formed hydrophobic pocket lined by nonpolar residues, G_{53} and L_{55}, and aromatic residues, F_{54}, F_{67}, and W_{68} (**Fig. 4.10d**). By contrast, the P_9 H_{66} maps to the IMC–protease interface. This interface in bacterial subtilisins displays a higher than average B-factors as seen from the crystal structure of the IMC:SbtE complex (76), and increasing these dynamics can decrease rates of autoprocessing (100). Varying solvent conditions modulate IMC binding to this interface, prolonging its release and hence the activation of the mature protease (110). This suggests that this interface is not solvent accessible. Based on our modeling analysis, we predicted that protonation of the H_{69} located in a hydrophobic pocket could have dramatic effects on the structure of the IMC, unlike protonation of the solvent-accessible H_{66}. Hence, H_{69} but not H_{66} as earlier hypothesized (114) could function as a potential inbuilt primary pH sensor within PCs. Using mutants that mimic the protonation state of histidine the role of His_{69} in the folding and activation of furin has been elucidated (26). These studies suggest that the protonation state of His_{69}, which is located in a solvent-accessible hydrophobic pocket, plays a critical role in regulating the secondary propeptide cleavage. Mutations that interfere with the His_{69} block propeptide excision, resulting in accumulation of profurin in the ER by a mechanism that requires the cytosolic sorting protein PACS-2 and COPI. Following propeptide excision, the furin•propeptide complex traffics to the mildly acidic TGN/endosomal system where protonation of His_{69} disrupts the solvent-accessible hydrophobic pocket to expose the P1/P6 Arg internal cleavage site His-Arg-Gly-Val-Thr-Lys-Arg_{75}↓, leading to release of the inhibitory propeptide and furin activation.

Given that the H_{69} – the pH sensor in furin – is conserved within other PCs it remains unclear why different PCs display unique pH sensitivity. For example, PC1, which requires a lower pH for its activation, becomes activated in the dense core secretory granules. We hypothesize that additional residues are likely involved in controlling the subtleties of pH-dependent activation of individual PCs. Further whether additional residues are involved for H_{69} to function as a pH sensor in furin also remains unknown.

9.1. Functional Advantages of Sequence-Encoded pH Sensors in Proteins

While post-translational modifications or cofactors can also regulate biological activity, one of the major advantages of using changes in proton concentrations for regulation and signaling is the potential for exceptionally efficacious spatial and temporal responses. Protons are small single subatomic particles that can diffuse rapidly through water to induce reversible chemical changes which result in considerable electrostatic perturbations, which in turn changes protein structure, dynamics, and interactions. The pH-driven changes in the affinity of hemoglobin for oxygen binding (Bohr's effect) is a classic example of a single-site, proton-induced allosteric regulation. In this case, the effect of pH is mediated by a His-Asp salt bridge which breaks when His gets deprotonated at neutral pH. The protonation of key histidine residues triggers interaction of the translocation domain of diphtheria toxin with the host cell membrane and causes flaviviruses to fuse into membranes. In addition to modulating activity of enzymes, viral entry, and transformation, exquisite pH-sensitive conformational switches are found to regulate activity of ion and water channels, affinity of proteins for their cognate binding partners, protein stability, solubility, and catalytic activity.

Despite established effects of small changes in proton concentrations on diverse cell functions, our understanding of how these changes affect proteins and macromolecular assemblies driving specific cell processes is limited. Although changes in solvent pH affect the ionization state of all weak acids and bases, and all cellular proteins contain amino acids with titratable groups, only select proteins appear to be bonafide pH sensors. New insights from protein structures and biomolecular simulations are beginning to reveal the structural basis for tight coupling between protonation state and protein conformation. However, our understanding of how physiological changes in pH affect protein conformations and macromolecular assemblies is limited.

9.2. Putative Mechanisms by Which pH Sensors Function

Based on our understanding of the activation of bacterial subtilisins we hypothesize that protonation of the intrinsic pH sensor of furin induces conformational changes that destabilize the IMC domain to initiate rapid activation. This hypothesis is based on the fact that the pH sensor residue in furin – H_{69} – is solvent accessible and is located within the loop that contains internal cleavage site conserved within all PCs. This cleavage site maps onto a loop between strands β3 and α2 and contributes toward a distinct solvent-exposed hydrophobic pocket on the surface of the IMC and is adjacent to the catalytic domain (**Fig. 4.10b, e**). The hydrophobic pocket is lined by residues G_{53}, L_{55}, F_{54}, F_{67}, and W_{68} (**Fig. 4.10e**). At a near-neutral pH of the ER, the mostly deprotonated H_{69} displays more hydrophobic properties that allow the imidazole side chain to be stabilized within the hydrophobic pocket. In this conformation the pH-sensitive loop tightly abuts the IMC and reduces the exposure of the internal cleavage site to trans-proteolysis. Moreover, the formation of this pocket appears to be essential for initiating the folding process because an $H_{69}K$ substitution does not result in a folded/processed complex (26). Upon entry into the TGN the imidazole side chain of H_{69} is exposed to a more acidic pH (~6.0) and gets protonated. The additional proton increases the polarity of the imidazole side chain forcing it out of the hydrophobic pocket (**Fig. 4.11**). This repulsion may cause a conformational change in the IMC domain, which can potentially destabilize the global conformation of the IMC domain and/or may increase the exposure of the internal cleavage site to trans-proteolysis. Biophysical and computational studies are currently underway to elaborate the mechanism by which the protonation of H_{69} facilitates activation and preliminary results suggest that (i) His_{69} protonation may induce local conformational changes and (ii) additional residues may be required to optimize the sensitivity of furin and other PCs to their compartment-specific activation (Shinde et al. unpublished data).

It is interesting to note that residue H_{69} in furin is absolutely conserved in all PCs. However, unlike furin, other PC family members become activated at different pHs. For example, similar to furin, PC1 is synthesized in the ER and traverses the TGN network but only becomes active in the mature dense secretory granules where the pH is approximately 5.5. If the conserved His residue is alone responsible for pH sensing, *why*

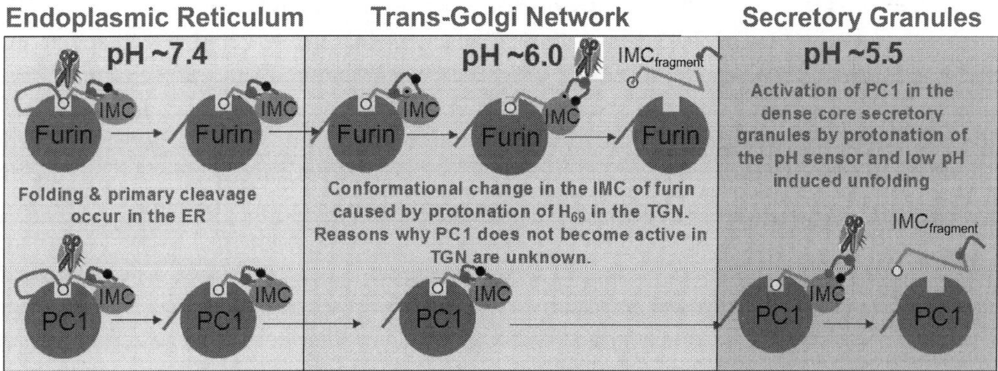

Fig. 4.11. A cartoon model depicting putative mechanism for activation of furin and PC1/PC3 through protonation of H_{69}. Both furin and PC1/PC3 undergo their first cleavage at a pair of conserved dibasic residues. This cleavage allows the non-covalent complexes of furin and PC1/PC3 to exit the ER into the TGN. Here, His_{69}, the pH sensor in furin, gets protonated and induces a conformational change within the pH-sensitive loop. This exposes the internal cleavage site at Arg_{75}, which can undergo proteolysis that releases inhibition. Although the residue corresponding to His_{69} in furin is conserved, PC1/PC3 does not undergo the second internal cleavage and suggests that additional residues must be involved in fine-tuning the pH sensitivity of individual PCs. Upon internal cleavage the inhibitory C-terminal region of the IMC is released and renders the catalytic site free for proteolysis.

does PC1 not become activated in the TGN? It is tempting to hypothesize that there are hitherto unknown determinants that are additionally required to allow different PCs to have different sensitivities for pH-dependent activation. For example, the determinants that enable the isolated IMC domains of furin and PC1 to display different conformational preferences may play a vital role in activation. Additional determinants could also be located within the P-domains that are found exclusively within eukaryotic PCs and which are completely absent in prokaryotic subtilases. This is consistent with the data that suggest that the P-domains play an important role in regulating stability, calcium dependence, and pH dependence of PCs. Alternately, there may be additional sequence determinants within IMCs that have yet to be identified. For example, IMCs have a higher preponderance of charges than their cognate catalytic domains. Moreover, the type of charges can vary significantly between different IMC domains, which may impart unique sensitivity to specific local environments within the IMC domain alone. Every amino acid has a unique p*K*a value. When one considers amino acids with charged side chains, aspartic acid and glutamic acid have carboxyl groups on their side chains and are fully ionized at pH 7.4 while arginine and lysine have amino groups as side chains and are fully protonated at pH 7.4. The p*K*a for histidine, which contains an imidazole ring for its side chain, is the closest to physiological pH and hence profoundly affects its chemical properties. Since every amino acid has a unique p*K*a value, a combination of specific charged and polar residues and their surface accessibility can dramatically affect the functional properties of the IMC domain through net charge at a particular pH.

The IMC domains of aqualysin and subtilisin, which are prokaryotic enzymes, have pIs that are 4.8 and 9.8, respectively. While these IMCs can be interchanged and the chimeric protease can fold correctly in vitro, they do display dramatically different affinities and therefore different protease activation profiles for their catalytic domains. Surprisingly, the IMC domains have lower affinity for their own catalytic domains, suggesting that the binding affinity of an IMC domain may be optimized for its cognate catalytic domain, so as to enable efficient spatial and temporal activation. The protease aqualysin was isolated from *Thermus aquaticus*, a strain that tolerates extreme temperatures and was first discovered in the Lower Geyser Basin of Yellowstone National Park. Subtilisin on the other hand was identified in *B. subtilis*, a bacterium commonly found in soil and thrives in much milder environmental conditions. In *B. subtilis* the protease is extracellular and is secreted under conditions of starvation to allow the bacterium to scavenge amino acids by breaking down decaying proteins from the soil. To activate the protease, the IMC of subtilisin has to respond to conditions in the soil, which are normally not extreme. On the other hand, aqualysin requires very extreme conditions of temperature and pH in order to become active; and unlike the intrinsically unstructured IMC of subtilisin, the IMC domain from aqualysin is well structured and more stable at room temperature. However, at 55–60°C aqualysin's IMC domain loses structure and can be released to facilitate rapid activation. Although the IMC domains of these proteases can be swapped and can facilitate correct folding at room temperature in vitro, it is likely that the subtilisin IMC would not function at the temperature where *T. aquaticus* survives. Similarly, while the aqualysin IMC domain folds subtilisin, the activation of the catalytic domain through IMC release and degradation requires drastically different conditions. This is because the aqualysin IMC is structured and binds subtilisin with an affinity that is more than 10-fold stronger.

Both aqualysin and subtilisin are ESPs. *How do they differ from ISPs?* This question can be answered by examining ESPs and ISPs from *Bacillus* sp. As mentioned earlier, ESPs are extracellular, kinetically stable, secreted proteases, while ISPs are thermodynamically stable intracellular proteases (111). It is important to note that the observed differences between folding pathways of ISPs and ESPs are not mere consequences of their cellular location because the intracellular or extracellular expression of ESPs does not yield an active protease in the absence of its cognate propeptide (30). While the energy barriers associated with kinetic stability increase the stability – and hence longevity – of the catalytic domain in harsh extracellular environments, similar barriers leading to high stability in ISPs may impede intracellular protein turnover. This is consistent with our finding that intracellular expression of ESPs (aqualysin 1 and subtilisin E) is detrimental to cell growth due to extensive proteolysis (unpublished data). This suggests that biological requirements, and not just specific conformations, dictate selection of folding pathways. We hypothesize that IMC-mediated folding of PCs has evolved to enable their precise spatiotemporal activation within a complex, compartment-specific environment within a eukaryotic cell (**Fig. 4.11**).

IMC-dependent folding mechanisms are also found within the cathepsins, which function within the acid environment of the lysosome (pH ~ 5). While the activation of lysosomal cathepsins, like PCs, is also specific and pH dependent, the precise mechanisms by which these proteases sense lysosomal pH are unknown. In some cases propeptides are known to regulate different types of cellular processes such as transport and localization, hierarchical organization, or oligomerization and regulation of protein activity–function by indirectly modulating protein conformations and can be termed as "post-translational modulators" of protein structure and function (29, 43).

10. Implication of pH Sensing for Biological Functions of PCs and Cathepsins

Changes in intracellular pH regulate numerous normal and pathological processes in the cell. Such changes in intracellular pH are permissive for growth factor-induced cell proliferation, cell cycle progression, and differentiation and are necessary for haptokinetic migration and amoeboid chemotaxis. Additionally, increases in cytosolic pH are a hallmark of transformed cells from different tissue origins and genetic backgrounds, making it a common characteristic of distinct cancers and possibly a common critical driving force for tumor progression. Since PCs are involved in the activation of matrix metalloproteases, growth factors, adhesion molecules, and which are crucial for cellular transformation, acquisition of the tumorigenic phenotype, and metastases formation, it is not surprising that they play major roles in tumor progression and malignancy. Experiments suggest that inhibition of PCs can alter the malignant phenotype of various tumor cells and can be used to target tumor angiogenesis. Measurement of pH inside tissue suggests that the microenvironment within tumors is generally more acidic than normal tissues. Major contributors to tumor acidity likely include the excessive production of lactic acid and ATP hydrolysis in hypoxic zones of tumors. Additional reduction in tumor pH may be achieved by administrating glucose (and/or insulin) and hydralazine, a drug which modifies the relative blood flow to tumors and normal tissues. *How would the increased acidification of tumors enhance the activation of PCs?* A more acidic pH is likely to cause premature activation of furin within the ER. This in turn may alter cellular homeostasis through premature, unregulated proteolysis. It is also known that vesicles shed by cancer cells mediate several tumor–host interactions, and that the tumor microenvironment influences the release and the activity of tumor-shed microvesicles. Such vesicles contain cathepsin-B, the activity of which was measured to be significantly increased at an acidic pH. The enhanced cathepsin-B activity stimulates both the activation of gelatinase and the invasiveness of endothelial cells observed at low pH values. As a consequence, the acidic microenvironment found in most solid tumors may contribute to protease-mediated proinvasive capabilities of tumor-shed vesicles.

In conclusion it is clear that the pH gradient within the secretory and endocytic pathway is essential for normal homeostasis (23). Cells have evolved multiple mechanisms to maintain this pH gradient and there is accumulating evidence that altered pH regulation is a hallmark of several disease states (127). Although there are tantalizing links between altered pH homeostasis and disease there is no definitive example where acidification is a primary cause of disease pathogenesis. However, this is likely to change through an increase in our understanding of how eukaryotic proteins may have evolved mechanisms to exploit such compartment-specific pH for their biological function.

Acknowledgements

Our apologies to colleagues whose work we did not cite because of space limitations. Special thanks go to Jimmy Dikeakos, David Radler, Laura Figoski, and Stephanie Dillon for insightful discussions, for review of the manuscript, and for editing. U.S. is supported by grants from the National Science Foundation (NSF-0746589) and the Oregon Nanoscience and Microtechnologies Institute and G.T. is supported by grant from the National Institutes of Health (DK37274 and CA151564).

References

1. Nakayama, K. (1997) Furin: A mammalian subtilisin/Kex2p-like endoprotease involved in processing of a wide variety of precursor proteins *Biochem J* **327**(Pt 3), 625–35.
2. Thomas, G. (2002) Furin at the cutting edge: From protein traffic to embryogenesis and disease *Nat Rev Mol Cell Biol* **3**, 753–66.
3. Fuller, R. S., Brake, A., and Thorner, J. (1989) Yeast prohormone processing enzyme (KEX2 gene product) is a Ca2+-dependent serine protease *Proc Natl Acad Sci USA* **86**, 1434–8.
4. Fuller, R. S., Brake, A. J., and Thorner, J. (1989) Intracellular targeting and structural conservation of a prohormone-processing endoprotease *Science* **246**, 482–6.
5. Fuller, R. S., Sterne, R. E., and Thorner, J. (1988) Enzymes required for yeast prohormone processing *Annu Rev Physiol* **50**, 345–62.
6. Thomas, G., Thorne, B. A., Thomas., L. et al. (1988) Yeast KEX2 endopeptidase correctly cleaves a neuroendocrine prohormone in mammalian cells *Science* **241**, 226–30.
7. Bergeron, F., Leduc, R., and Day, R. (2000) Subtilase-like pro-protein convertases: From molecular specificity to therapeutic applications *J Mol Endocrinol* **24**, 1–22.
8. Seidah, N. G., Benjannet, S., Wickham, L. et al. (2003) The secretory proprotein convertase neural apoptosis-regulated convertase 1 (NARC-1): Liver regeneration and neuronal differentiation *Proc Natl Acad Sci USA* **100**, 928–33.
9. Seidah, N. G., Mowla, S. J., Hamelin, J. et al. (1999) Mammalian subtilisin/kexin isozyme SKI-1: A widely expressed proprotein convertase with a unique cleavage specificity and cellular localization *Proc Natl Acad Sci USA* **96**, 1321–6.
10. Toure, B. B., Munzer, J. S., Basak, A. et al. (2000) Biosynthesis and enzymatic characterization of human SKI-1/S1P and the processing of its inhibitory prosegment *J Biol Chem* **275**, 2349–58.
11. Seidah, N. G., Mayer, G., Zaid, A. et al. (2008) The activation and physiological functions of the proprotein convertases *Int J Biochem Cell Biol* **40**, 1111–25.
12. Lambert, G., Charlton, F., Rye, K. A., and Piper, D. E. (2009) Molecular basis of PCSK9 function *Atherosclerosis* **203**, 1–7.
13. Siezen, R. J., and Leunissen, J. A. (1997) Subtilases: The superfamily of subtilisin-like serine proteases *Protein Sci* **6**, 501–23.
14. Rawlings, N. D., Barrett, A. J., and Bateman, A. (2009) MEROPS: The peptidase database *Nucleic Acids Res* **38**, D227–33.
15. Smith, E. L., Markland, F. S., Kasper, C. B., DeLange, R. J., Landon, M., and Evans, W. H. (1966) The complete amino acid sequence of two types of subtilisin, BPN' and Carlsberg *J Biol Chem* **241**, 5974–6.

16. Wright, C. S., Alden, R. A., and Kraut, J. (1969) Structure of subtilisin BPN' at 2.5 angstrom resolution *Nature* **221**, 235–42.
17. Bryan, P. N. (2000) Protein engineering of subtilisin *Biochim Biophys Acta* **1543**, 203–22.
18. Lipkind, G., Gong, Q., and Steiner, D. F. (1995) Molecular modeling of the substrate specificity of prohormone convertases SPC2 and SPC3 *J Biol Chem* **270**, 13277–84.
19. Oliva, A. A., Jr., Steiner, D. F., and Chan, S. J. (1995) Proprotein convertases in amphioxus: Predicted structure and expression of proteases SPC2 and SPC3 *Proc Natl Acad Sci USA* **92**, 3591–5.
20. Henrich, S., Cameron, A., Bourenkov, G. P. et al. (2003) The crystal structure of the proprotein processing proteinase furin explains its stringent specificity *Nat Struct Biol* **10**, 520–6.
21. Holyoak, T., Wilson, M. A., Fenn, T. D. et al. (2003) 2.4 A resolution crystal structure of the prototypical hormone-processing protease Kex2 in complex with an Ala-Lys-Arg boronic acid inhibitor *Biochemistry* **42**, 6709–18.
22. Henrich, S., Lindberg, I., Bode, W., and Than, M. E. (2005) Proprotein convertase models based on the crystal structures of furin and kexin: Explanation of their specificity *J Mol Biol* **345**, 211–27.
23. Demaurex, N. (2002) pH Homeostasis of cellular organelles *News Physiol Sci* **17**, 1–5.
24. Casey, J. R., Grinstein, S., and Orlowski, J. (2010) Sensors and regulators of intracellular pH *Nat Rev Mol Cell Biol* **11**, 50–61.
25. Anderson, E. D., Molloy, S. S., Jean, F., Fei, H., Shimamura, S., and Thomas, G. (2002) The ordered and compartment-specific autoproteolytic removal of the furin intramolecular chaperone is required for enzyme activation *J Biol Chem* **277**, 12879–90.
26. Feliciangeli, S. F., Thomas, L., Scott, G. K. et al. (2006) Identification of a pH sensor in the furin propeptide that regulates enzyme activation *J Biol Chem* **281**, 16108–16.
27. Ehrmann, M., and Clausen, T. (2004) Proteolysis as a regulatory mechanism *Annu Rev Genet* **38**, 709–24.
28. Anderson, E. D., VanSlyke, J. K., Thulin, C. D., Jean, F., and Thomas, G. (1997) Activation of the furin endoprotease is a multiple-step process: Requirements for acidification and internal propeptide cleavage *EMBO J* **16**, 1508–18.
29. Shinde, U., and Inouye, M. (2000) Intramolecular chaperones: Polypeptide extensions that modulate protein folding *Semin Cell Dev Biol* **11**, 35–44.
30. Ikemura, H., Takagi, H., and Inouye, M. (1987) Requirement of pro-sequence for the production of active subtilisin E in Escherichia coli *J Biol Chem* **262**, 7859–64.
31. Ikemura, H., and Inouye, M. (1988) In vitro processing of pro-subtilisin produced in Escherichia coli *J Biol Chem* **263**, 12959–63.
32. Zhu, X. L., Ohta, Y., Jordan, F., and Inouye, M. (1989) Pro-sequence of subtilisin can guide the refolding of denatured subtilisin in an intermolecular process *Nature* **339**, 483–4.
33. Shinde, U., Fu, X., and Inouye, M. (1999) A pathway for conformational diversity in proteins mediated by intramolecular chaperones *J Biol Chem* **274**, 15615–21.
34. Yabuta, Y., Takagi, H., Inouye, M., and Shinde, U. (2001) Folding pathway mediated by an intramolecular chaperone: Propeptide release modulates activation precision of pro-subtilisin *J Biol Chem* **276**, 44427–34.
35. Ohta, Y., Hojo, H., Aimoto, S. et al. (1991) Pro-peptide as an intramolecular chaperone: Renaturation of denatured subtilisin E with a synthetic pro-peptide [corrected] *Mol Microbiol* **5**, 1507–10.
36. Ohta, Y., and Inouye, M. (1990) Pro-subtilisin E: Purification and characterization of its autoprocessing to active subtilisin E in vitro *Mol Microbiol* **4**, 295–304.
37. Eder, J., and Fersht, A. R. (1995) Pro-sequence-assisted protein folding *Mol Microbiol* **16**, 609–14.
38. Shinde, U., and Inouye, M. (1993) Intramolecular chaperones and protein folding *Trends Biochem Sci* **18**, 442–6.
39. Ellis, R. J. (1993) The general concept of molecular chaperones *Philos Trans R Soc Lond B Biol Sci* **339**, 257–61.
40. Ellis, R. J. (2007) Protein misassembly: Macromolecular crowding and molecular chaperones *Adv Exp Med Biol* **594**, 1–13.
41. Ellis, R. J., and van der Vies, S. M. (1991) Molecular chaperones *Annu Rev Biochem* **60**, 321–47.
42. Shinde, U. P., Liu, J. J., and Inouye, M. (1997) Protein memory through altered folding mediated by intramolecular chaperones *Nature* **389**, 520–2.
43. Chen, Y. J., and Inouye., M. (2008) The intramolecular chaperone-mediated protein folding *Curr Opin Struct Biol* **18**, 765–70.
44. Silen, J. L., and Agard, D. A. (1989) The alpha-lytic protease pro-region does not require a physical linkage to activate the protease domain in vivo *Nature* **341**, 462–4.

45. Baker, D., Sohl, J. L., and Agard, D. A. (1992) A protein-folding reaction under kinetic control *Nature* **356**, 263–5.
46. Jaswal, S. S., Sohl, J. L., Davis, J. H., and Agard, D. A. (2002) Energetic landscape of alpha-lytic protease optimizes longevity through kinetic stability *Nature* **415**, 343–6.
47. Valls, L. A., Hunter, C. P., Rothman, J. H., and Stevens, T. H. (1987) Protein sorting in yeast: The localization determinant of yeast vacuolar carboxypeptidase Y resides in the propeptide *Cell* **48**, 887–97.
48. Ammerer, G., Hunter, C. P., Rothman, J. H., Saari, G. C., Valls, L. A., and Stevens, T. H. (1986) PEP4 gene of Saccharomyces cerevisiae encodes proteinase A, a vacuolar enzyme required for processing of vacuolar precursors *Mol Cell Biol* **6**, 2490–9.
49. Ramos, C., Winther, J. R., and Kielland-Brandt, M. C. (1994) Requirement of the propeptide for in vivo formation of active yeast carboxypeptidase Y *J Biol Chem* **269**, 7006–12.
50. Winther, J. R., and Sorensen, P. (1991) Propeptide of carboxypeptidase Y provides a chaperone-like function as well as inhibition of the enzymatic activity *Proc Natl Acad Sci USA* **88**, 9330–4.
51. Winther, J. R., Sorensen, P., and Kielland-Brandt, M. C. (1994) Refolding of a carboxypeptidase Y folding intermediate in vitro by low-affinity binding of the proregion *J Biol Chem* **269**, 22007–13.
52. van den Hazel, H. B., Kielland-Brandt, M. C., and Winther, J. R. (1993) The propeptide is required for in vivo formation of stable active yeast proteinase A and can function even when not covalently linked to the mature region *J Biol Chem* **268**, 18002–7.
53. Smith, S. M., and Gottesman, M. M. (1989) Activity and deletion analysis of recombinant human cathepsin L expressed in Escherichia coli *J Biol Chem* **264**, 20487–95.
54. Ogino, T., Kaji, T., Kawabata, M. et al. (1999) Function of the propeptide region in recombinant expression of active procathepsin L in Escherichia coli *J Biochem (Tokyo)* **126**, 78–83.
55. Wiederanders, B., Kaulmann, G., and Schilling, K. (2003) Functions of propeptide parts in cysteine proteases *Curr Protein Pept Sci* **4**, 309–26.
56. Hou, W. S., Bromme, D., Zhao, Y. et al. (1999) Characterization of novel cathepsin K mutations in the pro and mature polypeptide regions causing pycnodysostosis *J Clin Invest* **103**, 731–8.
57. Schulz, E. C., Neumann, P., Gerardy-Schahn, R., Sheldrick, G. M., and Ficner, R. (2010) Structure analysis of endosialidase NF at 0.98 A resolution *Acta Crystallogr D Biol Crystallogr* **66**, 176–80.
58. Schulz, E. C., Schwarzer, D., Frank, M. et al. (2010) Structural basis for the recognition and cleavage of polysialic acid by the bacteriophage K1F tailspike protein EndoNF *J Mol Biol* **397**, 341–51.
59. McIver, K. S., Kessler, E., and Ohman, D. E. (2004) Identification of residues in the Pseudomonas aeruginosa elastase propeptide required for chaperone and secretion activities *Microbiology* **150**, 3969–77.
60. Tang, B., Nirasawa, S., Kitaoka, M., Marie-Claire, C., and Hayashi, K. (2003) General function of N-terminal propeptide on assisting protein folding and inhibiting catalytic activity based on observations with a chimeric thermolysin-like protease *Biochem Biophys Res Commun* **301**, 1093–8.
61. Gray, T. E., Eder, J., Bycroft, M., Day, A. G., and Fersht, A. R. (1993) Refolding of barnase mutants and pro-barnase in the presence and absence of GroEL *EMBO J* **12**, 4145–50.
62. Suter, U., Heymach, J. V., Jr., and Shooter, E. M. (1991) Two conserved domains in the NGF propeptide are necessary and sufficient for the biosynthesis of correctly processed and biologically active NGF *EMBO J* **10**, 2395–400.
63. Thorne, B. A., and Plowman, G. D. (1994) The heparin-binding domain of amphiregulin necessitates the precursor pro-region for growth factor secretion *Mol Cell Biol* **14**, 1635–46.
64. Steiner, D. F. (2004) The proinsulin C-peptide–a multirole model *Exp Diabesity Res* **5**, 7–14.
65. Voorberg, J., Fontijn, R., Calafat, J., Janssen, H., van Mourik, J. A., and Pannekoek, H. (1993) Biogenesis of von Willebrand factor-containing organelles in heterologous transfected CV-1 cells *EMBO J* **12**, 749–58.
66. Weissman, J. S., and Kim, P. S. (1992) The pro region of BPTI facilitates folding *Cell* **71**, 841–51.
67. Morgunova, E., Tuuttila, A., Bergmann, U. et al. (1999) Structure of human pro-matrix metalloproteinase-2: Activation mechanism revealed *Science* **284**, 1667–70.
68. Baker, D., Shiau, A. K., and Agard, D. A. (1993) The role of pro regions in protein folding *Curr Opin Cell Biol* **5**, 966–70.
69. Baker, D., and Agard, D. A. (1994) Kinetics versus thermodynamics in protein folding *Biochemistry* **33**, 7505–9.

70. Shinde, U., and Inouye, M. (1994) The structural and functional organization of intramolecular chaperones: The N-terminal propeptides which mediate protein folding *J Biochem (Tokyo)* **115**, 629–36.
71. Kojima, S., Iwahara, A., and Yanai, H. (2005) Inhibitor-assisted refolding of protease: A protease inhibitor as an intramolecular chaperone *FEBS Lett* **579**, 4430–6.
72. Yabuta, Y., Subbian, E., Oiry, C., and Shinde, U. (2003) Folding pathway mediated by an intramolecular chaperone. A functional peptide chaperone designed using sequence databases *J Biol Chem* **278**, 15246–51.
73. Li, Y., Hu, Z., Jordan, F., and Inouye, M. (1995) Functional analysis of the propeptide of subtilisin E as an intramolecular chaperone for protein folding. Refolding and inhibitory abilities of propeptide mutants *J Biol Chem* **270**, 25127–32.
74. Kobayashi, T., and Inouye, M. (1992) Functional analysis of the intramolecular chaperone. Mutational hot spots in the subtilisin pro-peptide and a second-site suppressor mutation within the subtilisin molecule *J Mol Biol* **226**, 931–3.
75. Buevich, A. V., Shinde, U. P., Inouye, M., and Baum, J. (2001) Backbone dynamics of the natively unfolded pro-peptide of subtilisin by heteronuclear NMR relaxation studies *J Biomol NMR* **20**, 233–49.
76. Jain, S. C., Shinde, U., Li, Y., Inouye, M., and Berman, H. M. (1998) The crystal structure of an autoprocessed Ser221Cys-subtilisin E-propeptide complex at 2.0 A resolution *J Mol Biol* **284**, 137–44.
77. Shinde, U., and Inouye, M. (1995b) Folding pathway mediated by an intramolecular chaperone: Characterization of the structural changes in pro-subtilisin E coincident with autoprocessing *J Mol Biol* **252**, 25–30.
78. Shinde, U., Li, Y., Chatterjee, S., and Inouye, M. (1993) Folding pathway mediated by an intramolecular chaperone *Proc Natl Acad Sci USA* **90**, 6924–8.
79. Ruan, B., Hoskins, J., and Bryan, P. N. (1999) Rapid folding of calcium-free subtilisin by a stabilized pro-domain mutant *Biochemistry* **38**, 8562–71.
80. Ruan, B., Hoskins, J., Wang, L., and Bryan, P. N. (1998) Stabilizing the subtilisin BPN' pro-domain by phage display selection: How restrictive is the amino acid code for maximum protein stability? *Protein Sci* **7**, 2345–53.
81. Marie-Claire, C., Yabuta, Y., Suefuji, K., Matsuzawa, H., and Shinde, U. (2001) Folding pathway mediated by an intramolecular chaperone: The structural and functional characterization of the aqualysin I propeptide *J Mol Biol* **305**, 151–65.
82. Gallagher, T., Gilliland, G., Wang, L., and Bryan, P. (1995) The prosegment-subtilisin BPN' complex: Crystal structure of a specific 'foldase' *Structure* **3**, 907–14.
83. Radisky, E. S., King, D. S., Kwan, G., and Koshland, D. E., Jr. (2003) The role of the protein core in the inhibitory power of the classic serine protease inhibitor, chymotrypsin inhibitor 2 *Biochemistry* **42**, 6484–92.
84. Alexander, P. A., Ruan, B., and Bryan, P. N. (2001) Cation-dependent stability of subtilisin *Biochemistry* **40**, 10634–9.
85. Yabuta, Y., Subbian, E., Takagi, H., Shinde, U., and Inouye, M. (2002) Folding pathway mediated by an intramolecular chaperone: Dissecting conformational changes coincident with autoprocessing and the role of Ca(2+) in subtilisin maturation *J Biochem (Tokyo)* **131**, 31–7.
86. Inouye, M., Fu, X., and Shinde, U. (2001) Substrate-induced activation of a trapped IMC-mediated protein folding intermediate *Nat Struct Biol* **8**, 321–5.
87. Shinde, U., and Inouye, M. (1995a) Folding mediated by an intramolecular chaperone: Autoprocessing pathway of the precursor resolved via a substrate assisted catalysis mechanism *J Mol Biol* **247**, 390–5.
88. Bryan, P., Wang, L., Hoskins, J. et al. (1995) Catalysis of a protein folding reaction: Mechanistic implications of the 2.0 A structure of the subtilisin-prodomain complex *Biochemistry* **34**, 10310–18.
89. Li, Y., and Inouye, M. (1996) The mechanism of autoprocessing of the propeptide of prosubtilisin E: Intramolecular or intermolecular event? *J Mol Biol* **262**, 591–4.
90. Comellas-Bigler, M., Maskos, K., Huber, R., Oyama, H., Oda, K., and Bode, W. (2004) 1.2 A crystal structure of the serine carboxyl proteinase pro-kumamolisin; structure of an intact pro-subtilase *Structure (Camb)* **12**, 1313–23.
91. Eder, J., Rheinnecker, M., and Fersht, A. R. (1993) Folding of subtilisin BPN': Characterization of a folding intermediate *Biochemistry* **32**, 18–26.
92. Kuwajima, K. (1989) The molten globule state as a clue for understanding the folding and cooperativity of globular-protein structure *Proteins* **6**, 87–103.
93. Eder, J., Rheinnecker, M., and Fersht, A. R. (1993) Folding of subtilisin BPN': Role of the pro-sequence *J Mol Biol* **233**, 293–304.

94. Bryan, P., Alexander, P., Strausberg, S. et al. (1992) Energetics of folding subtilisin BPN' *Biochemistry* **31**, 4937–45.
95. Sohl, J. L., Jaswal, S. S., and Agard, D. A. (1998) Unfolded conformations of alpha-lytic protease are more stable than its native state *Nature* **395**, 817–19.
96. Alexander, P. A., Ruan, B., Strausberg, S. L., and Bryan, P. N. (2001) Stabilizing mutations and calcium-dependent stability of subtilisin *Biochemistry* **40**, 10640–4.
97. Almog, O., Gallagher, T., Tordova, M., Hoskins, J., Bryan, P., and Gilliland, G. L. (1998) Crystal structure of calcium-independent subtilisin BPN' with restored thermal stability folded without the prodomain *Proteins* **31**, 21–32.
98. Wang, L., Ruan, B., Ruvinov, S., and Bryan, P. N. (1998) Engineering the independent folding of the subtilisin BPN' pro-domain: Correlation of pro-domain stability with the rate of subtilisin folding *Biochemistry* **37**, 3165–71.
99. Ruvinov, S., Wang, L., Ruan, B. et al. (1997) Engineering the independent folding of the subtilisin BPN' prodomain: Analysis of two-state folding versus protein stability *Biochemistry* **36**, 10414–21.
100. Fu, X., Inouye, M., and Shinde, U. (2000) Folding pathway mediated by an intramolecular chaperone. The inhibitory and chaperone functions of the subtilisin propeptide are not obligatorily linked *J Biol Chem* **275**, 16871–8.
101. Shastry, M. C., and Udgaonkar, J. B. (1995) The folding mechanism of barstar: Evidence for multiple pathways and multiple intermediates *J Mol Biol* **247**, 1013–27.
102. Hayashi, T., Matsubara, M., Nohara, D., Kojima, S., Miura, K., and Sakai, T. (1994) Renaturation of the mature subtilisin BPN' immobilized on agarose beads *FEBS Lett* **350**, 109–12.
103. Matsubara, M., Kurimoto, E., Kojima, S., Miura, K., and Sakai, T. (1994) Achievement of renaturation of subtilisin BPN' by a novel procedure using organic salts and a digestible mutant of Streptomyces subtilisin inhibitor *FEBS Lett* **342**, 193–6.
104. Yabuta, Y., Subbian, E., Takagi, H., Shinde, U., and Inouye, M. (2002) Folding pathway mediated by an intramolecular chaperone: Dissecting conformational changes coincident with autoprocessing and the role of Ca(2+) in subtilisin maturation *J Biochem* **131**, 31–7.
105. Anfinsen, C. B. (1973) Principles that govern the folding of protein chains *Science* **181**, 223–30.
106. Karplus, M. (1997) The Levinthal paradox: Yesterday and today *Fold Des* **2**, S69–75.
107. Franke, A. E., Danley, D. E., Kaczmarek, F. S. et al. (1990) Expression of human plasminogen activator inhibitor type-1 (PAI-1) in Escherichia coli as a soluble protein comprised of active and latent forms. Isolation and crystallization of latent PAI-1 *Biochim Biophys Acta* **1037**, 16–23.
108. Huntington, J. A., Pannu, N. S., Hazes, B., Read, R. J., Lomas, D. A., and Carrell, R. W. (1999) A 2.6 A structure of a serpin polymer and implications for conformational disease *J Mol Biol* **293**, 449–55.
109. Huntington, J. A., Read, R. J., and Carrell, R. W. (2000) Structure of a serpin-protease complex shows inhibition by deformation *Nature* **407**, 923–6.
110. Subbian, E., Yabuta, Y., and Shinde, U. P. (2005) Folding pathway mediated by an intramolecular chaperone: Intrinsically unstructured propeptide modulates stochastic activation of subtilisin *J Mol Biol* **347**, 367–83.
111. Subbian, E., Yabuta, Y., and Shinde, U. (2004) Positive selection dictates the choice between kinetic and thermodynamic protein folding and stability in subtilases *Biochemistry* **43**, 14348–60.
112. Bryan, P. N. (2002) Prodomains and protein folding catalysis *Chem Rev* **102**, 4805–16.
113. Siezen, R. J. (1996) Subtilases: Subtilisin-like serine proteases *Adv Exp Med Biol* **379**, 75–93.
114. Tangrea, M. A., Bryan, P. N., Sari, N., and Orban, J. (2002) Solution structure of the pro-hormone convertase 1 pro-domain from Mus musculus *J Mol Biol* **320**, 801–12.
115. Tangrea, M. A., Alexander, P., Bryan, P. N., Eisenstein, E., Toedt, J., and Orban, J. (2001) Stability and global fold of the mouse prohormone convertase 1 pro-domain *Biochemistry* **40**, 5488–95.
116. Nour, N., Basak, A., Chretien, M., and Seidah, N. G. (2003) Structure-function analysis of the prosegment of the proprotein convertase PC5A *J Biol Chem* **278**, 2886–95.
117. Nour, N., Mayer, G., Mort, J. S. et al. (2005) The cysteine-rich domain of the secreted proprotein convertases PC5A and PACE4 functions as a cell surface anchor and interacts with tissue inhibitors of metalloproteinases *Mol Biol Cell* **16**, 5215–26.
118. Dikeakos, J. D., Lacombe, M. J., Mercure, C., Mireuta, M., and Reudelhuber, T. L. (2007) A hydrophobic patch in a charged alpha-helix is sufficient to target proteins to

dense core secretory granules *J Biol Chem* **282**, 1136–43.
119. Malfait, A. M., Arner, E. C., Song, R. H. et al. (2008) Proprotein convertase activation of aggrecanases in cartilage in situ *Arch Biochem Biophys* **478**, 43–51.
120. Jutras, I., Seidah, N. G., Reudelhuber, T. L., and Brechler, V. (1997) Two activation states of the prohormone convertase PC1 in the secretory pathway *J Biol Chem* **272**, 15184–8.
121. Leduc, R., Molloy, S. S., Thorne, B. A., and Thomas, G. (1992) Activation of human furin precursor processing endoprotease occurs by an intramolecular autoproteolytic cleavage *J Biol Chem* **267**, 14304–8.
122. Wan, L., Molloy, S. S., Thomas, L. et al. (1998) PACS-1 defines a novel gene family of cytosolic sorting proteins required for trans-Golgi network localization *Cell* **94**, 205–16.
123. Kucerova, H., and Chaloupka, J. (1995) Intracellular serine proteinase behaves as a heat-stress protein in nongrowing but as a cold-stress protein in growing populations of Bacillus megaterium *Curr Microbiol* **31**, 39–43.
124. Kucerova, H., Hlavacek, O., Vachova, L., Mlichova, S., and Chaloupka, J. (2001) Differences in the regulation of the intracellular Ca2+-dependent serine proteinase activity between Bacillus subtilis and B. megaterium *Curr Microbiol* **42**, 178–83.
125. Vachova, L. (1996) Activation of the intracellular Ca(2+)-dependent serine protease ISP1 of bacillus megaterium by purification or by high Ca2+ concentrations *Biochem Mol Biol Int* **40**, 947–54.
126. Vachova, L., Kucerova, H., Benesova, J., and Chaloupka, J. (1994) Heat and osmotic stress enhance the development of cytoplasmic serine proteinase activity in sporulating Bacillus megaterium *Biochem Mol Biol Int* **32**, 1049–57.
127. Weisz, O. A. (2003) Organelle acidification and disease *Traffic* **4**, 57–64.

Chapter 5

The Novel Role of Cathepsin L for Neuropeptide Production Illustrated by Research Strategies in Chemical Biology with Protease Gene Knockout and Expression

Lydiane Funkelstein and Vivian Hook

Abstract

Neuropeptides are essential for cell–cell communication in the nervous and endocrine systems. Production of active neuropeptides requires proteolytic processing of proneuropeptide precursors in secretory vesicles that produce, store, and release neuropeptides that regulate physiological functions. This review describes research strategies utilizing chemical biology combined with protease gene knockout and expression to demonstrate the key role of cathepsin L for production of neuropeptides in secretory vesicles. Cathepsin L was discovered using activity-based probes and mass spectrometry to identify proenkephalin cleaving activity as cathepsin L. Significantly, in vivo protease gene knockout and expression approaches illustrate the key role of cathepsin L for neuropeptide production. Notably, cathepsin L is colocalized with neuropeptide secretory vesicles, the major site of proteolytic processing of proneuropeptides to generate active neuropeptides. Cathepsin L participates in producing opioid neuropeptides consisting of enkephalin, β-endorphin, and dynorphin, as well as in generating the POMC-derived peptide hormones ACTH and α-MSH. In addition, NPY, CCK, and catestatin neuropeptides utilize cathepsin L for their biosynthesis. The role of cathepsin L for neuropeptide production indicates its unique biological role in secretory vesicles, which contrasts with its role in lysosomes for protein degradation. Interesting evaluations of protease gene knockout studies in mice that lack cathepsin L compared to the PC1/3 and PC2 (PC, prohormone convertase) indicate the significant role of cathepsin L in neuropeptide production. Thus, dual cathepsin L and prohormone convertase protease pathways participate in neuropeptide production. These recent new findings indicate cathepsin L as a novel 'proprotein convertase' for production of neuropeptides that mediate cell–cell communication in health and disease.

Key words: Neuropeptides, proteases, chemical biology, mass spectrometry, gene knockout, gene expression, immunofluorescence confocal microscopy, prohormone convertase, aminopeptidase, carboxypeptidase, neuroendocrine.

M. Mbikay, N.G. Seidah (eds.), *Proprotein Convertases*, Methods in Molecular Biology 768,
DOI 10.1007/978-1-61779-204-5_5, © Springer Science+Business Media, LLC 2011

1. Introduction

1.1. Neuropeptides for Cell–Cell Communication in the Nervous and Endocrine Systems

Neuropeptides in the nervous system are essential for activity-dependent neurotransmission of information among neurons, and peripheral peptides are required for endocrine regulation of physiological functions. Moreover, the nervous and endocrine systems communicate with one another via these peptide neurotransmitters and hormones, collectively known as neuropeptides. Knowledge of the biosynthetic mechanisms for the production of neuropeptides is critical for understanding cell–cell communication in neurotransmission and peptide hormone actions.

Production of neuropeptides requires proteolytic processing of their respective precursor proteins. This results in a multitude of distinct peptides with diverse physiological actions, such as enkephalin and opioid peptide regulation of analgesia (1, 2), ACTH induction of steroid synthesis (3), galanin involvement in cognition (4), neuropeptide Y participation in regulating feeding behavior (5, 6), and numerous other neuroendocrine functions (**Table 5.1**). The primary structures for prohormones indicate that neuropeptides within the precursors are typically flanked at

Table 5.1
Neuropeptides in the nervous and endocrine systems

Neuropeptide	Regulatory function
Enkephalin	Analgesia
β-Endorphin	Analgesia
Dynorphin	Analgesia
ACTH	Steroid production
α-MSH	Skin pigmentation
Insulin	Glucose metabolism
Glucagon	Glucose metabolism
Galanin	Cognition
NPY	Blood pressure (peripheral) and obesity (CNS)
Somatostatin	Growth regulation
Vasopressin	Water balance

Peptide neurotransmitters and hormones are collectively termed neuropeptides. Examples of several neuropeptides and their regulatory functions are listed. Abbreviations are adrenocorticotropic hormone (ACTH), α-melanocyte stimulating hormone (α-MSH), and neuropeptide Y (NPY).

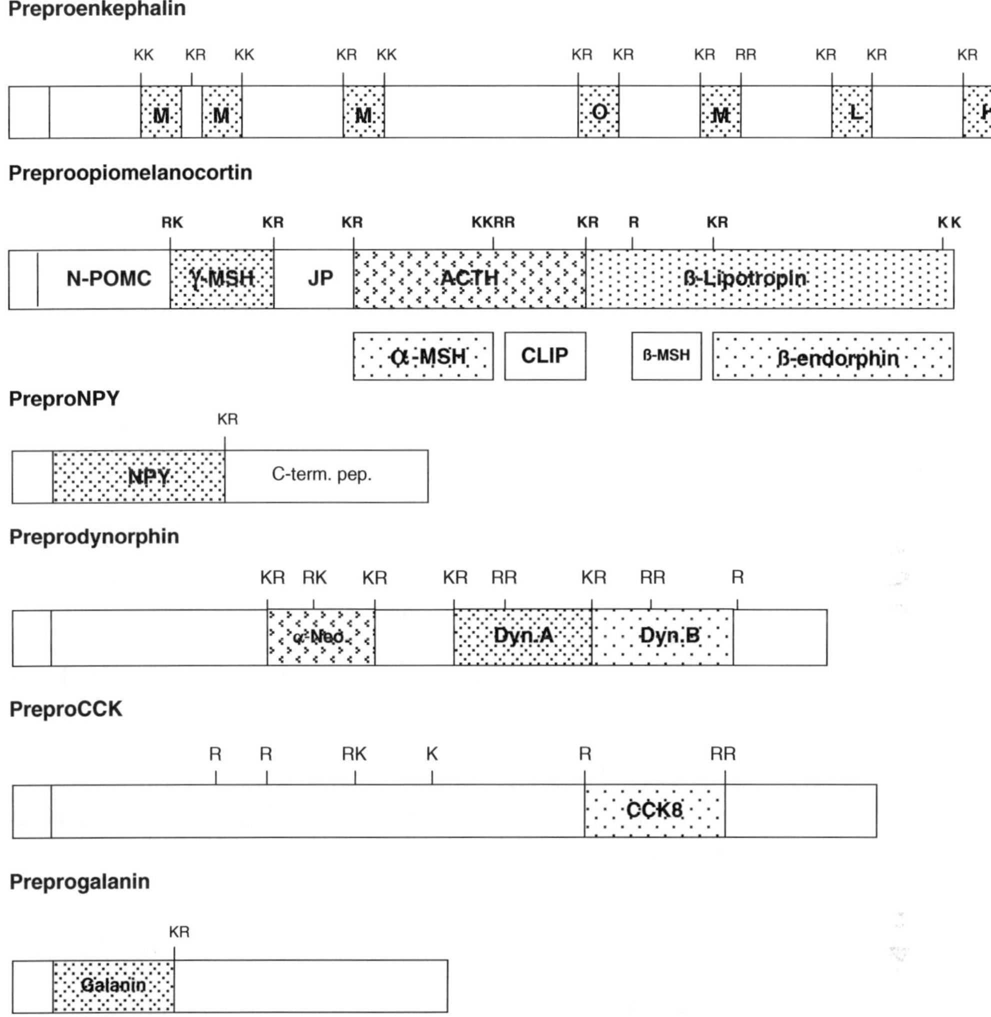

Fig. 5.1. Structural features of proneuropeptides for proteolytic processing. Prohormone precursor protein structures indicate that active peptide neurotransmitters and hormones are flanked by multibasic residues that represent sites of proteolytic processing to generate active neuropeptides. The precursor proteins are shown for preproenkephalin, preproopiomelanocortin, preproNPY (NPY, neuropeptide Y), preprodynorphin, preproCCK (CCK, cholecystokinin), and preprogalanin. The NH_2-terminal signal sequence is known to be cleaved by signal peptidases at the RER (rough endoplasmic reticulum) and the resultant prohormone undergoes trafficking to Golgi apparatus and packaged into secretory vesicles where prohormone processing occurs.

their NH_2- and COOH-termini by pairs of basic residues and sometimes by monobasic residues (7–9) (**Fig. 5.1**). These multibasic and monobasic residues provide sites of proteolytic processing that are cleaved to generate the active neuropeptides. Clearly, proteolysis represents key steps for the biosynthesis of essential peptide neurotransmitters and hormones.

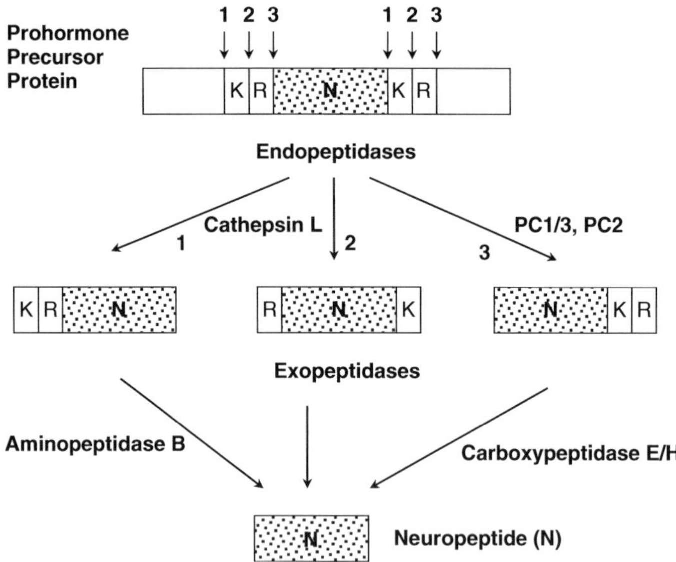

Fig. 5.2. Cathepsin L and prohormone convertase pathways for neuropeptide production. Proneuropeptides, also known as prohormones, typically contain active peptides flanked by paired basic residues. The dibasic processing sites undergo proteolytic cleavage at one of three sites (numbered 1, 2, and 3) which consist of cleavage at the NH_2- or COOH-terminal sides of the dibasic residues or between the dibasic residues. Peptide intermediates generated by cleavage at the NH_2-terminal side of the dibasic site, or between the dibasic residues, will then require Arg/Lys aminopeptidase, represented by aminopeptidase B, to remove basic residues at the NH_2-termini. Cleavage of proneuropeptides at the COOH-terminal side of dibasic residues then requires carboxypeptidase E to remove NH_2-terminal basic residues.

Proteolytic cleavage of prohormones may occur at one of three positions at paired basic processing sites. These cleavages may consist of processing at the COOH- and NH_2-termini of the dibasic residues or between the dibasic residues (**Fig. 5.2**). Resultant peptide intermediates require removal of basic residues from COOH- and/or NH_2-termini by carboxypeptidase and aminopeptidase enzymes, respectively.

1.2. Interdisciplinary Research Strategies Demonstrate Cathepsin L as a Significant and Novel Proprotein Convertase for Neuropeptide Production

Significantly, recent strategies utilizing chemical biology with protease gene knockout and expression demonstrate cathepsin L, with aminopeptidase B, as a newly identified protease pathway (7, 10–18) for production of active neuropeptides. Cathepsin L cleaves dibasic residue processing sites at their NH_2-termini and between the two basic residue sites of proneuropeptides (7, 10–16, 19). The peptide intermediates generated by cathepsin L subsequently require removal of NH-terminal and COOH-terminal amino acid extensions by aminopeptidase B and carboxypeptidase E, respectively, for production of the final neuropeptide. These findings complement ongoing studies in the field for prohormone processing by the prohormone convertases, consisting of PC1/3 and PC2 as well as related PC enzymes (7–9). Subsequent to PC activity, the final neuropeptides are generated by carboxypeptidase E (20, 21).

The novel biological role of cathepsin L for prohormone processing in regulated secretory vesicles is the focus of this review. The prominent role of cathepsin L in the production of active peptides contrasts with its previously known function as a lysosomal protease. These recent results demonstrate cathepsin L as a novel 'proprotein convertase' for production of neuropeptides. Thus, this review will focus on the interdisciplinary research strategies utilized to identify and demonstrate cathepsin L as a proneuropeptide processing for biosynthesis of neuropeptides in secretory vesicles.

2. Research Strategy Using Chemical Biology with Protease Gene Knockout and Expression Approaches Indicates Participation of Cathepsin L in the Production of the Enkephalin Neuropeptide

The biochemical strategy to elucidate the major proenkephalin cleaving activity in neuropeptide-containing secretory vesicles was to identify the protease subclass for the activity and identify the responsible enzyme protein by activity-probe labeling followed by mass spectrometry (10). Model neuropeptide-containing secretory vesicles isolated from sympathoadrenal chromaffin cells of the sympathetic nervous system were utilized for purification of the proenkephalin cleaving activity. The activity was substantially inhibited by selective inhibitors of cysteine proteases (20–22).

2.1. Identification of Cathepsin L in Proenkephalin Processing by Activity-Based Profiling and Mass Spectrometry

Chemical biology has developed sophisticated activity-based probes for identification of protease and enzyme families (23). Activity-based profiling of active cysteine proteases was instrumental for identification of the protease responsible for proenkephalin cleaving activity in chromaffin granules. The activity probe DCG-04, the biotinylated form of E64c that inhibits cysteine proteases, was utilized for specific affinity labeling of the 27 kDa protease enzyme of the proenkephalin cleaving enzyme activity (10). Two-dimensional gels resolved DCG-04 labeled proteins of $\sim$27 kDa, whose identification was indicated as cathepsin L by mass spectrometry of tryptic peptides (**Fig. 5.3**). These findings suggested a new biological function for cathepsin L in secretory vesicles for producing the enkephalin neuropeptide.

2.2. Cathepsin L Localization in Secretory Vesicles That Contain Enkephalin and Numerous Neuropeptides

The biochemical identification of cathepsin L in secretory vesicles suggested the novel localization of cathepsin L in this organelle. Confirmation of the localization of cathepsin L within secretory vesicles (chromaffin granules) was achieved by immunoelectron microscopy and by colocalization with enkephalin and NPY immunofluorescence confocal microscopy (**Fig. 5.4**) (10, 12). Cathepsin L was also found to undergo cosecretion with

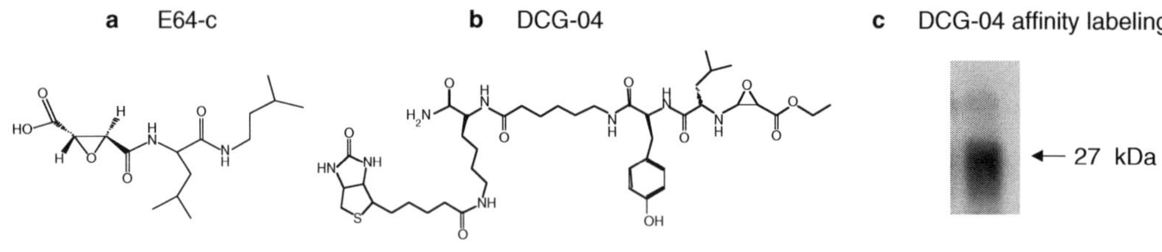

d Identification of Cathepsin L by Peptide Sequencing

^{1}MNPSFFLTVL CLGVASAAPK LDPNLDAHWH QWKATHRRLY GMNEEEWRRA VWEKNKKIID^{60}

^{61}LHNQEYSEGK HAFRMAMNAF GDMTNEEFRQ VMNGFQNQKH KKGKLFHEPL LVDVPKSVDW^{120}

^{121}TKKGYVTPVK NQGQCGSCWA FSATGALEGQ MFRKTGKLVS LSEQNLVDCS RAQGNQGCNG^{180}

^{181}GLMDNAFQYI KDNGGLDSEE SYPYLATDTN SCNYKPECSA ANDTGFVDIP QREKALMKAV^{240}

^{241}ATVGPISVAI DAGHTSFQFY KSGIYYDPDC SSKDLDHGVL VVGYGFEGTD SNNNKFWIVK^{300}

^{301}NSWGPEWGWN GYVKMAKDQN NHCGIATAAS YPTV^{334}

Fig. 5.3. Identification of proenkephalin cleaving activity in secretory vesicles as cathepsin L. (**a**) E64-c cysteine protease inhibitor. The cysteine protease inhibitor E64-c was found to be a potent inhibitor of the proenkephalin (PE) cleaving activity in secretory vesicles isolated from adrenal medullary chromaffin cells of the sympathetic nervous system. (**b**) Structure of DCG-04, an activity-based probe for cysteine proteases. The modified cysteine protease inhibitor DCG-04, resulting from biotinylation of E64-c, was utilized for affinity labeling of PE cleaving activity in secretory vesicles. (**c**) DCG-04 affinity labeling of cysteine protease activity in secretory vesicles. DCG-04 affinity labeling of purified PE cleaving activity reveals a 27 kDa protein band. This band was subjected to peptide sequencing by tryptic digestion and tandem mass spectrometry. (**d**) Identification of purified PE cleaving activity as cathepsin L by mass spectrometry for peptide sequencing. Peptides derived from tryptic digests of DCG-04 affinity-labeled 27 kDa proteins, sequenced by mass spectrometry, are illustrated as the *underlined* amino acid sequences of bovine cathepsin L.

enkephalin whose secretion is stimulated by activation of the regulated secretory pathway in these cells (10).

Cellular routing and trafficking of cathepsin L was demonstrated by coexpression of cathepsin L with proenkephalin in neuroendocrine PC12 cells (derived from rat adrenal medulla). Expression of cathepsin L resulted in its trafficking to secretory vesicles that contain enkephalin and neuropeptides (11). These findings indicated the novel location of cathepsin L in neuropeptide-containing secretory vesicles.

2.3. Cathepsin L Gene Knockout and Expression Result in Regulation of Enkephalin Neuropeptide Production from Proenkephalin

The in vivo role of cathepsin L in the production of enkephalin peptides was assessed in cathepsin L gene knockout mice. Brain levels of (Met)enkephalin (ME) were reduced by ~50% compared to wild-type control mice (**Fig. 5.5**) (10). Enkephalin was measured by radioimmunoassay that specifically detected ME and not proenkephalin. Brains contained a higher ratio of proenkephalin/ME, indicating retarded proenkephalin processing. Thus, the knockout results demonstrate the in vivo function of cathepsin L for enkephalin neuropeptide production.

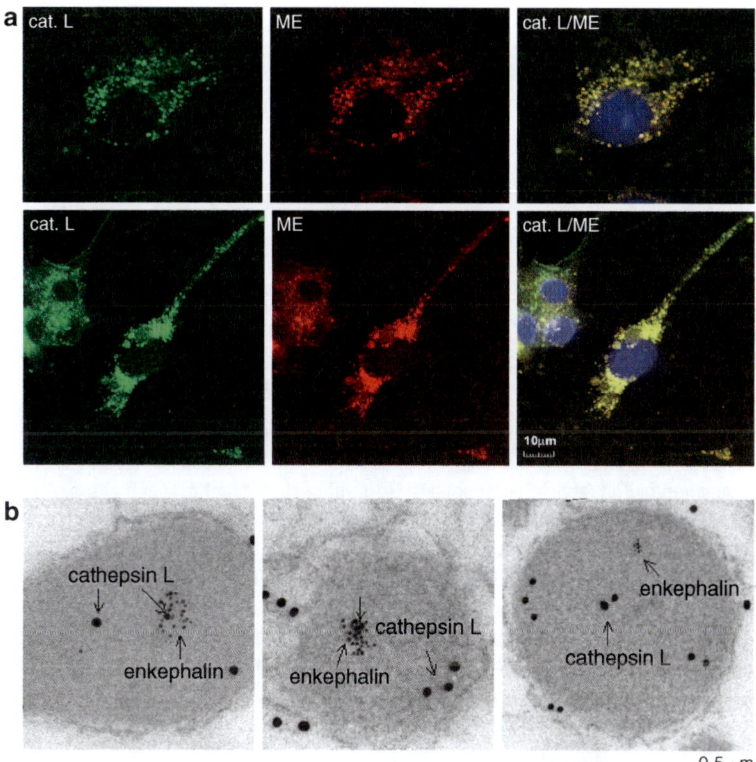

Fig. 5.4. Localization of cathepsin L within enkephalin-containing secretory vesicles of neuroendocrine chromaffin cells. (**a**) Cathepsin L colocalization with (Met)enkephalin (ME) by immunofluorescence confocal microscopy. Immunofluorescence localization of cathepsin L (cat. L) was assessed by anti-cathepsin L detected with anti-rabbit IgG-Alexa 488 (*green* fluorescence), and ME was detected with anti-ME and anti-mouse IgG-Alexa 594 (*red*). Colocalization is illustrated by overlay of the images, illustrated by *yellow* fluorescence. (**b**) Immunoelectron microscopy demonstrates colocalization of cathepsin L and (Met)enkephalin in secretory vesicles. Cathepsin L in secretory vesicles was indicated by anti-cathepsin L detected with 15 nm colloidal gold-conjugated anti-rabbit IgG, and ME was detected with anti-ME and 6 nm colloidal gold conjugated to anti-mouse IgG. The presence of both 15 and 6 nm gold particles within these vesicles demonstrated the in vivo colocalization of cathepsin L and ME.

Studies of cathepsin L expression showed that cathepsin L participates in cellular processing of proenkephalin into (Met)enkephalin in the regulated secretory pathway of PC12 neuroendocrine cells (**Fig. 5.5**) (11). Cathepsin L generated high molecular weight PE-derived intermediates (of ~23, 18–19, 8–9, and 4.5 kDa) that were similar to those in vivo in chromaffin granules assessed by Western blots. Such results demonstrated a cellular role for cathepsin L in the production of (Met)enkephalin in secretory vesicles.

2.4. Cleavage Specificity of Cathepsin L for Dibasic Prohormone Processing Sites

Cathepsin L cleaves at dibasic and monobasic prohormone processing sites. Production of (Met)enkephalin by cathepsin L was assessed with the enkephalin-containing peptide substrates BAM-22P and Peptide F, with identification of peptide products by

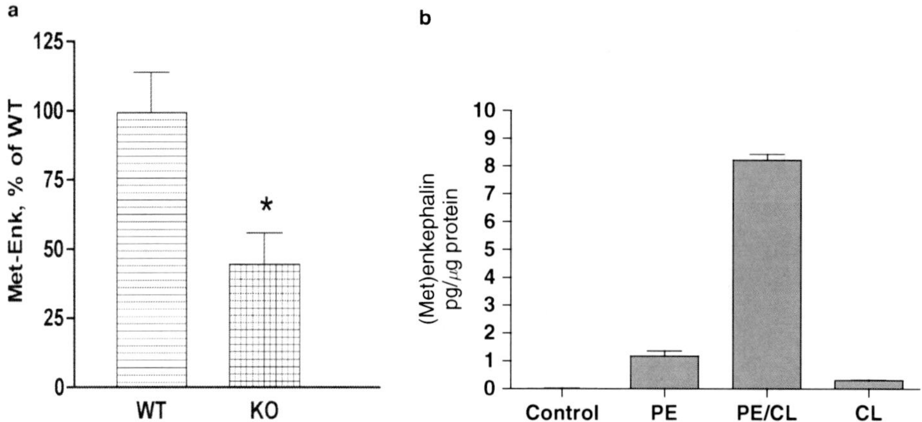

Fig. 5.5. Regulation of (Met)enkephalin in cathepsin L gene knockout mice. (**a**) (Met)enkephalin (ME) levels in brains of knockout mice. ME levels in extracts of brain tissue from cathepsin L gene knockout mice (–/–) and wild-type control mice (+/+) were measured by RIA, shown as the mean ± s.e.m., with $N = 10$ for each group. A significant decrease (*) in enkephalin levels in knockout mice was observed ($p < 0.013$, two-tailed *t*-test). (**b**) Elevated (Met)enkephalin production during cathepsin L expression. Elevation of cellular content of (Met)enkephalin in PC12 cells was observed after coexpression of cathepsin L (CL) and proenkephalin (PE). The radioimmunoassay (RIA) for (Met)enkephalin measures processed (Met)enkephalin since the RIA does not cross-react with PE. Controls included cells transfected with vector alone (no insert, control), PE alone, and cathepsin L (CL) alone. Experiments were conducted by transfection of triplicate wells of cells for each group, with RIA assay of (Met)enkephalin conducted in duplicate assays performed twice. Results are expressed as $x \pm$ s.e.m. (mean + standard error of the mean).

Peptide F:

↓↓ ↓

<u>YGGFM</u>KKMDELYPLEVEEEANGGEVLGKR<u>YGGFM</u>

BAM-22P:

↓↓ ↓

<u>YGGFM</u>RRVGRPEWWMDYQKRYG

Fig. 5.6. Cathepsin L cleaves dibasic and monobasic processing sites of enkephalin-containing peptide substrates, BAM-22P and Peptide F. (**a**) Cathepsin L cleavage of BAM-22P: analyses of peptide products by MALDI-TOF mass spectrometry (MS). BAM-22P was incubated with cathepsin L, and monoisotopic masses (MH^+, in Daltons) of peptide products are illustrated for mass peaks. The primary sequence of BAM-22P (residues 1–22) is shown, and *arrows* indicate cathepsin L cleavage sites. The *inset* shows that (Met)enkephalin (ME) was readily detected by electrospray MS, as well as by MALDI-TOF MS. (**b**) Cathepsin L cleavage of Peptide F: analyses of peptide products by MALDI-TOF MS. Cathepsin L cleavage products of Peptide F were identified by MALDI-TOF MS. Cathepsin L cleavage sites are illustrated by *arrows* above the Peptide F sequence.

MALDI-TOF mass spectrometry (10). Cathepsin L generated (Met)enkephalin by cleaving BAM-22P at the dibasic ↓Arg-↓Arg and monobasic ↓Arg sites (**Fig. 5.6**). Peptide F was cleaved by cathepsin L at dibasic ↓Lys-↓Lys and ↓Lys-Arg sites (**Fig. 5.6**). Moreover, cathepsin L processing of full-length ^{35}S-enkephalin precursor produced identical products as those present in vivo in adrenal medulla (Hook et al., unpublished observations). Further cleavage studies with peptide–MCA substrates containing dibasic cleavage sites illustrated that cathepsin L cleaves at the COOH-terminal side of the dibasic sites, as well as at the N-terminal side of basic residues (19). Thus, cathepsin L generates peptide intermediates with basic residue extensions at NH_2- and COOH-termini, which will then be removed by aminopeptidase B and carboxypeptidase E exopeptidases, respectively (*see* **Fig. 5.2**). These exopeptidases have been characterized and shown to participate in neuropeptide biosynthesis in secretory vesicles (7, 11, 17, 18, 20). The basic residue cleavage specificities of cathepsin L are appropriate for processing proprotein precursors into active neuropeptides.

2.5. Cathepsin L Meets the Criteria Expected of a Prohormone Processing Enzyme for Neuropeptide Biosynthesis

Cathepsin L meets the key criteria expected of proteases for processing proneuropeptides into peptide neurotransmitters and hormones. These criteria are as follows: (1) the protease must be present in the organelle site where production of the active peptide occurs, primarily in secretory vesicles; (2) the protease must possess the appropriate substrate cleavage specificity to generate the active peptide; and (3) protease inhibition or gene knockdown should reduce production of the active peptide. Fulfillment of these criteria demonstrates the novel biological role of cathepsin L in the biosynthesis of enkephalin and numerous neuropeptides, as described in the next section.

3. Prominent Function of Cathepsin L for Production of Numerous Neuropeptides

The significant role of cathepsin L for processing proenkephalin into the active enkephalin neuropeptide raised the question of the possible function of cathepsin L for biosynthesis of other neuropeptides. Continued investigation of cathepsin L in secretory vesicles demonstrated its prominent role in the biosynthesis of numerous neuropeptides represented by neuropeptide Y (NPY), POMC-derived peptide hormones consisting of ACTH, β-endorphin, and α-MSH, as well as dynorphin, CCK, and catestatin neuropeptides (12–16). Illustration of the biological function of cathepsin L for neuropeptide production has been demonstrated by protease gene knockout (summarized in **Table 5.2**) and protease gene expression combined with inhibitors and related approaches.

3.1. Neuropeptide Y (NPY)

NPY in the brain functions as a peptide neurotransmitter in the regulation of feeding behavior (5, 6). In the peripheral sympathetic nervous system, NPY is released and regulates blood pressure (24). Thus, NPY has central and peripheral effects

Table 5.2
Reduced levels of neuropeptides in cathepsin L knockout mice

Neuropeptide	Tissue	Wild type (%)	Cathepsin L KO (%)
Met-enkephalin	Brain cortex	100	44
NPY	Brain cortex	100	22
CCK8	Brain cortex	100	75
Dynorphin A	Brain cortex	100	25
Dynorphin B	Brain cortex	100	17
α-Neoendorphin	Brain cortex	100	10
ACTH	Pituitary	100	23
β-Endorphin	Pituitary	100	18
α-MSH	Pituitary	100	7

Results of tissue levels of several neuropeptides in brain cortex and pituitary in cathepsin L knockout mice compared to wild-type controls (100%) are illustrated (10, 12–15). Substantial decreases in levels of these neuropeptides occur in cathepsin L knockout mice compared to control wild-type mice. These data indicate a role of cathepsin L in neuropeptide production.

for regulating physiological functions. Novel findings show that cathepsin L participates as a key proteolytic enzyme for NPY production in secretory vesicles (12). Notably, NPY in cathepsin L knockout (KO) mice was substantially reduced in brain and adrenal medulla by 80 and 90%, respectively. Participation of cathepsin L in producing NPY predicts their colocalization in secretory vesicles, a primary site of NPY production. Indeed, cathepsin L is colocalized with NPY in brain cortical neurons and in chromaffin cells of adrenal medulla, demonstrated by immunofluorescence confocal microscopy. Immunoelectron microscopy confirmed the localization of cathepsin L with NPY in regulated secretory vesicles of chromaffin cells. Functional studies showed that coexpression of proNPY with cathepsin L in neuroendocrine PC12 cells resulted in increased production of NPY. Furthermore, in vitro processing indicated cathepsin L processing of proNPY at paired basic residues. These findings demonstrate a role for cathepsin L in the production of NPY from its proNPY precursor. These unique findings demonstrate the biological role of cathepsin L in the production of NPY.

3.2. ACTH, β-Endorphin, and α-MSH Derived from the Common POMC Prohormone

Cathepsin L participates in the biosynthesis of the pituitary hormones ACTH, β-endorphin, and α-MSH that are synthesized by proteolytic processing of their common POMC (proopiomelanocortin) precursor (13). These peptides each have distinct functions; ACTH stimulates glucocorticoid synthesis in adrenal cortex, β-endorphin is an endogenous opioid for analgesia, and α-MSH is involved in pigmentation of the skin. Key findings show that cathepsin L functions as a major proteolytic enzyme for production of POMC-derived peptide hormones in secretory vesicles.

Specifically, cathepsin L knockout (KO) mice showed major decreases in ACTH, β-endorphin, and α-MSH that were reduced to 23, 18, and 7% of wild-type controls (100%) in pituitary. These decreased peptide levels were accompanied by increased levels of POMC consistent with proteolysis of POMC by cathepsin L. Immunofluorescence microscopy showed colocalization of cathepsin L with β-endorphin and α-MSH in the intermediate pituitary and with ACTH in the anterior pituitary. In contrast, cathepsin L was only partially colocalized with the lysosomal marker lamp-1 in pituitary, consistent with its extralysosomal function in secretory vesicles. Expression of cathepsin L in pituitary AtT-20 cells resulted in increased ACTH and β-endorphin in the regulated secretory pathway. Furthermore, treatment of AtT-20 cells with CLIK-148, a specific inhibitor of cathepsin L, resulted in reduced production of ACTH and accumulation of POMC. These findings demonstrate a prominent role for cathepsin L in the production of ACTH, β-endorphin, and α-MSH peptide hormones in the regulated secretory pathway.

3.3. Dynorphin Neuropeptides

Dynorphin opioid neuropeptides mediate neurotransmission for analgesia and behavioral functions (25–27). Dynorphin A, dynorphin B, and α-neoendorphin are generated from prodynorphin by proteolytic processing. Recent studies demonstrate the significant role of the cysteine protease cathepsin L for producing dynorphins (15). Cathepsin L knockout mouse brains showed extensive decreases in dynorphin A, dynorphin B, and α-neoendorphin that were reduced by 75, 83, and 90%, respectively, compared to controls. Moreover, cathepsin L in brain cortical neurons was colocalized with dynorphins in secretory vesicles, the primary site of neuropeptide production. Cellular coexpression of cathepsin L with prodynorphin in PC12 cells resulted in increased production of dynorphins A and B. Comparative studies of PC1/3 and PC2 convertases showed that PC1/3 knockout mouse brains had a modest decrease in dynorphin A, and PC2 knockout mice showed a minor decrease in α-neoendorphin. Overall, these results demonstrate a prominent role for cathepsin L, jointly with PC1/3 and PC2, for production of dynorphins in brain.

3.4. Cholecystokinin (CCK)

Cholecystokinin (CCK) is a peptide neurotransmitter whose production requires proteolytic processing of the proCCK precursor to generate active CCK8 neuropeptide in brain. Recent investigation demonstrates the significant role of the cysteine protease cathepsin L for CCK8 production (14). In cathepsin L knockout (KO) mice, CCK8 levels were substantially reduced in brain cortex by an average of 75%. To evaluate the role of cathepsin L in producing CCK in the regulated secretory pathway of neuroendocrine cells, pituitary AtT-20 cells that stably produce CCK were treated with the specific cathepsin L inhibitor, CLIK-148. CLIK-148 inhibitor treatment resulted in decreased amounts of CCK secreted from the regulated secretory pathway of AtT-20 cells. CLIK-148 also reduced cellular levels of CCK9 (Arg-CCK8), consistent with CCK9 as an intermediate product of cathepsin L, shown by the decreased ratio of CCK9/CCK8. The decreased CCK9/CCK8 ratio also suggests a shift in the production to CCK8 over CCK9 during inhibition of cathepsin L. During reduction of the PC1/3 processing enzyme by siRNA, the ratio of CCK9/CCK8 was increased, suggesting a shift to the cathepsin L pathway for production of CCK9. The changes in ratios of CCK9 compared to CCK8 are consistent with dual roles of the cathepsin L protease pathway that includes aminopeptidase B to remove NH_2-terminal Arg or Lys and the PC1/3 protease pathway. These results suggest that cathepsin L functions as a major protease responsible for CCK8 production in mouse brain cortex and participates with PC1/3 for CCK8 production in pituitary cells.

3.5. Catestatin

The active catestatin peptide secreted from adrenal medulla of the sympathetic nervous system regulates blood pressure in stress (28, 29). Catestatin is generated from the precursor chromogranin A (CgA) by proteolytic processing (16). Notably, cathepsin L participates in catestatin formation. Endogenous cathepsin L colocalizes with CgA in the secretory vesicles of primary rat chromaffin cells. Transfection of PC12 cells with the cathepsin L cDNA resulted in its localization to secretory vesicles of PC12 cells. Cathepsin L cleaves CgA in vitro to catestatin-related peptides that show activity for inhibition of nicotine-induced catecholamine secretion from PC12 cells. These findings indicate that CgA can be utilized as a substrate for cathepsin L in the production of catestatin-like peptides.

4. Novel Secretory Vesicle Function of Cathepsin L in Contrast to Lysosomal Function

These recent studies indicate the unique localization of cathepsin L in secretory vesicles for its biological function of neuropeptide production. As described in this review, cathepsin L is colocalized with the neuropeptides enkephalin, NPY, β-endorphin, ACTH, α-MSH, dynorphins, and CCK (7, 10–15).

The novel secretory vesicle function of cathepsin L contrasts with its known role in lysosomes for protein degradation. Cathepsin L was first identified as a degrading protease localized in rat liver lysosomes (30) and in lysosomes of other types of cells and species (31, 32). Yet, in addition to its lysosomal function, additional studies have indicated cathepsin L in secretory vesicles of rat pituitary GH4C1 (33) and mouse NIH3T3 cell lines (34).

Comparison of the secretory vesicle localization of cathepsin L in secretory vesicles compared to lysosomes reveals differences in the portion of cathepsin L in these two organelles in different cell types. In bovine chromaffin cells of the sympathetic nervous system, cathepsin L is primarily colocalized with NPY and enkephalin (10, 12), with no localization with the lysosomal marker lamp-1 (35) (**Fig. 5.7**). This data suggests a primary

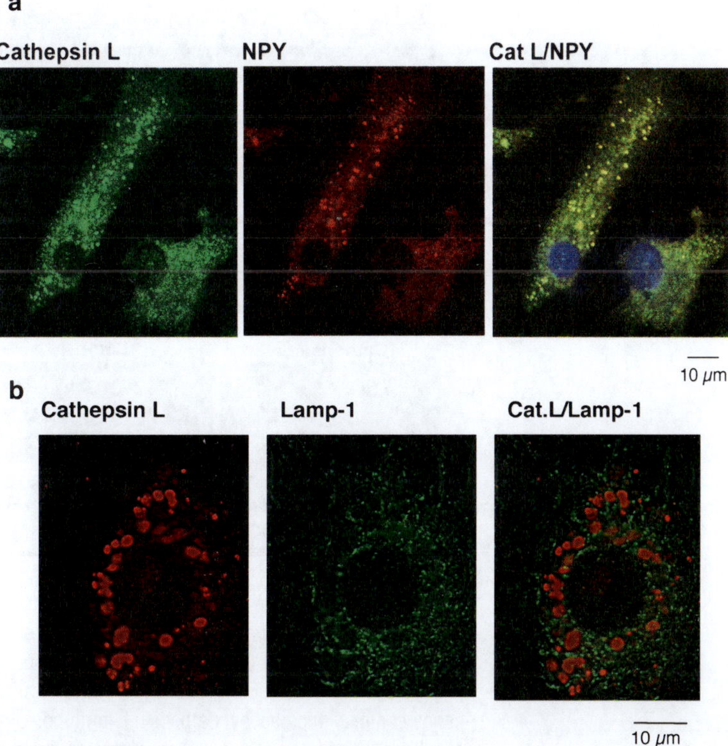

Fig. 5.7. In chromaffin cells of the sympathoadrenal nervous system, cathepsin L is primarily localized to neuropeptide-containing secretory vesicles compared to lysosomes. (**a**) Cathepsin L localization with NPY assessed by immunofluorescence confocal microscopy. Adrenal medullary chromaffin cells in primary culture were analyzed for colocalization of cathepsin L with NPY present in secretory vesicles by immunofluorescence confocal microscopy. Excellent colocalization of cathepsin L (*green* fluorescence) and NPY (*red* fluorescence) was observed in the merged image (*yellow* fluorescence), with the majority of cellular cathepsin L localized with NPY in secretory vesicles. (**b**) Evaluation of cathepsin L with the lysosomal marker lamp-1 by immunofluorescence confocal microscopy. Cathepsin L localization in chromaffin cells (*red* fluorescence) was compared to that of lamp-1 (*green* fluorescence), a lysosomal marker. Results show that cathepsin L in chromaffin cells shows little colocalization with lamp-1. These data indicate that cathepsin L in chromaffin cells is primarily localized with NPY and other neuropeptides (10) in secretory vesicles.

function of cathepsin L in secretory vesicles of chromaffin cells. But in mouse pituitary, while cathepsin L is colocalized with β-endorphin and α-MSH in secretory vesicles, a portion of cellular cathepsin L is colocalized with the lysosomal marker lamp-1 (13) (**Fig. 5.8**). Also, in mouse pituitary AtT-20 cells, cathepsin L is partially colocalized with ACTH in secretory vesicles and with

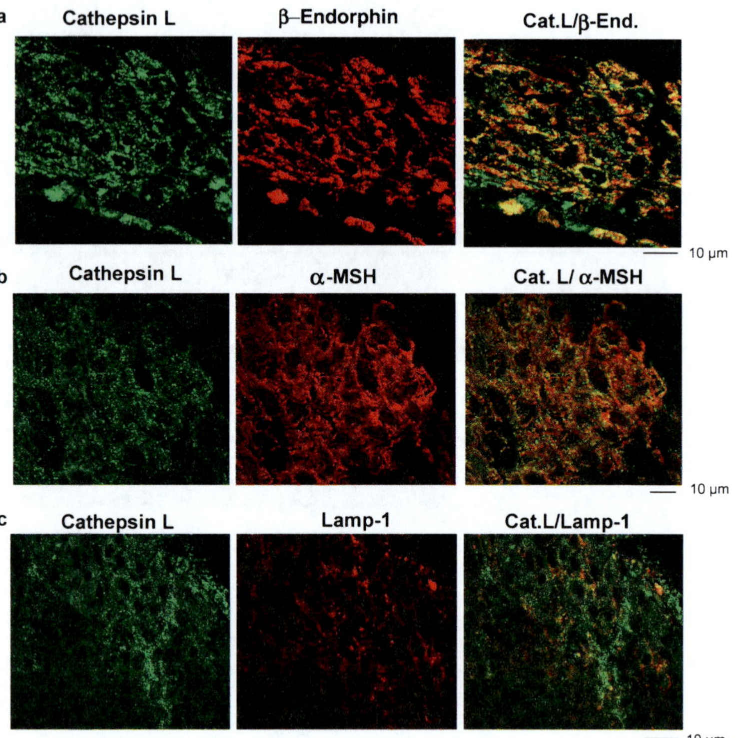

Fig. 5.8. In intermediate pituitary, cathepsin L is highly localized with β-endorphin and α-MSH peptide hormones and is partially localized with the lysosomal marker lamp-1. (**a**) Cathepsin L and β-endorphin colocalization. Colocalization of cathepsin L with β-endorphin in intermediate pituitary (mouse) was demonstrated by immunofluorescence confocal microscopy. Cathepsin L immunoreactivity (*green* fluorescence) showed excellent overlapping colocalization with β-endorphin (*red* fluorescence), shown by the *yellow* fluorescence of merged cathepsin L/β-endorphin fluorescent immunostaining. The majority of β-endorphin, contained in secretory vesicles, was colocalized with cathepsin L. (**b**) Cathepsin L and α-MSH colocalization. The overlapping colocalization of cathepsin L with α-MSH peptide hormone in intermediate pituitary was illustrated by immunofluorescence confocal microscopy. The majority of cathepsin L (*green* fluorescence) and α-MSH (*red* fluorescence) was colocalized, shown by the merged areas of *yellow* fluorescence. (**c**) Partial colocalization of cathepsin L and lamp-1 localization in mouse intermediate pituitary cells. Comparison of the cellular distribution of cathepsin L in intermediate lobe pituitary cells with that of the lysosomal marker lamp-1. Cathepsin L (*green* fluorescence) was partially colocalized with the lysosomal marker lamp-1 (*red* fluorescence). This subcellular distribution of cathepsin L is consistent with the localization of cathepsin L in secretory vesicles as well as in lysosomes. Cathepsin L localization to extralysosomal areas is consistent with its localization within secretory vesicles.

lamp-1 in lysosomes (13, 14). In the human pituitary, ACTH and β-endorphin were also colocalized with cathepsin L (36).

In addition, cathepsin L has been found in the nucleus of mouse and human cells for proteolysis of the histone H3 in the regulation of gene expression (37–39). Thus, cathepsin L has biological functions in organelles, i.e., secretory vesicles and nucleus, that differ from its lysosomal function.

Clearly, new findings indicate cathepsin L localization in several organelles, with notable function in several secretory vesicles for neuropeptide production as discussed in this review. These studies lead to the conclusion that cathepsin L not only functions in lysosomes for protein degradation but is also emerging in its biological functions in secretory vesicles for neuropeptide production.

5. Combined Roles of Cathepsin L and Prohormone Convertases PC1/3 and PC2 for Neuropeptide Production

The biological role of cathepsin L for production of neuropeptides, combined with function of the subtilisin-like prohormone convertases, indicates dual protease pathways for neuropeptide production (*see* **Fig. 5.2**). The proprotein convertase family consists of PC1/3, PC2, furin, PACE4, PC3, PC5/6, and PC7 for processing at basic residues (7, 9, 40, 41). Several related proteases that cleave at non-basic residues have been identified, consisting of the subtilisin/kexin-like isozyme-1 (SKI-1/SIP) and the neural apoptosis-regulated convertase-1 (PCSK9/NARC-1) (9, 40–42). These excellent reviews of the proprotein convertases have reported their roles in neuropeptide biosynthesis. Due to the limitations in space of this focused chapter on cathepsin L in secretory vesicles, readers are referred to these reviews on the PC enzymes.

6. Conclusion: Cathepsin L Represents a Distinct Protease Pathway, Combined with Prohormone Convertases, for Biosynthesis of Neuropeptides

Recent findings demonstrate the key biological role of cathepsin L in secretory vesicles for processing proneuropeptides into active peptide neuropeptides and hormones. The cathepsin L cysteine protease pathway participates with the prohormone convertase subtilisin-like protease pathway for neuropeptide biosynthesis (*see* **Fig. 5.2**). The presence of the two distinct mechanistic protease classes for proneuropeptide processing indicates cellular insurance for selection of protease pathways for production of specific neuropeptides. It will be of interest to understand regulatory mechanisms for selection of both or either processing pathways for production of particular neuropeptides in different neuroendocrine tissues. Indeed, endogenous protease inhibitors exist for each of these two distinct protease pathways. Cathepsin L in secretory vesicles is regulated by the endogenous inhibitors endopin 2 (a serpin inhibitor) (43), kunitz protease inhibitor form of APP (KPI-APP; (44)), and cystatin C (45). The PC1/3 and PC2 enzymes are regulated by endogenous inhibitors consisting of proSAAS (46) and 7B2 (47), respectively. The presence of different protease subclasses allows specific regulation by inhibitors.

Future knowledge of specific drug regulators for particular neuropeptides can lead to future translational research for small molecule regulation of cathepsin L and prohormone convertase pathways in the control of physiological functions. For example, regulation of opioid peptide production – enkephalin, β-endorphin, and dynorphin – may lead to new drugs for analgesia and pain relief. Specific small molecule control of hypothalamic NPY in the control of feeding behavior may lead to improvement in obese conditions. Regulation of hypothalamic CRF and pituitary ACTH production is important for the control of steroid biosynthesis in adrenal cortex for metabolic regulation. PC-related proteases have been implicated in sterol and lipid metabolism (48), tumor progression (49), atherosclerosis (50), and other physiological and disease conditions.

The newly identified biological function of cathepsin L for neuropeptide biosynthesis opens new avenues for targeting this protease in drug development for control of neuropeptides in health and disease. This research field is an exciting area for understanding neuropeptide regulation of neuroendocrine functions in health and disease.

Acknowledgments

The authors appreciate support of this research by grants to V. Hook from the National Institutes of Health. The authors also appreciate scientific advice by Dr. Shin-Rong Hwang and technical assistance by Mr. Thomas Toneff at the Skaggs School of Pharmacy and Pharmaceutical Sciences, University of California, San Diego, La Jolla, CA.

References

1. Law, P. Y., Wong, Y. H., and Loh, H. H. (2000) Molecular mechanisms and regulation of opioid receptor signaling *Annu Rev Pharmacol Toxicol* **40**, 389–430.
2. Snyder, S. H., and Pasternak, G. W. (2003) Historical review: Opioid receptors *Trends Pharmacol Sci* **24**, 198–205.
3. Frohman, L. A. (1995) Diseases of the anterior pituitary. In: *Endocrinology and Metabolism*, Third Edition, P. Felig, J. D. Baxter, and L. A. Frohman, eds. New York, NY: McGraw-Hill. Health Professions Division, pp. 293–7.
4. Steiner, R. A., Hohmann, J. G., Holmes, A., Wrenn, C. C., Cadd, G., Jureus, A., Clifton, D. K., Luo, M., Gutshall, M., Ma, S. Y., Mufson, E. J., and Crawley, J. N. (2001) Galanin transgenic mice display cognitive and neurochemical deficits characteristic of Alzheimer's disease *Proc Natl Acad Sci USA* **98**, 4184–9.
5. Gehlert, D. R. (1999) Role of hypothalamic neuropeptide Y in feeding and obesity *Neuropeptides* **33**, 329–38.
6. Wieland, H. A., Hamilton, B. S., Krist, B., and Doods, H. N. (2000) The role of NPY in metabolic homeostasis: Implications for obesity therapy *Expert Opin Investig Drugs* **9**, 1327–46.
7. Hook, V., Funkelstein, L., Lu, D., Bark, S., Wegrzyn, J., and Hwang, S. R. (2008) Proteases for processing proneuropeptides into peptide neurotransmitters and hormones *Annu Rev Pharmacol Toxicol* **48**, 393–423.
8. Steiner, D. F. (1998) The proprotein convertases *Curr Opin Chem Biol* **2**, 31–9.
9. Seidah, N. G., and Prat, A. (2002) Precursor convertases in the secretory pathway, cytosol and extracellular milieu *Essays Biochem* **38**, 79–94.
10. Yasothornsrikul, S., Greenbaum, D., Medzihradszky, K. F., Toneff, T., Bundey, R., Miller, R., Schilling, B., Petermann, I., Dehnert, J., Logvinova, A., Goldsmith, P., Neveu, J. M., Lane, W. S., Gibson, B., Reinheckel, T., Peters, C., Bogyo, M., and Hook, V. (2003) Cathepsin L in secretory vesicles functions as a prohormone-processing enzyme for production of the enkephalin peptide neurotransmitter *Proc Natl Acad Sci USA* **100**, 9590–5.
11. Hwang, S. R., Garza, C., Mosier, C., Toneff, T., Wunderlich, E., Goldsmith, P., and Hook, V. (2007) Cathepsin L expression is directed to secretory vesicles for enkephalin neuropeptide biosynthesis and secretion *J Biol Chem* **282**, 9556–63.
12. Funkelstein, L., Toneff, T., Hwang, S. R., Reinheckel, T., Peters, C., and Hook, V. (2008) Cathepsin L participates in the production of neuropeptide Y in secretory vesicles, demonstrated by protease gene knockout and expression *J Neurochem* **106**, 384–91.
13. Funkelstein, L., Toneff, T., Hwang, S. R., Beuschlein, F., Lichtenauer, U. D., Reinheckel, T., Peters, C., and Hook, V. (2008) Major role of cathepsin L for producing the peptide hormones ACTH, beta-endorphin, and alpha-MSH, illustrated by protease gene knockout and expression *J Biol Chem* **83**, 35652–9.
14. Beinfeld, M. C., Funkelstein, L., Foulon, T., Cadel, S., Kitagawa, K., Toneff, T., Reinheckel, T., Peters, C., and Hook, V. (2009) Cathepsin L plays a major role in cholecystokinin production in mouse brain and in pituitary AtT-20 cells: Protease gene knockout and inhibitor studies *Peptides* **30**, 1882–991.
15. Minokadeh, A., Funklestein, L., Toneff, T., Hwang, S. R., Reinheckel, T., Peters, C., Zadina, J., and Hook, V. (2010) Cathepsin L participates in dynorphin neuropeptide production in brain cortex, illustrated by protease gene knockout and expression *Mol Cell Neurosci* **43**, 98–107.
16. Biswas, N., Rodriquez-Flores, J. L., Courel, M., Gayen, J. R., Vaingankar, S. M., Mahata, M., Torpey, J. W., Taupenot, L., O'Connor, D. T., and Mahata, S. K. (2009) Cathepsin L colocalizes with chromogranin A I chromaffin vesicles to generate active peptides *Endocrinology* **150**, 3547–57.
17. Yasothornsrikul, S., Toneff, T., Hwang, S. R., and Hook, V. Y. H. (1998) Arginine and lysine aminopeptidase activities in chromaffin granules of bovine adrenal medulla: Relevance to prohormone processing *J Neurochem* **70**, 153–63.
18. Hwang, S. R., O'Neill, A., Bark, S., Foulon, T., and Hook, V. (2007) Secretory vesicle aminopeptidase B related to neuropeptide processing: Molecular identification and subcellular localization to enkephalin- and NPY-containing chromaffin granules *J Neurochem* **100**, 1340–50.
19. Azaryan, A. V., and Hook, V. Y. H. (1994) Unique cleavage specificity of 'prohormone thiol protease' related to proenkephalin processing *FEBS Lett* **341**, 197–202.
20. Fricker, L. D. (1988) Carboxypeptidase E *Annu Rev Physiol* **50**, 309–21.

21. Hook, V. Y. H., and Yasothornsrikul, S. (1998) Carboxypeptidase and aminopeptidase proteases in pro-neuropeptide processing. In: *Proteolytic and Cellular Mechanisms in Prohormone and Proprotein Processing*, V. Y. H. Hook, ed. Austin, TX: Landes Bioscience Publishers, pp. 121–40.
22. Yasothornsrikul, S., Aaron, W., Toneff, T., and Hook, V. Y. (1999) Evidence for the proenkephalin processing enzyme prohormone thiol protease (PTP) as a multicatalytic cysteine protease complex: Activation by glutathione localized to secretory vesicles *Biochemistry* **38**, 7421–30.
23. Cravatt, B. F., Wright, A. R., and Kozarich, J. W. (2008) Activity-based protein profiling: From enzyme chemistry to proteomic chemistry *Annu Rev Biochem* 77, 383–414.
24. Wiest, R., Jurzik, L., Herold, T., Straub, R. H., and Scholmerich, J. (2007) Role of NPY for vasoregulation in the splanchnic circulation during portal hypertension *Peptides* **28**, 396–404.
25. Akil, H., Owens, C., Gutstein, H., Taylor, L., Currran, E., and Watson, S. (1998) Endogenous opioids: Overview and current issues *Drug Alcohol Depend* **51**, 127–40.
26. Wang, Z., Gardell, L. W., Ossipov, M. H., Vanderah, T. W., Brennan, M. B., Hochgeschwender, U., Hruby, V. J., Malan, T. P., Lai, J., and Porreca, F. (2001) Pronociceptive actions of dynorphin maintain chronic neuropathic pain *J Neurosci* **21**, 1779–86.
27. Shippenberg, T. S., Zapata, A., and Chefer, V. I. (2007) Dynorphin and the pathophysiology of drug addiction *Pharmacol Therapeut* **116**, 306–21.
28. Mahata, S. K., Mahata, M., Fung, M. M., and O'Connor, D. T. (2010) Catestatin: A multifunctional peptide from chromogranin A *Regul Pept* [Epub ahead of print].
29. Vaingankar, S. M., Li, Y., Biswas, N., Gayen, J., Choksi, S., Rao, F., Zielger, M. G., Mahata, S. K., and O'Connor, D. T. (2010) Effects of chromogranin A deficiency and excess in vivo: Biphasic blood pressure and catecholamine responses *J Hypertens* [Epub ahead of print].
30. Ansorge, S., Kirschke, H., and Friedrich, K. (1977) Conversion of proinsulin into insulin by cathepsins B and L from rat liver lysosomes *Acta Biol Med Ger* **36**, 1723–7.
31. Yokota, S., Nishimura, Y., and Kato, K. (1988) Localization of cathepsin L in rat kidney revealed by immunoenzyme and immunogold techniques *Histochemistry* **90**, 277–83.
32. Ryvnyak, V. V., Ryvnyak, E. I., and Tudos, R. V. (2004) Electron histochemical localization of cathepsin L in the liver *Bull Exp Biol Med* **137**, 90–1.
33. Waguri, S., Sato, N., Watanabe, T., Ishidoh, K., Kominami, E., Sato, K., and Uchiyama, Y. (1995) Cysteine proteinases in GH4C1 cells, a rat pituitary tumor cell line, are secreted by the constitutive and regulated secretory pathways *Eur J Cell Biol* **67**, 308–18.
34. Collette, J., Bocock, J. P., Ahn, K., Chapman, R. L., Godbold, G., Yeyeodu, S., and Erickson, A. H. (2004) Biosynthesis and alternate targeting of the lysosomal cysteine protease cathepsin L *Int Rev Cytol* **241**, 1–51.
35. Fukuda, M. (1991) Lysosomal membrane glycoproteins. Structure, biosynthesis, and intracellular trafficking *J Biol Chem* **266**, 21327–30.
36. Hook, V., Funkelstein, L., Toneff, T., Mosier, C., and Hwang, S. R. (2009) Hyman pituitary contains dual cathepsin L and prohormone convertase processing pathway components involved in converting POMC into the peptide hormones ACTH, alpha-MSH, and beta-endorphin *Endocrine* **35**, 429–37.
37. Hiwasa, T., and Sakiyama, S. (1996) Nuclear localization of procathepsin L/MEP in ras-transformed mouse fibroblasts *Cancer Lett* **99**, 87–91.
38. Duncan, E. M., Muratore-Schroeder, T. L., Cook, R. G., Garcia, B. A., Shabanowitz, J., Hunt, D. F., and Allis, C. D. (2008) Cathepsin L proteolytically processes histone H3 during mouse embryonic stem cell differentiation *Cell* **135**, 284–94.
39. Hook, V., Funkelstein, L., Toneff, T., Mosier, C., and Hwang, S. R. (2009) Human pituitary contains dual cathepsin L and prohormone convertase processing pathway components involved in converting POMC into the peptide hormones ACTH, alpha-MSH, and beta-endorphin *Endocrine* **35**, 429–37.
40. Fugere, M., and Day, R. (2005) Cutting back on pro-protein convertases: The latest approaches to pharmacological inhibition *Trends Pharmacol Sci* **26**, 294–301.
41. Scamuffa, N., Calvo, F., Chretien, M., Seidah, N. G., and Khatib, A. M. (2006) Proprotein convertases: Lessons from knockouts *FASEB J* **20**, 1954–63.
42. Thomas, G. (2002) Furin at the cutting edge: From protein traffic to embryogenesis and disease *Nature Rev* **3**, 753–66.
43. Hwang, S. R., Stoka, V., Turk, V., and Hook, V. Y. (2005) The novel bovine serpin endopin 2C demonstrates selective

inhibition of the cysteine protease cathepsin L compared to the serine protease elastase, in cross-class inhibition *Biochemistry* **44**, 7757–67.

44. Hook, V. Y. H., Sei, C., Yasothornsrikul, S., Toneff, T., Kang, Y. -H., Efthimiopoulos, S. et al. (1999) The kunitz protease inhibitor form of the amyloid precursor protein (KPI/APP) inhibits the proneuropeptide processing enzyme prohormone thiol protease (PTP). Colocalization of KPI/APP and PTP in secretory vesicles *J Biol Chem* **274**, 3165–72.
45. Leonardi, A., Turk, B., and Turk, V. (1996) Inhibition of cathepsins L and S by tefins and cystatins *Bio Chem Hoppe Seyler* **377**, 319–21.
46. Basak, A., Koch, P., Dupelle, M., Fricker, L. D., Devi, L. A. et al. (2001) Inhibitory specificity and potency of proSAAS-derived peptides toward proprotein convertase 1 *J Biol Chem* **276**, 32720–8.
47. Fortenberry, Y., Liu, J., and Lindberg, I. (1999) The role of the 7B2 CT peptide in the inhibition of prohormone convertase 2 in endocrine cell lines *J Neurochem* **73**, 994–1003.
48. Seidah, N. G., Khatib, A. M., and Prat, A. (2006) The proprotein convertases and their implication in sterol and/or lipid metabolism *Biol Chem* **387**, 871–7.
49. Bassi, D. E., Fu, J., Lopez de Cicco, R., and Klein-Szanto, A. J. (2005) Proprotein convertases: "master switches" in the regulation of tumor growth and progression *Mol Carcinogen* **44**, 151–61.
50. Stawowy, P., and Fleck, E. (2005) Proprotein convertases furin and PC5: Targeting atherosclerosis and restenosis at multiple levels *J Molec Med* **83**, 865–75.

Chapter 6

In Vitro Assay for Protease Activity of Proprotein Convertase Subtilisin Kexins (PCSKs): An Overall Review of Existing and New Methodologies

Ajoy Basak, Andrew Chen, Swapan Majumdar, and Heather Palmer Smith

Abstract

The mammalian proprotein convertase subtilisin kexins (PCSKs) previously called proprotein or prohormone convertases (PCs) are a family of Ca^{+2}-dependent endoproteases in the subtilisin family. These proteolytic enzymes exert their many crucial physiological and biological functions in vivo via their ability to cleave larger inactive precursor proteins into their biologically active mature forms. This event takes place in a highly efficient and selective manner. Such actions of PCSKs either alone or in combination to cleave specific protein bonds are the hallmark events that not only define the normal functions and metabolism of the body but also may lead to a variety of diseases or disorders with associated conditions. These include among others, diabetes, obesity, cancer, cardiovascular diseases, reproduction abnormalities as well as viral bacterial infections. These conditions were the direct consequences of an enhanced level of enzymatic activity of one or more PCSKs except only PCSK9, whose protease activity in relation to its physiological substrate has yet to be characterized. Owing to this finding, a large number of research studies have been exclusively devoted to develop rapid, efficient and reliable in vitro methods for examining the protease activity of these enzymes. Several assays have been developed to monitor PCSK activity and these are widely used in chemical, biochemical, cellular and animal studies. This review will cover various methodologies and protocols that are currently available in the literature for PCSK activity assays. These include liquid phase methods using fluorogenic, chromogenic and intramolecularly quenched fluorescent substrates as well as a newly developed novel solid phase fluorescence method. This review will also highlight the usefulness of these methodologies and finally a comparative analysis has been made to examine their merits and demerits with some key examples.

Key words: Proprotein convertase subtilisin kexins (PCSKs), protease assay, fluorogenic peptides, chromogenic peptides, solid phase assay, intramolecularly quenched fluorogenic peptides, protein processing.

M. Mbikay, N.G. Seidah (eds.), *Proprotein Convertases*, Methods in Molecular Biology 768,
DOI 10.1007/978-1-61779-204-5_6, © Springer Science+Business Media, LLC 2011

1. Introduction

Proprotein convertase subtilisin kexins (PCSKs) also known as proprotein or prohormone convertases (PCs) are a family of Ca^{+2}-dependent endoproteases of bacterial subtilisin and yeast kexin types (1–3). These proteolytic enzymes selectively cleave larger functionally inactive protein precursors into their smaller active forms. This hypothesis of precursor cleavage known as the "prohormone theory" was first put forward simultaneously more than four decades ago by Steiner (4) and Chrétien-Li (5). Subsequently the theory was well established by a wide variety of in vitro, ex vivo and in vivo studies involving cell lines, tissue and animal studies (6, 7). So far nine members of this family have been discovered which perform many important and key proteolytic tasks for maintaining the regular function, metabolism, growth and other physiological events in the body. Any deviation, dysfunction, or imbalance of these actions may result in serious disease and disorder conditions. Proproteins processed by PCs include prohormones, proneuropeptides, surface proteins, growth factor precursors, adhesion molecules, receptors, enzymes, viral glycoproteins and bacterial toxins. The nine PC members are PC1/PC3 (PCSK1) (8, 9), PC2 (PCSK2) (8), furin/PACE (PCSK3) (10, 11), PC4 (PCSK4) (12, 13), PC5/PC6 (PCSK5) (14, 15), PACE4 (PCSK6) (16), PC7/PC8/LPC (PCSK7) (17–19), SKI-1/S1P (PCSK8) (20, 21) and the most recent NARC-1 (PCSK9) (22). Based on their cleavage pattern and substrate specificity, PCSKs can be classified into three categories. (i) *Kexin type*: PCSKs 1–7 belong to this class which cleave selected protein bonds at the carboxy-terminal of a basic amino acid residue mostly Arg characterized by the presence of the motif Arg/Lys/His-X_n-Arg/Lys/X-Arg↓, where X = any amino acid except the sensitive Cys residue and $n = 1, 3,$ or 5 (1–3). (ii) *Pyrolysin type*: So far PCSK8 is the only member of this type which cleaves specific protein bonds at the carboxy-terminal of a non-basic preferably hydrophobic amino acid within the consensus sequence Arg/Lys/His-X-Φ-Ψ/Φ↓, where Φ = alkyl side chain containing hydrophobic amino acid such as Leu/Ile and Ψ = small amino acid such as Gly/Ala (1–3). (iii) *Proteinase K type*: This includes PCSK9 which cleaves only its own prodomain at Val-Phe-Ala-Gln↓Ser-Ile-Pro site. So far no additional information is available in the literature about the protease activity and physiological or synthetic substrates of this enzyme. A large number of research studies have been conducted on the protease activity in vitro as well as in vivo, substrate specificity and natural protein substrates of PCSK enzymes ever since the discovery of its first member in 1990. All the studies revealed the crucial role of protease activity of these enzymes in proprotein maturation and in normal health as well as various diseases. The findings were further confirmed by studies involving knockout and transgenic mice as well as gene overexpression studies (23–26). As a result, PCSKs are considered as potential targets for intervention of many diseases. This also includes PCSK9 even though its protease activity was not so well demonstrated (27). Except for PCSK9, all other PCSKs exhibit their biological functions via their proteolytic activity which does not seem to play any significant role for PCSK9's ability to degrade LDL and other receptors that are required for cholesterol clearance from plasma (28). Owing to these findings, development of an efficient in vitro protease activity assay of PCSK enzymes is considered as extremely important since monitoring such activity may provide important information about the disease condition or disorder state during its progression. In addition it may provide possible strategies and targets for intervention of such diseases. Therefore, this review is directed towards understanding this important aspect of PCSK research and summarizes all the current methods available for the assay of PCSK activity while introducing a new and efficient solid phase methodology for the first time.

2. Protease Activities of PCSK Enzymes and Their Implications

A large number of in vitro, ex vivo and in vivo studies confirmed the crucial role of proteolytic activity of PCSK enzymes in the activation of various protein precursors into their mature and functionally active forms. A direct correlation between the level of PCSK protease activity and the extent of disease progression or severity has been well demonstrated by various cell culture, mouse model and biochemical studies (25). The diseases and conditions linked to enhanced PCSK activity include cancer (23, 24), hypertension (29), diabetes (30), fertilization defects (31), restricted placental growth (32), viral infections (33), abnormal bone and cartilage development (34, 35), high cholesterol and lipid synthesis (36, 37) and bacterial diseases (38). These clinical implications emphasize the need to develop robust, rapid and sensitive assays for the protease activity of PCSK enzymes which has drawn particular attention from the researchers in the field. It may be stressed that it is the enzymatic activity and not the protein or mRNA level of PCSK that is of the utmost significance in terms of the state of the disease or condition. It might be possible that the PCSK protein content may remain the same or even diminish; yet, the extent of maturation of the associated precursor protein may in fact increase leading to advanced disease progression.

So far a number of methodologies have been reported which can efficiently monitor PCSK enzymatic activity in vitro in a rapid and efficient manner.

3. In Vitro Assay of PCSK Activity

In the case of other proteases, there are two types of in vitro methods available for assaying enzymatic activity. These are liquid phase and solid phase based. While in the case of the former, the assay is performed in solution with a specific substrate in a single-phase system, the latter is conducted in a biphasic system where the enzyme, present in solution, is allowed to react with a substrate that is immobilized on an insoluble solid matrix such as resin. In either case, the substrate contains a unique functional tag which only upon cleavage by the enzyme is released into the medium and is detected and measured by its special physical property. In most cases the released free tag possesses either fluorescence or an absorbance property.

3.1. Liquid Phase Method

Liquid phase methods may be of three types based on the physical property of the released functional moiety. It can be colorimetric, fluorometric or chromometric (spectrophotometric). However, only the latter two methods have been found to be most effective and accordingly the substrate required for them is either fluorogenic or chromogenic, respectively. Both these types of substrates have been widely used for PCSK activity assay.

3.2. Fluorogenic Methods

In this method, a fluorogenic peptide containing a fluorescent moiety usually a coumarin derivative located at the C-terminus immediately post to the cleavage site is used as the substrate of the enzyme. The most common types of fluorescent groups used for PCSK assay are AMC (7-amino-4-methyl coumarin) (39) and AFC (7-amino-4-trifluoromethyl coumarin) (40). When uncleaved, the fluorescence intensity of the attached functional moiety in the peptide is significantly suppressed owing to the amide bond formation between the peptide terminal carboxyl group and coumarin's 7-amino group. This resulted in a low availability of the electron pair for sharing with the other group. However, following the cleavage of the peptide bond between the coumarin group and the amino acid next to it, the highly fluorescent free AMC or AFC group is released into the medium leading to an increase in fluorescence intensity. In the past several other fluorescent groups such as aminoisophthalic ethyl ester (AIE) (41), chloromethyl amino coumarin

Table 6.1
List of chromogenic and fluorogenic compounds with their abbreviations and chemical structures used for enzyme assay

Type of compound	Structure of detector	Name and abbreviation of detector function	Wavelengths (λ) for detection of released UV and fluorescence (nm)
Chromogenic	O_2N … NH_2	Para-nitro aniline (pNA)[a]	$\lambda_{max} = 405$
	NH_2	β-Naphthyl amine (bNA)[a]	$\lambda_{max} = 340$
Fluorogenic	NH_2	β-Naphthyl amine (bNA)[b]	$\lambda_{ex} = 335$ $\lambda_{em} = 410$
	NH_2, OMe	4-Methoxy-2-naphthyl amine (MNA)[a]	$\lambda_{ex} = 340$ $\lambda_{em} = 425$
	H_2N, CO_2Me, CO_2Me	5-Aminoisophthalic dimethyl ester (AIE)[a,c]	$\lambda_{ex} = 350$ $\lambda_{em} = 440$
	H_2N, O, O, CH_3	7-Amino-4-methyl coumarin (AMC)[a,c]	$\lambda_{ex} = 370$ $\lambda_{em} = 460$
	H_2N, O, O, CH_2Cl	4-Chloromethyl-7-amino coumarin (CMAC)[a,c]	$\lambda_{ex} = 370$ $\lambda_{em} = 450$
	H_2N, O, O, CF_3	7-Amino-4-trifluoromethyl coumarin (AFC)[a,c]	$\lambda_{ex} = 490$ $\lambda_{em} = 520$

[a]Beynon and Bond (103)
[b]Reinharz and Roth (104)
[c]Craik et al. (105)

(CMAC) (42) as well as 7-methoxy-4-(aminomethyl)coumarin (MAMC) and its putative O-demethylated metabolite 7-hydroxy-4-(aminomethyl)coumarin (HAMC) (43) (**Table 6.1**) have been used with limited success. This is primarily because of their poor electron donor capacity and low sensitivity and solubility problems. So far only the peptidyl coumarin derivatives AMC and AFC have been found to be effective for in vitro detection and kinetic analysis of protease activity of not only PCSKs but also enzymes of other classes as well. It is interesting to note that β-Naphthyl amine function can be used as either chromogenic or fluorogenic species.

3.2.1. Peptidyl-MCA Substrates

Several peptidyl-MCA (4-methyl-7-amino coumarinamide) substrates have been developed and successfully used for monitoring and studying in vitro the protease activity of all PCSK enzymes except PCSK9. The sequence of the peptide used varies and depends on the recognition motif of the enzyme under investigation. The two most commonly used substrates of kexin-type PCSKs are Boc-RVRR-MCA (44) and pE (pyroglutamic)-RTKR-MCA (45). These substrates are common to all kexin-type PCSK enzymes with no particular preference for any specific member, although they appear to be more potent for furin. They are now widely accepted as general PCSK substrate of kexin type. A list of various peptidyl substrates so far tested for the activity of this type of PCSK enzymes is shown in **Tables 6.2** and **6.3**. So far attempts to find AMC or AFC substrate with selectivity towards any individual member of PCSK did not succeed owing to their overlapping substrate specificities. Recently Pasquato et al. (46) tested several peptide-MCA derivatives as possible substrates for detecting and monitoring protease activity in vitro for PCSK8 or SKI-1. They reported that a heptapeptidyl-MCA Suc (Succinoyl)-YISRRLL-MCA, containing P7-Tyr, P4-Arg, P2-Leu and also P1-Leu, is by far the most selective and potent substrate of this convertase. Consequently it is the recommended substrate for assaying PCSK8 or SKI-1 activity (VAL-508-539-IRCM, Gestion Univalor, Limited Partnership, web: www.univalor.ca).

Table 6.2
List of peptidyl-MCA substrates for activity assay of kexin-type PCSK or PC enzymes

No.	Substrate	Enzymes suitable for assay	Reference
1.	Boc-R-V-R-R-MCA	Furin, PC5, PC7	(106, 107)
2.	pE-R-T-K-R-MCA	PC1, PC2, Furin, PC5, PC7	(108–110)
3.	R-Q-R-R-MCA	PC4	(111)
4.	R-E-K-R-MCA	PC4 (weak)	(112)
5.	R-K-K-R-MCA	Furin, PC4	(108)
6.	R-S-K-R-MCA	PC1	(113)
7.	K-S-K-R-MCA	PC1 (weak)	(113)
8.	Y-E-K-E-R-S-K-R-MCA	PC1	(113)
9.	Ac-K-T-K-Q-L-R-MCA	PACE4, PC1, PC4	(111)
10.	Ac-R-S-K-R-MCA	PC1, Furin	(113)
11.	Ac-S-K-R-MCA	PC1 (very weak)	(113)
12.	Ac-R-E-K-R-MCA	PC1	(113)
13.	Ac-R-F-A-R-MCA	PC1	(113)
14.	Ac-R-P-K-R-MCA	PC1	(113)
15.	Ac-R-K-K-R-MCA	PC1, Furin	(113)
16.	Ac-R-A-R-Y-R-R-MCA	Furin	(114)

Boc, *t*-Butyloxy carbonyl; pE, pyroglutamic acid; Ac, Acetyl

Table 6.3
List of peptidyl-MCA substrates for enzyme activity assay of pyrolysin-type PCSK8/SKI-1/S1P convertase

No.	Substrate	Potency of cleavage	Reference
1.	Succ-I-Y-I-S-R-R-L-L-MCA	++++++++++	(46)
2.	Succ-Y-I-S-R-R-L-L-MCA	+++++	
3.	Succ-F-I-S-R-R-L-L-MCA	++++	
4.	Succ-I-I-S-R-R-L-L-MCA	+++	
5.	Succ-A-I-S-R-R-L-L-MCA	++	
6.	Succ-V-I-S-R-R-L-L-MCA	++	
7.	Succ-S-I-S-R-R-L-L-MCA	+	
8.	Succ-I-S-R-R-L-L-MCA	+	

Succ, Succinoyl group
Number of + signs denote the efficiency of cleavage of peptide↓MCA bond by the enzyme. The amino acid sequence was derived from the SKI-1 processing site of surface glycoprotein (GPC) of Lassa virus. The amino acids shown ***underlined*** are mutant variants

3.2.2. Chemical Synthesis of Peptidyl-MCA Derivatives

Peptidyl-MCA derivatives have been chemically prepared in two stages. First the peptide with the desired sequence having a terminal free carboxyl group and all amino acid side chain functions well protected is prepared. This can be achieved by liquid phase chemistry in a stepwise manner (47) or more efficiently by solid phase Fmoc chemistry using 2-chlorotrityl resin as described previously (48). The protected peptide, obtained following purification if needed by silica gel column chromatography (49), is then coupled to 7-amino-4-methyl coumarin using HATU or TBTU as a coupling agent (50). Finally the coupled peptide was completely deprotected by treatment with Reagent B (88% trifluoroacetic acid + 2% triisopropyl silane + 5% phenol + 5% water as described in (51)). The final product was purified by RP-HPLC, lyophilized and fully characterized by mass and NMR spectroscopy.

3.2.3. Protocol for PCSK Assay Using Peptidyl-MCA Substrate

The assay is performed in a well plate (black-coloured opaque flat bottom 96-well format, Millipore) by taking a small aliquot (typically 5 or 10 μl) of the enzyme sample and incubating with shaking at 25 or 37°C in a total volume of either 100 or 50 μl in a buffer medium consisting of 25 mM Tris + 25 mM Mes + 2.5 mM $CaCl_2$, pH 7.4 (buffer A), in the presence of a fluorogenic substrate namely either Boc-RVRR-MCA or pERTKR-MCA (50 or 100 μM final concentration) dissolved in dimethyl sulfoxide (DMSO). The fluorescence released due to the formation of free AMC at any time point is recorded in a spectrofluorometer using the excitation and emission wavelengths fixed at 370 and 460 nm, respectively (52). Typically the procedure is as follows: To a single well, add 10 μl of pERTKR-MCA (500 μM stock concentration in DMSO), followed by 80 μl of assay buffer (buffer A) and 10 μl of enzyme sample. In an improved protocol, one can use a buffer containing 50 mM Tris–HCl (pH 7.5), 1% Triton X-100, 10% glycerol and a cocktail of protease inhibitors (aprotinin, 1 mM, PMSF, 1 mM and benzamidine, 1 mM) (53). After each half-hour the fluorescence intensity was measured in a stop time assay. The extent of enzyme reaction depends on the activity or the amount of fluorescence released per hour. A control experiment is run in parallel under identical conditions without the enzyme added. The raw fluorescence unit (RFU) value obtained per hour is converted into nmol AMC released per hour using a standard curve connecting ΔRFU/h with nmol AMC. In addition to this stop time or end time assay, online progression curve showing fluorescence release in real time can also be used for detecting PCSK activity. In the latter case, the enzyme activity is assessed by the initial slope of the linear progression curve. Higher the slope means higher the enzymatic activity.

3.2.4. Peptidyl-AFC Substrates

Except two reports (54, 55), nothing is known about the use of peptidyl-AFC for PCSK assay. Despite the fact that the peptide-AFC has been widely used for monitoring activity of proteases of other families such as the caspases, caspase-3/7/14 (56) as well as trypsin, elastase (57), kallikreins (58) and cathepsins (59, 60), so far only one peptidyl-AFC derivative, Z (Ac)-Arg-Glu-Lys-Arg-AFC (Z = carbobenzoxy, Ac = acetyl), has been successfully used to detect and monitor in vitro the catalytic activity of kexin-type PCSK enzymes such as PACE4 and furin (54, 55).

3.2.5. Protocol for PCSK Assay Using Peptide-AFC Substrate

A similar protocol was also used for this assay with the substrate Z (Ac)-Arg-Glu-Lys-Arg-AFC. Usually a final concentration of 40 μM of the substrate is used and the released fluorescence is measured with excitation and emission wavelengths fixed at 400 and 505 nm, respectively (53, 54). Other concentration levels of the substrate can also be used depending on the activity of the enzyme sample. Apart from AFC and AMC, 4-trifluoromethyumbelliferyl derivative of peptides was also tried to monitor protease activity but with limited success (61).

3.2.6. Comparison Between AFC and MCA Substrates

The fluorescence of free AFC had a remarkable property useful in the assay of proteases present in biological fluids and tissue samples. The acylated derivatives of both AMC and AFC possess a blue fluorescence. However, the liberated free AMC has a slightly shifted, higher blue fluorescence compared to free liberated AFC whose fluorescence is shifted into the green region of the spectrum. This led to an increased level of spectrum difference

(Stokes shift) between the emission peak of peptidyl-AFC and the emission peak of the free AFC. This allows it to achieve lower detection limits of protease activity compared to using AMC derivatives. In fact the quantum efficiency of free AFC is nearly 40-fold higher than that of 4-methoxy-2-naphthylamine and 100 times that of 5-aminoisophthalic acid dimethyl ester (62). Moreover when applied to lysed cells samples, the peptide-MCA substrate showed a high background. The longer wavelength spectra of AFC (Ex/Em=405/535 nm vs. 370/460 nm for AMC) provided a greater sensitivity and less interference from the media (63). The use of peptide-AFC provided a homogenous assay with a wide linear range and high signal/background ratio. This assay is ideal for measuring enzyme activity in cells grown in 96-well plates (63).

3.2.7. Advantages and Limitations of Fluorogenic Methods

Currently fluorometric assay has been found to be most useful and practical not only for PCSKs but also for other proteases. The method is rapid and convenient for high sample throughput and most importantly it is very sensitive in terms of detection and for quantitative as well as kinetic measurements of protease activity of various enzymes including PCSKs. These methods are highly reliable and consistent. Although fluorometric assay is widely utilized, the technique does have some limitations. For example, the peptide substrates containing the highly hydrophobic aromatic fluorescent groups often have poor solubility in aqueous system making competition assays difficult to perform. Besides, the bulky fluorophore moiety may sterically alter enzyme–substrate recognition, resulting in misleading or diminished rates of catalysis. Results from monitoring peptide hydrolysis solely by fluorescence are also susceptible to artificially high rates of proteolysis since detection lacks the ability to differentiate between target and non-specific protease activity. Furthermore, fluorometric assays used to identify potential inhibitors are prone to artefacts with fluorescent compounds. Despite these limitations, they are still considered as the most widely used method for assaying the protease activity (64).

3.3. Chromogenic Method

This method is based on the liberation of a chromophoric group from its peptidyl derivative following cleavage by the protease under study. Chromophoric groups are those which exhibit strong absorption maximum at a specific wavelength in the UV–visible range, making them easily detectable following their release after proteolytic cleavage. The most common and potent chromogenic moieties are para nitro aniline and β-naphthyl amine (**Table 6.1**) (65), and consequently peptidyl-para-nitro anilides/β-naphthyl amides have been used for assay of protease activity including those of PCSKs (66). Usually such methods are much less sensitive and precise than the AMC- or

AFC-based fluorogenic substrates. They also suffer from poor solubility in aqueous medium.

3.4. Intramolecularly Quenched Fluorogenic (IQF) Method

This technique was introduced in the early 1990s for the purpose of detecting the protease activity of PCSK and other proteases in a quick, efficient and selective manner (67). It has also been used for making peptide libraries in order to study the substrate specificity of proteases (68). This technique is based on what is known as intramolecular quenching of fluorescence (IQF) of an electron donor function by an electron acceptor function both present within a peptide chain, one at the N-terminus and the other at the C-terminus of the sequence. The only criterion is that the sequence must contain a protease cleavage site and the two functions must be separated by a spatial distance of 8–10 Å for maximum quenching effect. Initially such peptide molecules exhibit a low level of fluorescence intensity owing to quenching effect within the molecule. Upon cleavage of the peptide bond by the protease, there is an immediate loss of quenching effect resulting in liberation of fluorescence intensity (67). This phenomenon can be compared with "fluorescence resonance energy transfer" (FRET) event (68, 69). FRET is the transfer of energy in the excited state from the initially excited donor (D) to an acceptor (A). The donor molecules typically emit at shorter wavelengths that overlap with the absorption of acceptors. The process is a distance-dependent interaction between the electronic excited states of two molecules without emission of a photon. Thus FRET is really the result of long-range dipole–dipole interactions between the donor and acceptor molecules. An excited donor molecule has several routes to release its captured energy during returning to the ground state. The excited state energy can be dissipated to the environment (as light or heat) or transferred directly to a second acceptor molecule, sending the acceptor to an excited state. The latter process is called FRET. This methodology is used to specifically study interaction and binding events. For example, RNA binding assay has been developed using FRET peptide containing 5-FAM/TAMRA-Mal as acceptor and donor functions respectively (70). Unlike in FRET, there is a complete cleavage of a peptide bond within the IQF segment by the protease under investigation, leading to a physical separation of the two groups and an abrupt enhancement of fluorescence intensity. So far in the literature several pairs of fluorescence and quenching compounds have been successfully used. These are shown as follows.

(i) *Abz/Tyx* (*Abz* = 2-amino benzoic acid, *Tyx* = 3-nitro tyrosine, Ex/Em=320/420 nm) (71); (ii) *Abz/Dnp* (*Dnp* = 2,4-dinitro phenyl, Ex/Em=320/420 nm) (72); (iii) *Dabcyl/Edans* (*Dabcyl* =4,4′-[dimethylamino phenyl]azo benzoic acid; *Edans* = 5-(2-aminoethyl)amino naphthalene-1-sulfonic acid, Ex/Em=

340/490 nm) (73); (iv) *Abz/EDDnp* (*EDDnp* = (2,4-dinitrophenyl)ethylene diamine) (74); (v) *DaciaC/Dnp* (*DaciaC* = S-(*N*-[4-methyl-7-*d*imethyl*a*mino-*c*oumar*i*n-3-yl]-carbox*a*midomethyl)-cysteine) (75); (vi) *L-Amp/Dnp* (*L-Amp* = L-2-amino-3-[7-methoxy-4-coumaryl]propionic acid) (76); (vii) *Mca/Dnp* (*Mca* = 7-methoxy coumarin 4-yl, Ex/Em=328/393 nm) (77); (viii) *Dns/pNP* (Dns [Dansyl] = 5-[dimethylamino]naphthalene-1-sulfonyl; pNP = para-nitrophenyl, Ex/Em=313/418 nm) (78); (ix) HiLyte Fluor^{TM} 488/(QXL^{TM} 520) Ex/Em 488/520 nm (79); (x) *QXL520/5-FAM* (79), Ex/Em=490/520 nm (80).

QXL 520 is a non-fluorescent dye that acts as a quencher to 5-FAM or HiLyte Fluor^{TM} 488 (http://www.genengnews.com/articles/chtitem.aspx?tid=2243&chid=2). The QXL 520 quencher offers several advantages. Its absorption spectrum overlaps with nearly the entire emission spectrum range of 5-FAM, thereby providing efficient quenching. This fluorescent/quencher pair has been recently used to develop sensitive assays for caspases, MMPs, secretases, HIV protease, HCV (hepatitis C virus) protease and others (80, 81). So far, several efficient fluorogenic IQF–peptide substrates have been developed to detect the protease activity of PCSK enzymes in vitro. These are mostly based on Abz/Tyx, Abz/Dnp, Abz/EDDnp, AMC/Dnp and Dabcyl/Edans pairs, and a few of them are listed in **Table 6.4**. The continuous fluorometric assays for endopeptidases based on IQF substrates were originally developed by Yaron et al. for the enzyme thimet (82).

3.5. SDS-PAGE Autoradiograph (In-Gel Activity Staining) Method

SDS-PAGE gel with autoradiography using radioactive or fluorescent labelled peptide substrates has also been used to measure protease activity particularly in tissues and ex vivo or in vivo conditions (83, 84). Typically an SDS gel is prepared using the normal buffer condition containing the labelled peptide substrate such as a peptide-MCA. This allows a homogenous coating of the substrate all over the gel. The enzyme sample is then loaded on the gel under usual conditions. The released AMC due to the action of the enzyme at specific spot can be captured using a film. The presence of a positive band will provide information about the activity and the molecular size of the enzyme under investigation while the intensity measures the level of protease activity. We have applied this method to detect and measure the proteolytic activity of recombinant PCSK4 enzyme (85).

3.6. RP-HPLC Method

Reverse-phase high performance liquid chromatography (RP-HPLC) has also been used in the past for monitoring protease activity in vitro of various enzymes including PCSKs. Typically a short peptide (10–20 mer) containing the enzyme-specific cleavage site is first selected. It is then digested with the

Table 6.4
List of some fluorogenic IQF peptides used for detection and measurement of PCSK protease activity. The chemical structures of each pair of fluorescent and quencher groups commonly used in IQF are shown in individual box below

No.	Substrate	Most potent PCSK enzyme(s)	Reference
1.	**Abz**-R-I-Y-I-S-R-R-L-L ↓ T-F-T-**Tyx**-A	SKI-1	(115)
2.	**Dabcyl**-R-G-V-V-N-A ↓ S-S-R-L-A-**Edans**	SKI-1	(73)
3.	**Abz**-P-A-K-S-A-R ↓ S-V-R-**Tyx**-A	PC4	(111)
4.	**Abz**-R-N-T-P-R-E-R-R-R-K-K-R ↓ G-L-**Tyx**-A	Furin, PC5, PC7	(71)
5.	**Abz**-V-P-R-M-E-K-R- ↓ Y-G-G-F-M-Q-**EDDnp**	PC2	(116)
6.	**Dnp**-F-A-Q-S-I-P-K-**AMC**	PCSK9	(27)

H-Abz-OH / H-Dnp
H-Abz-OH / H-EDDnp
H-MCA / H-Dnp
H-AMP-OH / H-Dnp
H-Abz-OH / H-Tyx-OH
Dabcyl-OH / H-Edans
H-DaciaC-OH / H-Dnp
5-FAM-OH

enzyme under study for various time intervals in an appropriate buffer, pH and Ca^{+2} ion concentration. The efficiency of such methods depends on HPLC separation of cleaved peptide fragments from its undigested peptide. These methods are usually very slow, time consuming and require multiple steps of operations. Moreover, it is much less sensitive due to the use of UV detection system for the HPLC peaks. This may be improved by using diode ray detection system. Alternatively fluorescence detector can be used for tryptophan-containing or fluorescence-labelled peptide substrates (54).

3.7. Solid Phase Method

Owing to several limitations associated with liquid phase methods that include slow pace of the reaction, occasional use of multiple steps, expensive reagents, lack of solubility of the substrate and possibility of rapid loss of enzyme activity via degradation during the assay run, solid phase methodologies are more appealing as they are less expensive and more sensitive. Although some preliminary studies have been made including one in our laboratory to develop such methodologies for protease assay, nothing has yet been firmly established. There is currently a great deal of research interest to develop such a method especially for the assay of PCSK activity. Earlier efforts have been made to use nano-particles, gold particles, magnetic sensors, radioactive probes (I^{125}, H^3, C^{14}, etc.) to develop a solid phase assay for protease activity (86–97). For example, an immobilized radio-labelled peptide such as AffiGel-10-Gly-Gly-Gly-Gly-Val-Ser-Gln-Asn-Tyr↓Pro-Ile-Val-Gln-[^{3}H]Gly-OH has been used to measure HIV protease activity based on amount of radioactivity released into the medium following cleavage at Tyr↓Pro site (93). Biochemical applications of these methods to detect and measure specific protease activity in crude biological samples have yet to be firmly demonstrated. So far nothing is known in the literature about solid phase assay for PCSK activity. This prompted us to develop our concept in the field.

3.7.1. A New Solid Phase Assay for PCSK Activity

Recently, we succeeded in designing a new solid phase approach for rapid detection and quantitative measurement of PCSK activity in a cost-effective manner. Our concept is based on the attachment of a highly sensitive fluorescent moiety to the N-terminus of a suitable peptide sequence which is then covalently attached via a suitable spacer to a solid matrix through its C-terminus. When such a fluorescent peptide attached to resin is treated with an active protease in a suitable buffer under optimum pH and temperature, it is expected that due to the cleavage of a peptide bond by the protease, an N-terminal fluorescent peptide fragment will be released into the incubation medium. *The intensity of this liberated fluorescence will be a direct measure of the enzymatic activity of the protease.* The extent of hydrolysing activity of the enzyme is directly associated with the released fluorescence intensity and will depend on the time of incubation, presence of cofactors, regulators, binding partner proteins and the sample size. We propose that this methodology can be easily extended to detect protease activity of any endoproteolytic enzyme, provided a suitable peptide motif selectively recognized by the enzyme under study is available. The basic principle of the methodology including the various steps involved is depicted schematically in **Fig. 6.1**. The presence of two spacer arms (linkers) on either end of the peptide segment is meant to provide more accessibility and exposure of the peptide sequence to the enzyme used (88).

3.7.2. Design, Synthesis and Evaluation of Immobilized Fluorescent Peptide (IFP)-Based Assay

The success of the above proposed fluorogenic solid phase assay depends on the choice of (i) fluorescence moiety, (ii) the peptide sequence, (iii) the solid matrix or resin, (iv) the method of attachment of the fluorescent tag to the peptide, (v) structural stability of the fluorescence moiety under various pH, light and temperature conditions and finally (vi) the strengths of the bonds

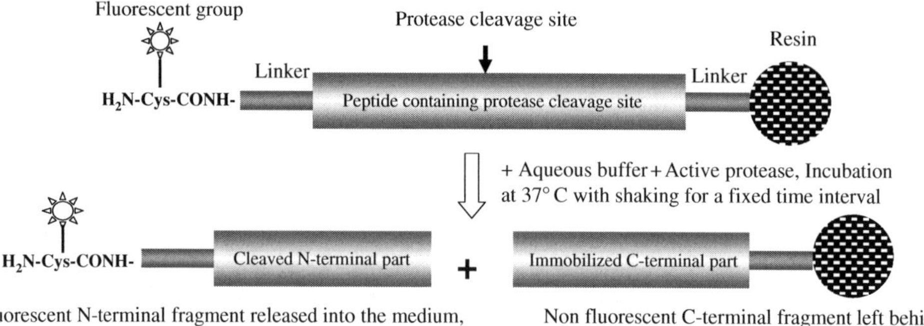

Fig. 6.1. General schematic diagram showing the methodology for novel solid phase fluorometric assay for protease activity.

connecting the peptide chain to the resin and to the fluorescent group. Meeting all these conditions may lead to an effective in vitro solid phase method for detection of protease activity. In our initial design for a solid phase substrate, we selected Texas Red dye (chemical structure shown in **Fig. 6.2**) as our choice of fluorescence moiety because of its extremely high sensitivity, relative stability and commercial availability in various useful activated forms that can be employed for efficient coupling to peptides via a free Cys-SH or terminal NH_2 groups. A Cys residue with a free thiol group is preferred most for above conjugation because of its high reactivity. In fact, the commercially available Texas Red C_2 maleimide which already contains a two-carbon atom linker and a reactive maleimide function is well suited for conjugation to the free SH group of a Cys-containing peptide. This Michael type of addition reaction occurs very efficiently leading to a high yield of coupled product (**Fig. 6.2**).

In our efforts to develop a rapid solid phase assay for PCSK4 enzyme based on the above strategy, we selected the proIGF-2-

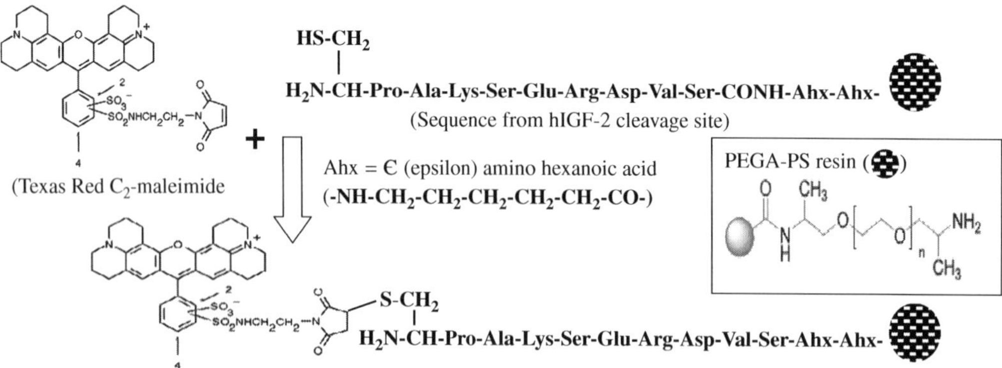

Fig. 6.2. General scheme showing the preparation of immobilized fluorescent IGF-2 peptide using amino-PEGA-PS resin.

(a potent PC4 substrate) derived sequence *Pro-Ala-Lys-Ser-Glu-Arg↓Asp-Val-Ser* encompassing its PC4 processing site as shown by a vertical arrow. An additional Cys residue is added to the N-terminus of this sequence for attachment of Texas Red dye, while an additional linker consisting of two units of the unnatural amino acid, Ahx (epsilon amino hexanoic acid), is appended to the C-terminus of the peptide prior to the solid matrix. For choice of resin, we opted for micro-porous Fmoc-amino-PEGA-PS (*p*oly*e*thylene *g*lycol poly*a*crylamide *p*oly*s*tyrene) resin, available from Novabiochem, San Diego, USA (for structure, *see* **Fig. 6.2** inset) (98). This type of resin is useful for permanently anchoring a peptide on solid matrix. Our design led to the building of the immobilized fluorescent peptide-1 (IFP-1) (**Fig. 6.3**).

A typical synthesis of such a solid phase immobilized peptide involves a few simple steps. First a 10 mM solution of Texas Red C_2 maleimide or Alexa Fluor 680 C_2 maleimide (Invitrogen Life Science, USA, MW=728) is prepared by dissolving 3.12 mg in DMSO (0.43 ml). In parallel, using the solid phase Fmoc chemistry the desired peptide was built on Fmoc-amino-PEGA-PS resin (440 mg, substitution 120 mmol Fmoc/g) with all side chain functional groups of amino acids fully protected but not the

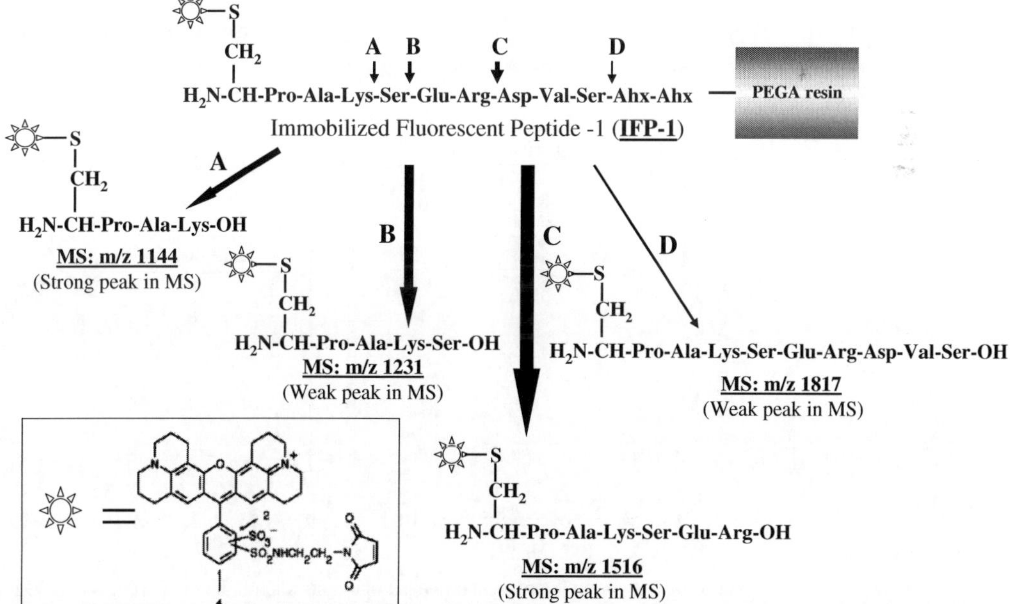

Fig. 6.3. Validation of the proposed solid phase fluorometric assay using commercial bovine trypsin which cleaves the fluorescent peptide on the resin releasing various fragments into the medium. The figure shows the cleavage sites based on the mass spectra of the products released into the medium. Primarily the major cleavage occurs at site C and then at site A. This is followed by further cleavages at sites B and D (both minor) upon prolonged incubation. PEGA resin = amino polyethylene glycol polyacrylamide polystyrene resin; MS = mass spectrum.

terminal amino group. The peptide resin (100 μl wet volume) was then allowed to swell in water (pH 6.8, 100 μl) for 30 min and treated with the above Texas Red solution (200 μl, 1.22-fold molar excess). The suspension is then stirred with a magnetic stirring bar at 37°C for 2 h and finally washed extensively (10× 1 ml) in order to remove all uncoupled fluorescent dye. The washing is considered complete when the fluorescence intensity reading measured at $\lambda_{ex} = 591$ nm and $\lambda_{em} = 608$ nm remained low and steady.

4. Results of the Solid Phase Assay

Before trying on recombinant PCSK4 activity (85), we tested our IFP-1 resin on the activity of a standard enzyme such as bovine trypsin (Sigma Chemical Co, USA). Trypsin is particularly suited for this study, since it has a very well-defined specificity, as it cleaves only the peptide bonds in which the carbonyl group is contributed either by an Arg or a Lys residue. Therefore, we expect IFP-1 which contains Arg and Lys residues to be cleaved by trypsin and thereby release fluorescent N-terminal peptide fragment into the medium. This can be measured using a fluorometer instrument and should reflect the enzyme activity of trypsin. With the amount of resin fixed, the intensity of released fluorescence should consistently increase

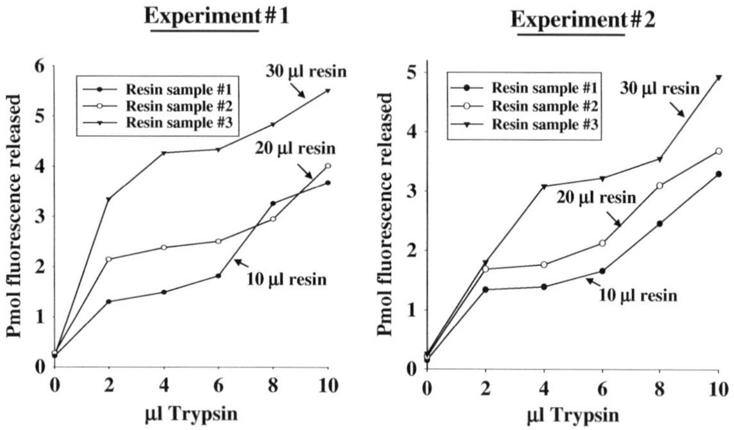

Fig. 6.4. Monitoring fluorescence release upon digestion of IFP-1 (~10, 20 and 30 ml wet volume) after 1 h treatment with increasing amounts (2, 4, 6, 8 and 10 ml) of standard bovine trypsin solution (1 mg/ml) as described in the text. Two sets of experiments (#1 and 2) were performed. The fluorescence released into the supernatant was measured using excitation and emission wavelengths fixed at 591 and 608 nm, respectively. The raw fluorescence unit (RFU) was converted into amount (pmol or picomole) of fluorescent peptide generated using a standard curve separately created.

with the amount of added enzyme. In fact our results shown in **Fig. 6.4** confirmed these expectations as one can notice increasing level of fluorescence intensity in the medium for three individual samples of IFP-1 resin in two sets of experiments (left and right panels, **Fig. 6.4**). The fluorescence released per unit of time is first measured as raw fluorescence unit (RFU) which is then transformed into pmol peptide cleaved using standard curve previously generated. This method can be useful for detailed kinetic analysis of the enzyme as long as there is a single site of cleavage. It is also interesting to mention that for each resin sample in either experiment, the RFU value increases also with time (*not shown in the figure*). Next, the question arises about the site(s) of cleavage(s). This is revealed by recording

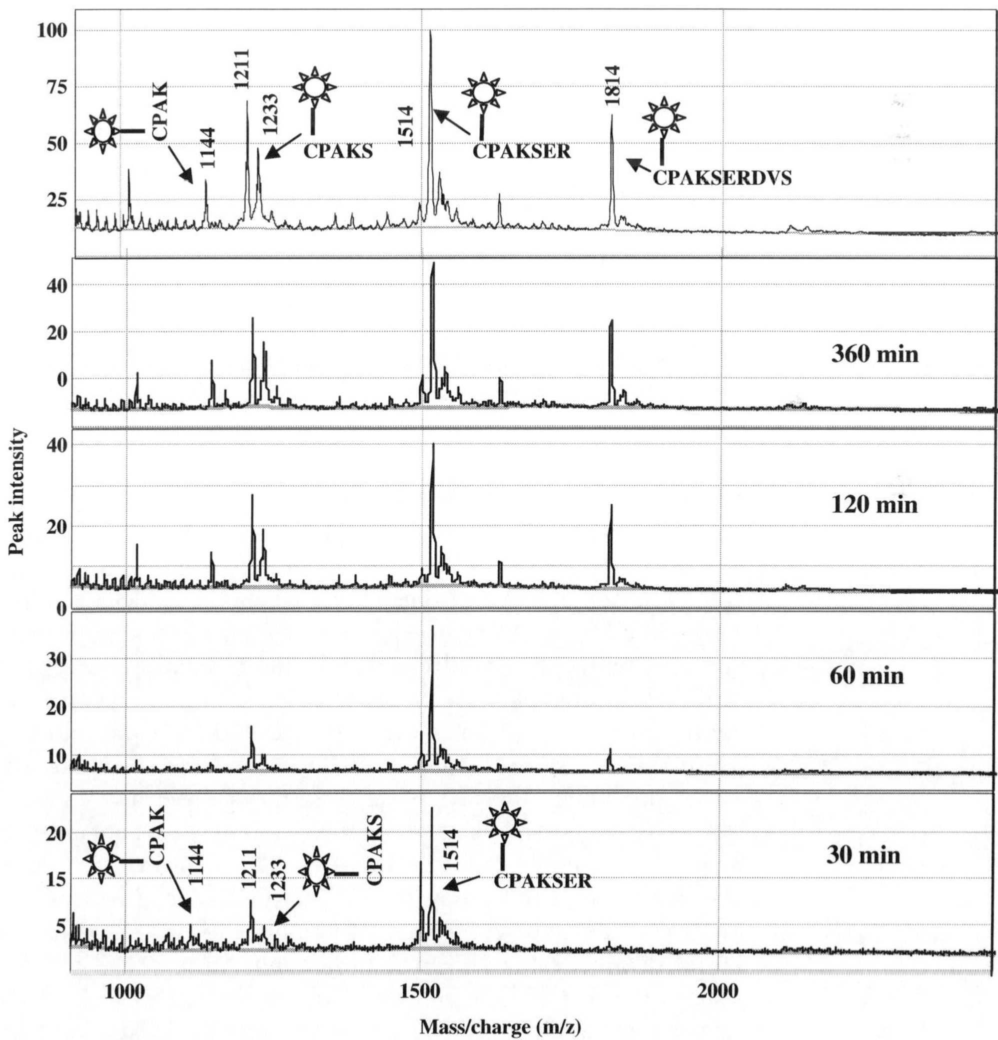

Fig. 6.5. Surface-enhanced laser desorption ionization time of flight mass spectra of fluorescent peptide fragment released into the media following digestion of IFP-1 for various time periods with commercial trypsin.

the mass spectra of the media collected at various time points. **Figure 6.5** shows the SELDI-TOF (surface-enhanced laser desorption time of flight) mass spectra results of the crude digested product(s) released into the medium after 30 min reaction. The data indicated a major cleavage at Arg↓Asp site, with minor cleavages at Lys↓Ser↓Glu positions. This is confirmed by the presence of peaks at *m/z* 1514 (very strong), 1233 (weak) and 1144 (extremely weak) which are attributed to the cleaved fluorescent fragments, Texas Red-C-P-A-K-S-E-R-OH (calculated MW=1516), Texas Red-C-P-A-K-S-OH (calculated MW=1231) and Texas Red-C-P-A-K-OH (calculated MW=1144), respectively. The results showing post-Arg and Lys cleavages are consistent with the behaviour of trypsin. However, the cleavage after the Ser residue is unexpected. We rationalize this by proposing that it originates from the cleavage by α-chymotrypsin, a common contaminant of trypsin (99). Upon prolonged digestion, these peaks become more intense meaning an increased level of cleavages but in addition, another peak at *m/z* 1814 was also observed due to the formation of Texas Red-C-P-A-K-S-E-R-D-V-S-OH (calculated MW = 1817). Again this peak may be the result of cleavage by the contaminant α-chymotrypsin in trypsin. Overall, our data suggest that IFP-1 can be used as a solid phase substrate for trypsin to detect its activity as long as the digestion time is limited to less than 30 min or less. Upon a longer period of incubation, one may notice multiple cleavages including those mediated by the contaminant α-chymotrypsin.

In parallel to the above study, another fluorescent peptide, IFP-2, was prepared and immobilized on PEGA-PS resin in order to confirm the efficacy and applicability of our methodology and also to develop a solid phase assay for PCSK8/SKI-1 activity. IFP-2 contains a short tetrapeptide (RRLL) recognition motif for the enzyme. Since in our previous peptide model we noted multiple cleavages particularly upon prolonged incubation with the enzyme trypsin, we decided this time to employ a sequence as short as possible yet containing the minimum recognition motif of the enzyme. Thus the sequence RRLL↓G was selected. Here a small amino acid Gly at P1′ position was chosen because of its preference for SKI-1. As before, we have also inserted two Ahx units as linkers on either end of this pentapeptide sequence as illustrated in **Fig. 6.5**. The protocol for using this resin for SKI-1 assay is similar to that described above. In brief, IFP-2 resin (10 μl wet volume + 10 μl water) is suspended at 37°C in 25 mM ammonium carbonate + 2.5 mM $CaCl_2$, pH 7.4 (15 μl), followed by incubation with recSKI-1 enzyme (typically a 5 μl sample, activity 2 U/μg protein, where U = pmol of peptide, Abz-Y-I-S-I-R-R-L↓L-T-F-T-Tyx-A cleaved per min) (73, 100). The fluorescence released into

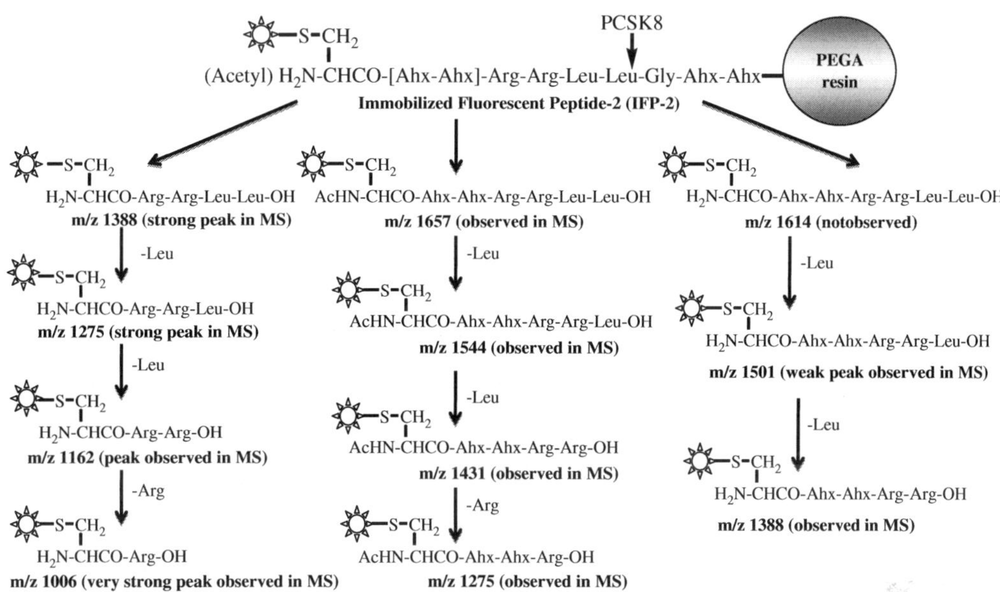

Fig. 6.6. Fluorescent-based solid phase assay of PCSK8 enzyme using its minimum tetrapeptide (RRLL) recognition motif. The fluorescent peptide was synthesized by solid phase peptide chemistry on amino-PEGA-PS resin as a mixture of free as well as N-terminal acetylated peptide with and without the two N-terminal Ahx linkers (indicated by *third bracket*). Treatment with rec-PCSKL8/SKI-1 enzyme led to the formation of highly fluorescent N-terminal fragments (established by mass spectral data) indicating cleavage at RRLL↓ site by the enzyme. The progress of the cleavage can be monitored by measuring the fluorescence intensity released into the reaction medium.

the medium after every 30 min was measured by taking 5 μl aliquots of supernatant following centrifugation. In addition a 1 μl aliquot was also removed for mass spectrum analysis. This way one can correlate enzyme activity and released fluorescence intensity and also determine the site of the cleavage(s). The methodology and the cleavage details based on mass spectral data are shown in **Figs. 6.6** and **6.7**, respectively, on a 12 h digest of the supernatant collected. The data confirm that SKI-1 cleaves the immobilized peptide at the expected RRLL↓G site leading to peaks at *m/z* 1388, 1431, 1501 (weak), 1544 and 1657 for the peptides dye-C-R-R-L-L-OH (calculated MW=1386), Ac-dye-C-Ahx-Ahx-R-R-OH (calculated MW=1428), dye-C-Ahx-Ahx-R-R-L-OH (calculated MW=1499), Ac-dye-Ahx-Ahx-C-R-R-L-OH (calculated MW=1541) and Ac-dye-C-Ahx-Ahx-R-R-L-L-OH (calculated MW=1654), respectively. The data also indicate that our synthetic immobilized fluorescent peptide was not clean and actually consisted of a mixture of three peptides namely dye-C-Ahx-Ahx-R-R-L-L-G-Ahx-Ahx-resin, Ac-dye-C-Ahx-Ahx-R-R-L-L-G-Ahx-Ahx-resin and dye-C-R-R-L-L-G-Ahx-Ahx-resin. This was due to an inefficient coupling of Ahx and partial acetylation of the terminal amino group during the capping process with acetic anhydride. This issue will be addressed and corrected

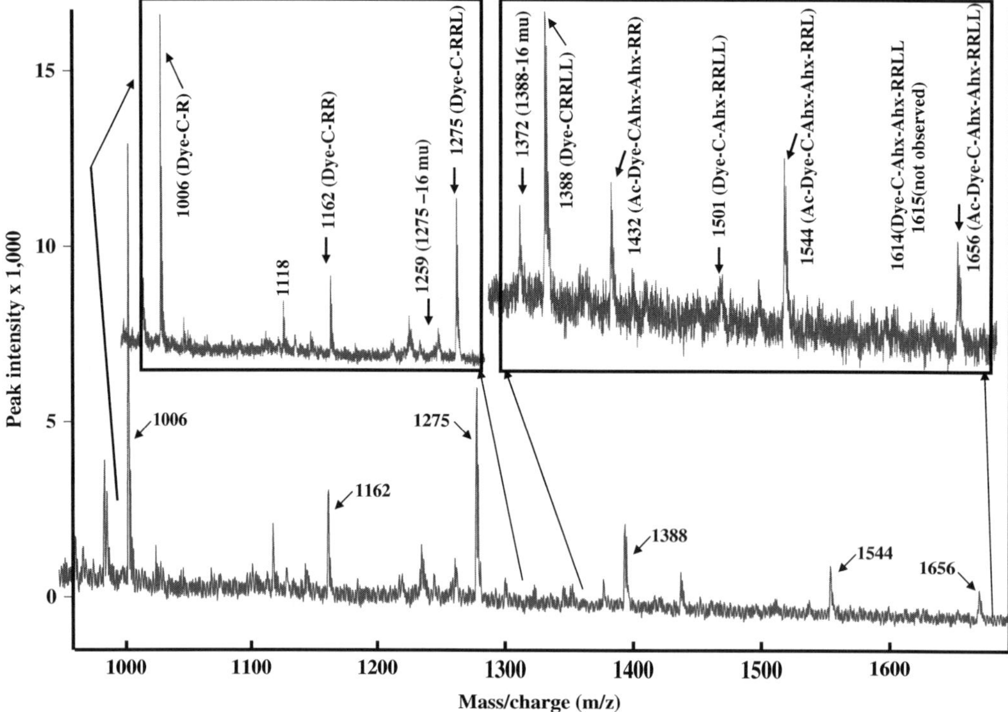

Fig. 6.7. Matrix-assisted laser desorption time of flight mass spectra of fluorescent peptide fragments released into the medium following 12 h digestion of IFP-2 by PCSK8/SKI-1 enzyme.

in our future work. However, all the peptides on the solid matrix contain an identical "RRLL↓G" cleavage site which serves as an efficient common substrate of SKI-1. Our initial data suggest that the above solid phase fluorogenic substrate method can be successfully used to detect and measure the protease activity of SKI-1, although it needs further improvement and perfection in terms of clean synthesis of immobilized substrate which will ultimately lead to a novel solid phase assay for not only SKI-1 but also other members of PCSK family.

5. Conclusion

So far, several in vitro liquid phase methods have been developed to detect, measure and kinetically analyse the protease activity of PCSK enzymes. In most cases, these are based on the use of peptidyl-MCA substrates such as pERTKR-MCA and Boc-RVRR-MCA for kexin-type PCSK enzymes and Suc-I-Y-I-S-R-R-L-L-MCA or Suc-RRLL-MCA for pyrolysin-type PCSK8 enzyme. In addition peptidyl-AFC such as REKR-AFC has also been used successfully to follow protease activity of PCSKs of kexin type. Despite some limitations, these fluorogenic peptidyl derivatives continue to be the most rapid, efficient and high-throughput agents for PCSK activity assay. Besides AMC and AFC, development of more sensitive fluorogenic functional moiety may remain as a future challenge for research. This is particularly crucial since a slight regulation of PCSK activity under physiological condition may trigger significant alterations in normal metabolic pathways leading to disease or disorder states. Our new proposed solid phase fluorogenic method presented here is promising in this respect although it needs to be further improved to address the issue of multiple cleavages and selectivity. To our knowledge, this is a first report of a solid phase assay for PCSK activity. The advantage of this method is that it offers rapidity, simplicity, sensitivity and high degree of selectivity. One immediate application of this method would be to use it for early detection of a disease where PCSK activity is directly linked. These may include cancer, diabetes, obesity, fertilization defects and viral/bacterial infections. It may also be useful for monitoring disease progression. A typical case may be the monitoring of progression of foetal growth restriction or intrauterine growth restriction (IUGR) and pre-eclampsia condition during pregnancy (101, 102). The crucial role of PCSK4-mediated processing of pro-IGF2 in these medical conditions has been well demonstrated (31, 32).

Overall, both solid and liquid phase methods can be useful for monitoring and characterizing catalytic activity of PCSKs in vitro as well as in ex vivo cellular and animal models. The availability of various knockout, knockdown and transgenic animals with different levels of PCSK expressions makes these assay methods more useful as research and clinical tools. Further research particularly on the proposed solid phase approach will lead to more powerful and efficient avenue for curbing human illnesses and monitoring their intervention with drug or other types of therapies.

Acknowledgements

The authors would like to thank Alex Duchene, a co-op student, for carrying out some of the initial work involving the fluorescent solid phase assay method. The authors are thankful to Canadian Institutes of Health Research (CIHR) for CANADA-HOPE scholarship grant (AB and SM) and Team grant program (MOP-69093) as well as Center for Catalysis Research and Innovation,

U Ottawa (AB) for financial assistance. The above funders for this study had no role in study design, data collection and analysis, decision to publish or preparation of the manuscript.

References

1. Seidah, N. G., and Prat, A. (2002) Precursor convertases in the secretory pathway, cytosol and extracellular milieu *Essays Biochem* **38**, 79–94.
2. Seidah, N. G., and Chrétien, M. (1999) Proprotein and prohormone convertases: A family of subtilases generating diverse bioactive polypeptides *Brain Res* **848**, 45–62.
3. Steiner, D. F. (1998) The proprotein convertases *Curr Opin Chem Biol* **2**, 31–9.
4. Steiner, D. F. (1967) Evidence for a precursor in the biosynthesis of insulin *Trans NY Acad Sci* **30**, 60–8.
5. Chrétien, M., and Li, C. H. (1967) Isolation, purification, and characterization of gamma-lipotropic hormone from sheep pituitary glands *Can J Biochem* **45**, 1163–74.
6. Kovac, S., Shulkes, A., and Baldwin, G. S. (2009) Peptide processing and biology in human disease *Curr Opin Endocrinol Diabetes Obes* **16**, 79–85.
7. Seidah, N. G., Mayer, G., Zaid, A., Rousselet, E., Nassoury, N. et al. (2008) The activation and physiological functions of the proprotein convertases *Int J Biochem Cell Biol* **40**, 1111–25.
8. Seidah, N. G., Gaspar, L., Mion, P., Marcinkiewicz, M., Mbikay, M., and Chrétien, M. (1990) cDNA sequence of two distinct pituitary proteins homologous to Kex2 and furin gene products: Tissue-specific mRNAs encoding candidates for prohormone processing proteinases *DNA Cell Biol* **9**, 415–24.
9. Smeekens, S. P., and Steiner, D. F. (1990) Identification of a human insulinoma cDNA encoding a novel mammalian protein structurally related to the yeast dibasic processing protease Kex2 *J Biol Chem* **265**, 2997–3000.
10. Kiefer, M. C., Tucker, J. E., Joh, R., Landsberg, K. E., Saltman, D., and Barr, P. J. (1991) Identification of a second human subtilisin-like protease gene in the fes/fps region of chromosome *DNA Cell Biol* **10**, 757–69.
11. van de Ven, W. J., Voorberg, J., Fontijn, R., Pannekoek, H., van den Ouweland, A. M., van Duijnhoven, H. L., Roebroek, A. J., and Siezen, R. J. (1990) Furin is a subtilisin-like proprotein processing enzyme in higher eukaryotes *Mol Biol Rep* **14**, 265–75.
12. Seidah, N. G., Day, R., Hamelin, J., Gaspar, A., Collard, M. W., and Chrétien, M. (1992) Testicular expression of PC4 in the rat: Molecular diversity of a novel germ cell-specific Kex2/subtilisin-like proprotein convertase *Mol Endocrinol* **6**, 1559–70.
13. Nakayama, K., Kim, W. S., Torii, S., Hosaka, M., Nakagawa, T., Ikemizu, J., Baba, T., and Murakami, K. (1992) Identification of the fourth member of the mammalian endoprotease family homologous to the yeast Kex2 protease. Its testis-specific expression *J Biol Chem* **267**, 5897–900.
14. Lusson, J., Vieau, D., Hamelin, J., Day, R., Chrétien, M., and Seidah, N. G. (1993) cDNA structure of the mouse and rat subtilisin/kexin-like PC5: A candidate proprotein convertase expressed in endocrine and nonendocrine cells *Proc Natl Acad Sci USA* **90**, 6691–5.
15. Miranda, L., Wolf, J., Pichuantes, S., Duke, R., and Franzusoff, A. (1996) Isolation of the human PC6 gene encoding the putative host protease for HIV-1 gp160 processing in CD4+ T lymphocytes *Proc Natl Acad Sci USA* **93**, 7695–700.
16. Rehemtulla, A., Barr, P. J., Rhodes, C. J., and Kaufman, R. J. (1993) PACE4 is a member of the mammalian propeptidase family that has overlapping but not identical substrate specificity to PACE *Biochemistry* **32**, 11586–90.
17. Bruzzaniti, A., Goodge, K., Jay, P., Taviaux, S. A., Lam, M. H., Berta, P., Martin, T. J., Moseley, J. M., and Gillespie, M. T. (1996) PC8 [corrected], a new member of the convertase family *Biochem J* **314**, 727–31.
18. Meerabux, J., Yaspo, M. L., Roebroek, A. J., Van de Ven, W. J., Lister, T. A., and Young, B. D. (1996) A new member of the proprotein convertase gene family (LPC) is located at a chromosome translocation breakpoint in lymphomas *Cancer Res* **56**, 448–51.
19. Seidah, N. G., Hamelin, J., Mamarbachi, M., Dong, W., Tardos, H., Mbikay, M., Chrétien, M., and Day, R. (1996) cDNA structure, tissue distribution, and chromosomal localization of rat PC7, a novel mammalian proprotein convertase closest to yeast kexin-like proteinases *Proc Natl Acad Sci USA* **93**, 3388–93.

20. Seidah, N. G., Hamelin, J., Mamarbachi, A. M., Benjannet, S., Toure, B. B., Basak, A., Munzer, J. S., Marcinkiewicz, M., Zhong, M., Barale, J.-C., Lazure, C., Murphy, R. A., Chrétien, M., and Marcinkiewicz, M. (1999) Mammalian subtilisin/kexin isozyme SKI-1: A widely expressed proprotein convertase with a unique cleavage specificity and cellular localization *Proc Natl Acad Sci USA* **96**, 1321–6.
21. Cheng, D., Espenshade, P. J., Slaughter, C. A., Jaen, J. C., Brown, M. S., and Goldstein, J. L. (1999) Secreted site-1 protease cleaves peptides corresponding to luminal loop of sterol regulatory element-binding proteins *J Biol Chem* **274**, 22805–12.
22. Seidah, N. G., Benjannet, S., Wickham, L., Marcinkiewicz, J., Bélanger Jasmin, S., Stifani, S., Basak, A., Prat, A., and Chrétien, M. (2003) The secretory proprotein convertase neural apoptosis-regulated convertase 1 (NARC-1): Liver regeneration and neuronal differentiation *Proc Natl Acad Sci USA* **100**, 928–33.
23. Khatib, A. M., Siegfried, G., Chrétien, M., Metrakos, P., and Seidah, N. G. (2002) Proprotein convertases in tumor progression and malignancy: Novel targets in cancer therapy *Am J Pathol* **160**, 1921–35.
24. Bassi, D. E., Fu, J., Lopez de Cicco, R., and Klein-Szanto, A. J. (2005) Proprotein convertases: "master switches" in the regulation of tumor growth and progression *Mol Carcinog* **44**, 151–61.
25. Chrétien, M., Seidah, N. G., Basak, A., and Mbikay, M. (2008) Proprotein convertases as therapeutic targets *Expert Opin Ther Targets* **12**, 1289–300.
26. Scamuffa, N., Calvo, F., Chrétien, M., Seidah, N. G., and Khatib, A. M. (2006) Proprotein convertases: Lessons from knockouts *FASEB J* **20**, 1954–63.
27. Kourimate, S., Chétiveaux, M., Jarnoux, A. L., Lalanne, F., and Costet, P. (2009) Cellular and secreted pro-protein convertase subtilisin/kexin type 9 catalytic activity in hepatocytes *Atherosclerosis* **206**, 134–40.
28. Mousavi, S. A., Berge, K. E., and Leren, T. P. (2009) The unique role of proprotein convertase subtilisin/kexin 9 in cholesterol homeostasis *J Intern Med* **266**, 507–19.
29. Stawowy, P., Blaschke, F., Kilimnik, A., Goetze, S., Kallisch, H., Chrétien, M., Marcinkiewicz, M., Fleck, E., and Graf, K. (2002) Proprotein convertase PC5 regulation by PDGF-BB involves PI3-kinase/p70s6-kinase activation in vascular smooth muscle cells *Hypertension* **39**, 399–404.
30. Ugleholdt, R., Poulsen, M. L., Holst, P. J., Irminger, J. C., Orskov, C., Pedersen, J., Rosenkilde, M. M., Zhu, X., Steiner, D. F., and Holst, J. J. (2006) Prohormone convertase 1/3 is essential for processing of the glucose-dependent insulinotropic polypeptide precursor *J Biol Chem* **281**, 11050–7.
31. Gyamera-Acheampong, C., and Mbikay, M. (2009) Proprotein convertase subtilisin/kexin type 4 in mammalian fertility: A review *Hum Reprod Update* **15**, 237–47.
32. Qiu, Q., Basak, A., Mbikay, M., Tsang, B. K., and Gruslin, A. (2005) Role of pro-IGF-II processing by proprotein convertase 4 in human placental development *Proc Natl Acad Sci USA* **102**, 11047–52.
33. Shiryaev, S. A., Remacle, A. G., Ratnikov, B. I., Nelson, N. A., Savinov, A. Y., Wei, G., Bottini, M., Rega, M. F., Parent, A., Desjardins, R., Fugere, M., Day, R., Sabet, M., Pellecchia, M., Liddington, R. C., Smith, J. W., Mustelin, T., Guiney, D. G., Lebl, M., and Strongin, A. Y. (2007) Targeting host cell furin proprotein convertases as a therapeutic strategy against bacterial toxins and viral pathogens *J Biol Chem* **282**, 20847–53.
34. Patra, D., Xing, X., Davies, S., Bryan, J., Franz, C., Hunziker, E. B., and Sandell, L. J. (2007) Site-1 protease is essential for endochondral bone formation in mice *J Cell Biol* **179**, 687–700.
35. Constam, D. B., Calfon, M., and Robertson, E. J. (1996) SPC4, SPC6, and the novel protease SPC7 are coexpressed with bone morphogenetic proteins at distinct sites during embryogenesis *J Cell Biol* **134**, 181–91.
36. Seidah, N. G., Khatib, A. M., and Prat, A. (2006) The proprotein convertases and their implication in sterol and/or lipid metabolism *Biol Chem* **387**, 871–7.
37. Seidah, N. G., and Prat, A. (2007) The proprotein convertases are potential targets in the treatment of dyslipidemia *J Mol Med* **85**, 685–96.
38. Gordon, V. M., and Leppla, S. H. (1994) Proteolytic activation of bacterial toxins: Role of bacterial and host cell proteases *Infect Immun* **62**, 333–40.
39. Jasinski, J. P., and Woudenberg, R. C. (1994) 7-Amino-4-methylcoumarin *Acta Crystallogr* **C50**, 1954–6.
40. Bissell, E. R., Mitchell, A. R., and Smith, R. E. (1980) Synthesis and chemistry of 7-amino-4-(trifluoromethyl) coumarin and its amino acid and peptide derivatives *J Org Chem* **45**, 2283–7.

41. Izquierdo, C., and Burguillo, J. (1989) Synthetic substrates for thrombin *Int J Biochem* **21**, 579–92.
42. Sedding, D. G., Homann, M., Seay, U., Tillmanns, H., Preissner, K. T., and Braun-Dullaeus, R. C. (2008) Calpain counteracts mechanosensitive apoptosis of vascular smooth muscle cells in vitro and in vivo *FASEB J* **22**, 579–89.
43. Onderwater, R. C., Venhorst, J., Commandeur, J. N., and Vermeulen, N. P. (1999) Design, Synthesis, and characterization of 7-methoxy-4-(aminomethyl) coumarin as a novel and selective cytochrome P450 2D6 substrate suitable for high-throughput screening *Chem Res Toxicol* **12**, 555–9.
44. Ledgerwood, E. C., Brennan, S. O., and George, P. M. (1997) Endoproteases other than furin have a role in hepatic proprotein processing *IUBMB Life* **42**, 1131–42.
45. Lopez-Perez, E., Seidah, N. G., and Checler, F. (1999) Proprotein convertase activity contributes to the processing of the Alzheimer's β-amyloid precursor protein in human cells: Evidence for a role of the prohormone convertase PC7 in the constitutive α-secretase pathway *J Neurochem* **73**, 2056–62.
46. Pasquato, A., Pullikotil, P., Asselin, M. C., Vacatello, M., Paolillo, L., Ghezzo, F., Basso, F., Di Bello, C., Dettin, M., and Seidah, N. G. (2006) The proprotein convertase SKI-1/S1P. In vitro analysis of Lassa virus glycoprotein-derived substrates and *ex vivo* validation of irreversible peptide inhibitors *J Biol Chem* **281**, 23471–81.
47. Bayer, E., and Mutter, M. (1972) Liquid phase synthesis of peptides *Nature* **237**, 512–13.
48. Basak, S., Stewart, N. A., Chrétien, M., and Basak, A. (2004) Aminoethyl benzenesulfonyl fluoride and its hexapeptide (Ac-VFRSLK) conjugate are both in vitro inhibitors of subtilisin kexin isozyme-1 *FEBS Lett* **573**, 186–94.
49. Scott, R. P. W. (1993) *Silica Gel and Bonded Phases. Their Production, Properties and Use in LC.* Wiley, Chichester, Separation Science Series.
50. Stewart, J. M., and Young, J. D. (1984) *Solid Phase Peptide Synthesis* (2nd ed.). Pierce Chemical Company, Rockford, IL.
51. Technical bulletin (1998, May) "Cleavage, Deprotection, and Isolation of Peptides after Fmoc Synthesis". PE-Biosystems web site: http://www3.appliedbiosystems.com/cms/groups/psm_marketing/documents/generaldocuments/cms_040654.pdf.
52. Thomas, G. (2002) Furin at the cutting edge: From protein traffic to embryogenesis and disease *Nat Rev Mol Cell Biol* **3**, 753–66.
53. Tanga, S. S., Zhangc, J. H., Liud, H. X., and Li, H. Z. (2009) PC2/CPE-mediated proprotein processing in tumor cells and its differentiated cells or tissues *Mol Cell Endocrinol* **303**, 43–9.
54. Hall, T., Kam, F., Min, F., Liu, M., Zobel, J. F., Marino, M. H., Malfait, A. M., Tortorella, M. D., and Tomasselli, A. G. (2007) A high performance liquid chromatography assay for monitoring proprotein convertase activity *J Chromatogr A* **1148**, 46–54.
55. Malfait, A. M., Arner, E. C., Song, R. H., Alston, J. T., Markosyan, S., Staten, M., Yang, Z., Griggs, D. W., and Tortorella, M. D. (2008) Proprotein convertase activation of aggrecanases in cartilage in situ *Arch Biochem Biophys* **478**, 43–51.
56. Park, K., Kuechle, M. K., Choe, Y., Craik, C. S., Lawrence, O. Y., and Presland, R. B. (2006) Expression and characterization of constitutively active human caspase-14 *Biochem Biophys Res Commun* **347**, 941–8.
57. Cox, S. W., Cho, K., Eley, B. M., and Smith, R. E. (2006) A simple, combined fluorogenic and chromogenic method for the assay of proteases in gingival crevicular fluid *J Periodontal Res* **25**, 164–71.
58. St. Leger, R. J., Joshi, L., Bidochka, M. J., Rizzo, N. W., and Roberts, D. W. (1996) Biochemical characterization and ultrastructural localization of two extracellular trypsins produced by *Metarhizium anisopliae* in infected insect cuticles *Appl Environ Microbiol* **62**, 1257–64.
59. Cox, S. W., and Eley, B. M. J. (1989) Detection of cathepsin B and L, elastase, tryptase, trypsin, and dipeptidyl peptidase IV-like activities in crevicular fluid from gingivitis and periodontitis patients with peptidyl derivatives of 7-amino-4-trifluoromethyl coumarin *Periodontal Res* **24**, 353–1.
60. Lojda, Z. (1996) The use of substrates with 7-amino-3-trifluoromethylcoumarine (AFC) leaving group in the localization of protease activities in situ *Acta Histochem* **98**, 215–28.
61. Patent inventors: Chelsky, Daniel (Moylan, PA) Burbaum (Cranbury, NJ): http://www.freepatentsonline.com/5856083.html.
62. Miller, J. N. (2008) Advances in Fluorescence Enzyme Detection Methods. In *Standardization and Quality Assurance in Fluorescence Measurements II*, Springer Series on Fluorescence. Springer, Berlin, Heidelberg.
63. Foley, J. D., Rosenbaum, H., and Griep, A. E. (2004) Temporal regulation of VEID-7-amino-4-trifluoromethylcoumarin cleav-

age activity and caspase-6 correlates with organelle loss during lens development *J Biol Chem* **279**, 32142–50., and video journal: Cupp-Enyard, C. (2009), Use of the protease fluorescent detection kit to determine protease activity *J Vis Exp* **30**; http://www.jove.com/index/Details.stp?ID=1514.

64. Albani, J. R. (2007) *Principles and Applications of Fluorescence Spectroscopy*. Wiley-Blackwell, Oxford.
65. Scully, M. F., and Kakkar, V. V. (Eds) (1977) Chromogenic peptide substrates: chemistry and clinical usage/International Symposium on Chromogenic Substrates, King's College Hospital Medical School.
66. Sharma, N., Liu, S., Tang, L., Irwin, J., Meng, G., and Rancourt, D. E. (2006) Implantation serine proteinases heterodimerize and are critical in hatching and implantation *BMC Dev Biol* **6**, 61–5.
67. Gore, M. G. (Ed) (2000) *Spectrophotometry and Spectrofluorometry: A Practical Approach*. Oxford University Press, Oxford.
68. Clegg, R. M. (1995) Fluorescence resonance energy transfer *Curr Opin Biotechnol* **6**, 103–10.
69. Cotrin, S. S., Puzer, L., de Souza, W. A., Juliano, J. L., Carmona, A. K., and Juliano, M. A. (2004) Positional-scanning combinatorial libraries of fluorescence resonance energy transfer peptides to define substrate specificity of carboxydipeptidases: Assays with human cathepsin *Anal Biochem* **335**, 244–52.
70. Mitobe, J., Morita-Ishihara, T., Ishihama, A., and Watanabe, H. (2009) Involvement of RNA-binding protein Hfq in the osmotic-response regulation of *invE* gene expression in *Shigella sonne BMC Microbiol* **9**, 110–16.
71. Basak, A., Zhong, M., Munzer, J. S., Chrétien, M., and Seidah, N. G. (2001) Implication of the proprotein convertases Furin, PC5 and PC7 in the cleavage of surface glycoproteins of Hong Kong, Ebola and Respiratory Syncytial viruses – a comparative analysis using fluorogenic peptides *Biochem J* **353**, 537–45.
72. Carmona, A. K., Schwager, S. L., Juliano, M. A., Juliano, L., and Sturrock, E. D. (2006) A continuous fluorescence resonance energy transfer angiotensin I-converting enzyme assay *Nat Protocol* **1**, 1971–6.
73. Basak, S., Mohottalage, D., and Basak, A. (2006) Multibranch and Pseudopeptide approach for design of novel inhibitors of Subtilisin Kexin Isozyme-1 *Prot Pept Lett* **13**, 863–76.
74. Oliveira, M. C. F., Hirata, I. Y., Chagas, J. R., Boschcov, P., Gomes, R. A. S., Figueiredo, A. F. S., and Juliano, L. (1992) Intramolecularly quenched fluorogenic peptide substrates for human rennin *Anal Biochem* **203**, 39–46.
75. James, J., Schmidt, J., and Robert, G. (2003) Fluorigenic substrates for the protease activities of botulinum neurotoxins, serotypes A, B, and F *Appl Environ Microbiol* **69**, 297–303.
76. Paschalidou, K., Neumann, U., Gerhartz, B., and Tzougraki, C. (2004) Highly sensitive intramolecularly quenched fluorogenic substrates for renin based on the combination of L-2-amino-3-(7-methoxy-4-coumaryl) propionic acid with 2,4-dinitrophenyl groups at various positions *Biochem J* **382**, 1031–8.
77. Kennelly, P. J. (2001) Protein phosphatases – a phylogenetic perspective *Chem Rev* **101**, 2291–312.
78. Goudreau, N., Guis, C., Soleilhac, J. M., and Roques, B. P. (1994) Dns-Gly-(*p*-NO_2)Phe-βAla, a specific fluorogenic substrate for neutral endopeptidase 24.11 *Anal Biochem* **219**, 87–95.
79. Rakhmanova, V., Po, P., and Meyer, R. Technical bulletin, A sensitive fluorimetric assay for detection of ß-secretase activity http://www.biocompare.com/Articles/ApplicationNote/1593/A-SensitiveFluorimetric-Assay-For-Detection-Of-Secretase-Activity.html.
80. Diwu, Z., Xiang, Q., He, J., Zhang, J., Wang, H., Tang, Y., and Hong, A. (2005) Novel QXL-based protease substrates and their applications in drug discovery (http://www.anaspec.com/resources/publications.asp)
81. Yu, X., Sainz, B., Jr., and Uprichard, S. L. (2009) Development of a cell-based hepatitis C virus infection fluorescent resonance energy transfer assay for high-throughput antiviral compound screening *Antimicrob Agents Chemother* **53**, 4311–19.
82. Yaron, A., Carmel, A., and Katchalski-Kazir, E. (1979) Intramolecularly quenched fluorogenic substrates for hydrolytic enzymes *Anal Biochem* **95**, 228–35.
83. Tonna, E. A., Aronson, R. B., and Pavelec, M. (1974) Autoradiographic assessment of proteolytic enzyme action using skeletal connective tissue matrices *Connect Tissue Res* **2**, 183–91.
84. Kaberdin, V. R., and McDowall, K. J. (2003) Expanding the use of zymography by the chemical linkage of small, defined substrates to the gel matrix *Genome Res* **13**, 1961–5.
85. Basak, A., Shervani, N. J., Mbikay, M., and Kolajova, M. (2008) Recombinant proprotein convertase 4 (PC4) from *Leishmania tarentolae* expression system: Purifica-

tion, biochemical study and inhibitor design *Protein Expr Purif* **60**, 117–26.
86. Varani, J., Johnson, K., and Kaplan, J. (1980) Development of a solid-phase assay for measurement of proteolytic enzyme activity *Anal Biochem* **107**, 377–84.
87. Guarise, C., Pasquato, L., De Filippis, V., and Scrimin, P. (2006) Gold nanoparticles-based protease assay *Proc Natl Acad Sci USA* **103**, 3978–82.
88. St Hilaire, P. M., Willert, M., Juliano, M. A., Juliano, L., and Meldal, M. (1999) Fluorescence-quenched solid phase combinatorial libraries in the characterization of cysteine protease substrate specificity *J Comb Chem* **1**, 509–23.
89. Eisenthal, R., and Danson, M. J. (Eds) (2002) *Enzyme Assays: A Practical Approach* (2nd ed.). Oxford University Press, Oxford.
90. Reymond, J. L. (Ed) (2006) *Enzyme Assays.* Wiley-VCH, Weinheim.
91. Peláez, C., Mejía, A., and Planas, A. (2004) Development of a solid phase kinetic assay for determination of enzyme activities during composting *Process Biochem* **39**, 971–5.
92. Weissleder, R., Zhao, M., Josephson, L., and Tang, T. (2003) Magnetic sensors for protease assays *Angew Chem Int Ed* **42**, 1375–8.
93. Wondrak, E. W., Copeland, T. D., Louis, J. M., and Oroszlan, S. (1990) A solid phase assay for the protease of human immunodeficiency virus *Anal Biochem* **188**, 82–5.
94. Jean, F., Basak, A., Chrétien, M., and Lazure, C. (1991) Detection of endopeptidase activity and analysis of cleavage specificity using a radiometric solid-phase enzymatic assay *Anal Biochem* **194**, 399–406.
95. Janelle, L., Fields, L., Hideaki, N., and Fields, G. B. (2004) Development of a solid-phase assay for analysis of matrix metalloproteinase activity *J Biomol Tech* **15**, 305–16.
96. Meldal, M. (1998) *Introduction to Combinatorial Solid-Phase Assays for Enzyme Activity and Inhibition Methods in Molecular Biology.* Humana, Totowa, NJ, pp. 51–57.
97. Fearrari, S., Marin, O., Pagano, M. A., Meggio, F., Hess, D., and Shemerly, M. E. (2005) Aurora – A site specificity: Study with synthetic peptide substrates *Biochem J* **390**, 293–302.
98. Kuramochi, K., Miyano, Y., Enomoto, Y., Takeuchi, R., Ishi, K., Takakusagi, Y., Saitoh, T., Fukudome, K., Manita, D., Takeda, Y., Kobayashi, S., Sakaguchi, K., and Sugawara, F. (2008) Identification of small molecule binding molecules by affinity purification using a specific ligand immobilized on PEGA resin *Bioconjug Chem* **19**, 2417–26.
99. Kostka, V., and Carpenter, F. H. (1964) Inhibition of chymotrypsin activity in crystalline trypsin preparations *J Biol Chem* **239**, 2417–26.
100. Majumdar, S., Chen, A., Palmer-Smith, H., and Basak, A. (2011) Novel Circular, Cyclic and Acyclic Ψ(CH(2)O) Containing Peptide Inhibitors of SKI-1/S1P: Synthesis, Kinetic and Biochemical Evaluations. *Curr Med Chem.* **18**, 2770–82.
101. Ilekis, J. V. (2007) Review article: Preeclampsia: A pressing problem: An executive summary of a national institute of child health and human development *Workshop Reprod Sci* **14**, 508–23.
102. Holeman, K. (2003) Fetal growth restriction and consequences for the offspring in animal models *J Soc Gynecol Invest* **10**, 392–9.
103. Beynon, R. J., and Bond, J. S. (2001) *Proteolytic Enzymes. A Practical Approach* (2nd ed.). Oxford University Press, Oxford.
104. Reinharz, A., and Roth M. (1969) Studies on Renin with synthetic substrates *Eurp J Biochem* 7, 334–9.
105. Craik, C. S., Page, M. J., and Madison, E. L. (2011) Proteases as Therapeutics *Biochem J* **435**, 1–16.
106. Molloy, S. S., Bresnahan, P. A., Leppla, S. H., Klimpel, K. R., and Thomas, G. (1992) Human furin is a calcium-dependent serine endoprotease that recognizes the sequence Arg-X-X-Arg and efficiently cleaves anthrax toxin protective antigen *J Biol Chem* **267**, 16396–402.
107. Komiyama, T., Coppola, J. M., Larsen, M. J., van Dort, M. E., Ross, B. D., Day, R., Rehemtulla, A., and Fuller, R. S. (2009) Inhibition of furin/proprotein convertase-catalyzed surface and intracellular processing by small molecules *J Biol Chem* **284**, 15729–38.
108. Munzer, J. S., Basak, A., Zhong, M., Mamarbachi, A., Hamelin, J., Savaria, D., Lazure, C., Hendy, G. N., Benjannet, S., Chrétien, M., and Seidah, N. G. (1997) In vitro characterization of the novel proprotein convertase PC7 *J Biol Chem* **272**, 19672–81.
109. Lee, S. N., Prodhomme, E., and Lindberg, I. (2004) Prohormone convertase 1 (PC1) processing and sorting: Effect of PC1 propeptide and proSAAS *J Endocrinol* **182**, 353–64.
110. Cornwall, G. A., Cameron, A., Lindberg, I., Hardy, D. M., Cormier, N., and Hsia, N. (2003) The cystatin-related epididymal spermatogenic protein inhibits the serine protease prohormone convertase *Endocrinology* **144**(3), 901–8.
111. Basak, S., Chrétien, M., Mbikay, M., and Basak, A. (2004) In vitro elucidation of

substrate specificity and bioassay of proprotein convertase 4 using intramolecularly quenched fluorogenic peptides *Biochem J* **380**, 505–14.

112. Basak, A., Touré, B. B., Lazure, C., Mbikay, M., Chrétien, M., and Seidah, N. G. (1999) Enzymic characterization in vitro of recombinant proprotein convertase PC4 *Biochem J* **343**, 29–37.

113. Jean, F., Boudreault, A., Basak, A., Seidah, N. G., and Lazure, C. (1995) Fluorescent peptidyl substrates as an aid in studying the substrate specificity of human prohormone convertase PC1 and human furin and designing a potent irreversible inhibitor *J Biol Chem* **270**, 19225–31.

114. Komiyama, T., Swanson, J. A., and Fuller, R. S. (2005) Protection from anthrax toxin-mediated killing of macrophages by the combined effects of furin inhibitors and Chloroquine *Antimicrob Agents Chemother* **49**, 3875–82.

115. Basak, A., Chrétien, M., and Seidah, N. G. (2002) A rapid fluorometric assay for the proteolytic activity of SKI-1/S1P based on the surface glycoprotein of the hemorrhagic fever Lassa virus *FEBS Let* **514**, 333–9.

116. Johanning, K., Juliano, M. A., Juliano, L., Lazure, C., Lamango, N. S., Steiner, D. F., and Lindberg, I. (1998) Specificity of prohormone convertase 2 on proenkephalin and proenkephalin-related substrates *J Biol Chem* **273**, 22672–80.

Chapter 7

Inhibitor Screening of Proprotein Convertases Using Positional Scanning Libraries

Iris Lindberg and Jon R. Appel

Abstract

Proprotein convertases represent an important class of biosynthetic enzymes that are increasingly viewed as targets for therapeutic approaches to infection, cancer, and potentially endocrine disorders. The identification of potent inhibitors can be accomplished by screening synthetic combinatorial libraries containing thousands of small molecules to millions of peptides. In this chapter, the screening of positional scanning libraries is described for the identification of PC1/3 and PC2 inhibitors.

Key words: Prohormone convertase 1/3, PC1/3, PC1, PC2, enzyme inhibitor, synthetic combinatorial libraries, positional scanning, mixture-based libraries.

1. Introduction

The proprotein convertases, discovered over 20 years ago, represent a family of eukaryotic subtilases with restricted tryptic-like specificity. These enzymes are involved in a wide variety of physiological processes, from endocrine control to synthesis of growth factors and neuropeptides. Proprotein convertases are typically divided into two classes, those with restricted tissue expression, such as the neuroendocrine enzymes prohormone convertases 1/3 and 2, and those with broad expression, such as furin, responsible for the proteolytic maturation of a large number of circulating molecules. Furin is also involved in pathogenic processes such as viral and bacterial infection, where pathogenic organisms take advantage of host processing machinery, and in the development of oncogenic processes due to the need for furin in metastatic events. The development of furin inhibitors is a

M. Mbikay, N.G. Seidah (eds.), *Proprotein Convertases*, Methods in Molecular Biology 768,
DOI 10.1007/978-1-61779-204-5_7,

rapidly moving field, and a variety of different inhibitors ranging from peptides to small molecules have been described (reviewed in (1)). While a potent, cell-permeable, small molecule proprotein convertase inhibitor has not yet been achieved, much progress has recently been made toward this goal (2–5). Development of specific prohormone convertase inhibitors is also expected to yield therapeutically useful molecules (6). For example, blocking the synthesis of glucagon – a PC2-specific event – could result in better glycemic control, whereas blocking the synthesis of ACTH, largely under the control of PC1/3, could be helpful in pituitary disease. As described below, we have taken a combinatorial library approach to the identification of convertase inhibitors.

Combinatorial library synthesis and screening methods, which enable the rapid identification of highly active compounds, have revolutionized basic research and drug discovery. A number of different combinatorial approaches based on the principles of solid phase synthesis have been used to generate enormous molecular diversities, including peptides, peptidomimetics, small organic molecules, and heterocyclic compounds (7–10). The main advantage of the various combinatorial approaches compared to traditional drug synthesis and screening is the fact that very large numbers of compounds can be simultaneously synthesized and rapidly screened in biological assays.

Positional scanning synthetic combinatorial libraries are composed of positional libraries or sublibraries, in which each diversity position is defined with a single building block, while the remaining positions are composed of mixtures of building blocks (11). Each positional sublibrary represents the same collection of individual compounds. Further, mixture-based libraries, when arranged in a positional scanning format, provide extensive structure–activity information in any given assay. As a simple means to illustrate the positional scanning library concept, a tripeptide combinatorial library is illustrated in **Fig. 7.1**. Three different amino acids are incorporated at each of the three diversity positions, resulting in 27 (3^3) individual peptides. When the same peptides are arranged as a positional scanning library, only 9 peptide mixtures (3 amino acids × 3 positions) need to be synthesized. Each of the three positional sublibraries, OXX, XOX, and XXO, contains the same diversity of peptides, but differ only in the location of the position defined with a single amino acid. The O positions represent one of the four amino acids while the remaining two positions are mixtures (X) of the same four amino acids. Shown below each mixture are the 9 peptides (3^2) that make up that mixture. In this example, assume that alanine-arginine-threonine, or ART, is the only tripeptide in this library that is recognized by a given receptor. Since each positional sublibrary contains the same diversity of peptides, the ART tripeptide

Tripeptide library

Position 1	OXX	Position 2	XOX	Position 3	XXO
	1 AXX		4 XAX		7 XXA
	2 RXX		5 XRX		8 XXR
	3 TXX		6 XTX		9 XXT

AXX	RXX	TXX	XAX	XRX	XTX	XXA	XXR	XXT
AAA	RAA	TAA	AAA	ARA	ATA	AAA	AAR	AAT
AAR	RAR	TAR	AAR	ARR	ATR	ARA	ARR	ART
AAT	RAT	TAT	AAT	ART	ATT	ATA	ATR	ATT
ARA	RRA	TRA	RAA	RRA	RTA	RAA	RAR	RAT
ARR	RRR	TRR	RAR	RRR	RTR	RRA	RRR	RRT
ART	RRT	TRT	RAT	RRT	RTT	RTA	RTR	RTT
ATA	RTA	TTA	TAA	TRA	TTA	TAA	TAR	TAT
ATR	RTR	TTR	TAR	TRR	TTR	TRA	TRR	TRT
ATT	RTT	TTT	TAT	TRT	TTT	TTA	TTR	TTT

Fig. 7.1. Conceptual illustration of a tripeptide positional scanning library. O, defined functionality; X, mixture of functionalities.

(outlined in each sublibrary in **Fig. 7.1**) is present in all three positional sublibraries. Thus, the only mixtures with activity are AXX, XRX, and XXT because the ART tripeptide is present only in those mixtures. The combination of these amino acids in their respective positions yields the tripeptide ART, which would then be synthesized and tested for its activity against this receptor. It should be noted that the activity observed for each of the three mixtures (AXX, XRX, and XXT) is due to the presence of the tripeptide ART *within* each mixture and not due to the individual amino acids (A, R, and T) that occupy the defined positions. In more complex libraries, more than one mixture is often found to exert activity at each position. Selection of the building blocks for the synthesis of individual compounds is based first on activity and then on differences in the chemical character of the building block.

Although the above example is a simple representation of the arrangement and use of a positional scanning library, the concepts described here apply to all types of mixture-based libraries having defined and mixture positions, using either amino acids (D or L) or scaffolds bearing various chemical substituents at different positions. For example, a hexapeptide library using 20 amino acids represents a total of 4.9×10^7 (20×19^5; cysteine is omitted in the mixture positions) individual peptides (**Fig. 7.2**). This can be formatted into a positional scanning library of 120 mixtures (20 amino acids $\times$ 6 positions). This library will be used to describe our identification of a precise hexapeptide inhibitor of PC1/3 (12), although similar principles apply to the identification of a nonpeptide inhibitor of PC2 (6).

Hexapeptide library									
Position 1	O	X	X	X	X	X	→	20 Mixtures	1. AXXXXX ↓ 20.YXXXXX
Position 2	X	O	X	X	X	X		"	
Position 3	X	X	O	X	X	X		"	
Position 4	X	X	X	O	X	X		"	
Position 5	X	X	X	X	O	X		"	
Position 6	X	X	X	X	X	O		"	
			Total # mixtures					120 Mixtures	
			Total # peptides					$20 \times 19^5 = 4.9 \times 10^7$	

Fig. 7.2. Representation of hexapeptide positional scanning library. O, one of 20 L-amino acids in defined position; X, equimolar mixture of 19 L-amino acids (cysteine omitted).

2. Materials

1. Positional scanning libraries are available on a collaborative basis from the Torrey Pines Institute for Molecular Studies. The synthesis of various peptide and nonpeptide libraries has been described (7, 13). In the following protocol a hexapeptide positional scanning library of 120 mixtures is used. Concentrations used range from 1 to 5 mg/ml in water, and the library can be stored at 4°C for several weeks while in use or –20°C when not in use.
2. Purified recombinant PC enzymes are prepared from the conditioned medium of overexpressing CHO cells (14, 15). ProPC2 is obtained as a zymogen and should be activated by dilution in pH 5.0 reaction buffer prior to use; activation proceeds rapidly, even on ice, upon lowering the pH. PC1/3, obtained as an active enzyme, is prone to autocatalytic conversion to smaller enzyme products which may have differential susceptibility to inhibitors. Intermediate stock solutions should be prepared in reaction buffer (it must contain detergent and BSA to avoid losses by adsorption), stored on ice, and used within hours; stock solutions of both enzymes are relatively stable to three cycles of freezing and thawing.
3. The fluorogenic substrate used is pERTKR-aminomethylcoumarin, obtained from Peptides International, Lexington, KY. Stock is made up as 20 mM solution of peptide in DMSO; net peptide weight is usually 75% (the remainder being salts and water).

4. 96-well polypropylene round-bottomed microtiter plates (Costar 3365).
5. Multichannel pipettors capable of delivering 5–40 μl in octuplicate and plastic reservoirs to use with these pipettors.
6. The PC2 assay buffer contains 100 mM sodium acetate, pH 5.0, 2 mM $CaCl_2$, 0.2% *n*-octylglucoside (RPI), 0.1% NaN_3, and 0.1 mg/ml crystalline fraction V BSA (Roche). For PC1/3 the assay pH is adjusted to 5.5.
7. A fluorometer plate reader, either filter or monochromator based, with the ability to perform kinetic reads (accumulation of product over time). We read our plates from the top since our plates are opaque.

3. Methods

3.1. General Considerations

Screening many plates is an operation that must be carefully timed, as most fluorometers will only read a single plate at a time. We stagger the start of the enzyme assays such that we can obtain 25-min kinetic reads of each new plate every 30 min. Depending on the speed of the equipment and operator, this timing may need to be varied.

We use polypropylene round-bottomed plates for two reasons. One, the round-bottomed format cuts the amount of reactants needed in half vs a flat-bottomed plate (50 vs 100 μl). Second, convertases are extremely sticky, and proteins tend to stick less tightly to polypropylene than to polystyrene. However, flat-bottomed plates will certainly work.

3.2. Experiment Preparation

1. Enzyme preparation: Prepare sufficient enzyme (diluted in reaction buffer, 40 μl per well) for the number of plates to be assayed within the next 2–3 h. Depending on the sensitivity of the instrument to be used, pg to ng/well concentrations are used for PC2; PC1 reactions will require 20–50× as much enzyme due to its lower specific activity. For maximum sensitivity in inhibitor detection, the amount of enzyme used should be kept as low as possible while still generating adequate reproducibility and signal-to-noise ratio for the uninhibited control. Remember to prepare sufficient enzyme such that the bottom of the multichannel pipettor reservoir is covered (about 1 ml extra). *Example*: for four 96-well plates which will require 40 μl of diluted enzyme, prepare 0.04 ml × 100 wells × 4 plates plus 1 ml = 17 ml.
2. Substrate preparation: Dilute the standard substrate, pERTKR-aminomethylcoumarin, to 0.5 mM in water. Use

5–10 μl per well (0.05–0.1 mM final concentration). Again, sufficient substrate must be prepared to accommodate the reservoir. It is possible to use substrate amounts lower than this (substrate is the most costly ingredient in the screen); this may increase sensitivity to competitive inhibitors, but the reproducibility and signal-to-noise ratio must first be checked.

3. Library preparation: Most positional scanning libraries are supplied in high concentrations (1–10 mg/ml), dissolved either in water, formamide, or DMSO. Peptide libraries containing hydrophobic amino acids in the defined positions are particularly likely to have various amounts of insoluble material. It is extremely important that the tubes be well mixed prior to pipetting (see below).

3.3. Enzyme Inhibition Assay

1. Using an eight-channel pipettor (*see* **Note 1**), pipette 40 μl of enzyme solution into all wells of a 96-well plate on a bed of wet ice except the end columns (these will become background controls).
2. Pipette 5 μl of well-mixed inhibitor (i.e., library) solution to duplicate rows starting from the second row. Pipette vehicle (the solution the library arrives in) into the control rows. Remember to mix tubes well prior to each row of pipetting, as the inhibitor mixtures often consist of particulate suspensions. We accomplish mixing using a dedicated plate shaker, a Glas-Col Pulse Vortexer Mixer, and shake the plate such that a homogeneous suspension is achieved prior to pipetting *each row* of samples. The presence of a homogeneous suspension should be verified by visual inspection prior to pipetting; samples settle remarkably quickly.
3. Preincubate plate for 30 min at room temperature to permit the inhibitor to bind the enzyme; note the start time (*see* **Note 2**). Preheat the fluorometer by running a mock plate.
4. Start the reaction by adding 5 μl of diluted substrate. There should be no bubbles visible. If this is a problem, briefly spin down the plates to break the bubbles. Note the time the reaction was started.
5. Place the plate in the fluorometer chamber and measure fluorescence at timed intervals. The fluorometer (380 nm excitation, 480 nm emission) should be pre-warmed and programmed to take two to three multiple reads over each minute, reading each plate once a minute repetitively for at least 25 min. After an initial period of warming, the rate of release of the fluorescent product aminomethylcoumarin will be directly proportional to the rate of the reaction. The rate of hydrolysis of inhibited wells (relative to control samples

lacking inhibitors) is often – though not always – linear and can be used as a direct measure of the relative inhibition by each library mixture. It is helpful to inspect the progress curves during the reaction so that you can see whether the inhibited rates are indeed linear and if they exhibit a lower slope than the non-inhibited rates (indicative of competitive inhibition) or are non-linear (likely indicating noncompetitive inhibition). This progress curve feature is available on most kinetic fluorometers. If rates are linear, calculate the average rates for each well and print out both the curves and the rates for each plate (or copy to a spreadsheet for later entry into a graphics program). If inhibited rates are not linear, use the maximum rate for each well as a measure of its inhibited velocity, but make sure at least nine time points are averaged to get this maximum (*see* **Note 3**), and note the time of preincubation as well as of the assay as both of these parameters will impact the maximum rate. In either case, duplicates should agree to within 15%; if they do not, adjust assay parameters until this precision level is obtained.

6. Immediately after beginning the first read, pipette enzyme and inhibitor into the second plate for the second preincubation. Note the time. Add substrate after 30 min at room temperature, at which time the first plate should be finished with data collection and the second plate can be inserted into the fluorometer. Immediately pipette the third plate and continue until all of the data have been collected.

3.4. Optimizing and Confirming Inhibition

A good understanding of the various assay parameters, such as signal-to-noise ratio, variability, and sensitivity, is required for the successful identification of active compounds from the library. The most important parameter to control for an assay system is the variability. To optimize assays one can vary preincubation time with inhibitor; enzyme, substrate, and inhibitor concentrations; and salt concentration in the buffer – salt, for example, can radically affect the potency of charged inhibitors. The use of repeated experiments and averaged data ensures the accurate identification of individual compounds having significant activity. For inhibitor variation, depending on the amount of inhibition originally observed, use half to 1/20th of the amount of inhibitor originally pipetted; dilutions (in water) can be prepared in parallel plates (remember to mix the original library tubes well and pipette as a homogeneous suspension!). Our experience indicates that it is unlikely that dilutions greater than 1/10 will exhibit much inhibition (*see* **Note 4**).

3.5. Positional Scanning Library Deconvolution

In order to identify the most active individual compounds from the positional scanning library, the individual peptides that correspond to the combination of the amino acids defined in the most active mixtures at each position are synthesized and tested. For practical purposes, the number of amino acids selected from each position that will be used to synthesize the individual peptides should be minimized as the number of compounds to be made rises exponentially (*see* **Note 5**). For example, if two amino acids were selected from each position of a hexapeptide library, one would need to synthesize 64 peptides (2^6).

Successful deconvolution of active individual compounds from mixture-based libraries is dependent on reproducible screening data and clear dose–response activities of the most active mixtures. In most cases, dose–response curves can be determined for the most active mixtures, and activities based on calculated IC_{50} values are used to select the building blocks that will be included in the synthesis of individual compounds (*see* **Note 6**). Some positions may exhibit more distinct residue preferences than others; this is due to binding requirements. For example, the P3 position is not nearly as discriminating as the P1 position for PC1/3 (12).

Upon synthesis and testing of individual peptides derived from the screening of positional scanning libraries, one should be able to confirm the activity of the selected mixtures. It is important to note that it is the *activity of the individual peptides* within a mixture, and not the amino acid in the defined position of the mixture, that results in the observed inhibitory activity. If no inhibitory activity is found within the set of individual peptides, then either additional peptides need to be prepared that include amino acids that were not initially selected or an iteration can be prepared using the most active mixture as a starting point. Upon iterating an active mixture and defining additional positions of the library with specific amino acids, the resulting mixtures become progressively less complex since they contain fewer different peptides, and this should result in an increase in inhibitory activity relative to the original mixtures from the library.

A representative screen of the hexapeptide positional scanning library with PC1/3 is shown in **Fig. 7.3**. To identify the hexapeptides from this library that inhibit PC1/3, the amino acids defined in the most active mixtures at each position were selected to make a set of individual peptides. In this example, 12 hexapeptides were made based on the combination from selecting leucine in position 1; lysine, leucine, methionine, and tyrosine in position 2; arginine in position 3; histidine, threonine, and valine at position 4; and lysine and arginine at positions 5 and 6, respectively. The activities of these peptides are shown in **Fig. 7.4**, in which Ac-LLRVKR-NH2 was found to be the most active inhibitor of PC1/3 (12). Two years later, this exact hexapeptide was found within a natural inhibitor of this enzyme (16).

A strategy termed biometrical analysis has been developed to systematically compare the results obtained from screening a peptide library composed of millions of sequences with the

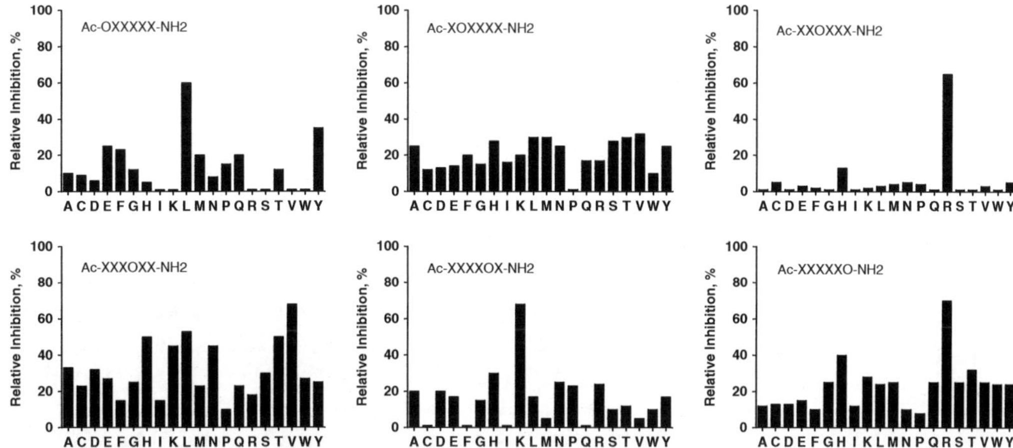

Fig. 7.3. Inhibition of PC1/3 activity by a hexapeptide positional scanning library (final concentration 1 mg/ml).

Position	1	2	3	4	5	6
Amino acid	L	K,L,M,Y	R	H,T,V	K	R

	Peptide	PC1/3 % Inhibition at 100 ng/ml	PC1/3 Ki (nM)
1	Ac-LKRHKR-NH2	11	>100
2	Ac-LKRTKR-NH2	19	>100
3	Ac-LKRVKR-NH2	59	5.7
4	Ac-LLRHKR-NH2	18	60
5	Ac-LLRTKR-NH2	34	16
6	Ac-LLRVKR-NH2	75	3.2
7	Ac-LMRHKR-NH2	10	>100
8	Ac-LMRTKR-NH2	25	>100
9	Ac-LMRVKR-NH2	61	4.9
10	Ac-LYRHKR-NH2	5	>100
11	Ac-LYRTKR-NH2	22	>100
12	Ac-LYRVKR-NH2	51	6.5

Fig. 7.4. Amino acids selected to synthesize individual peptides and the inhibitory activities of the hexapeptides against PC1/3.

millions of sequences within protein databases. This approach is based on the assumption that each amino acid in a peptide epitope or ligand in other protein/protein interaction contributes independently and additively to recognition or, in general terms, strength of interaction. Hence, the stimulatory value of each amino acid in a given position can be added to that of the amino acid in the next position throughout the length of the peptide. Consequently, a stimulatory score can be calculated for each peptide. A scoring matrix is generated by transforming the screening data of each of the 20 amino acids defined in each position of the library. Individual peptides are given a score calculated by adding the individual activities of the amino acids for the length of the library. This matrix is then used to score all the overlapping peptides of a given length in the protein databases and thus identify the sequences with the highest score. From this list a number of sequences with the highest scores can be synthesized and tested to confirm the activities (17, 18). We have used the biometrical analysis for T-cell clones of known and unknown specificities. When T-cell ligands are known, they are found within the set of peptides with the highest scores. This approach has also been successfully used for the identification of T-cell ligands for a CD4+ T-cell clone from a patient with Lyme disease. Both microbial epitopes from proteins in *Borrelia burgdorferi* (the infectious bacteria of Lyme disease) and candidate autoantigens were identified (17). These data further support the idea that combinatorial screening methods, combined with novel algorithms, have the potential to identify naturally occurring molecules.

4. Notes

1. A multichannel pipette (or a robotic pipetting station) is essential for pipetting libraries onto microtiter plates; this minimizes pipetting errors and improves assay reproducibility.
2. Some inhibitors display time-dependent inhibition; for these compounds, a much better signal will be observed using longer preincubation periods.
3. Data analysis: It is extremely important to have a good non-inhibited control value since all other values depend on this value. To obtain this, use a sufficient number of wells (usually between 3 and 5), which achieves less than 10% variability. For final data reduction, subtract the no-enzyme background from all wells, average the non-inhibited controls, and plot experimental data as the percent inhibition (not activity) and as a bar graph with either amino acid or tube number as the *X* axis. Any convenient graphics program can be used; we use Prism. Some mixtures may stimulate; this may not be an artifact, and these data should also be collected.

4. Library and enzyme concentration should be adjusted in order to perform the assay at the optimal sensitivity to detect inhibition by mixtures. For example, if there is no inhibition from the library, repeat the screening at a higher library concentration. Conversely, if high inhibition (>80%) is observed for all the mixtures of the library, repeat the screening at a lower library concentration. A two- to fourfold change in library concentration should result in a differentiating screening profile, in which there are mixtures that still inhibit as well as mixtures with little activity.
5. Often building blocks of similar chemical character will yield similar activities at a given position. This may indicate that a number of analogs of the same compound are responsible for the observed activity. Similar building blocks can be excluded from selection to reduce the number of final compounds needed to be synthesized. However, one can later synthesize analogs of the most active individual compounds using the building blocks that were originally excluded.
6. In a number of examples of library screening data, the distinction between active and inactive mixtures is more difficult to determine, because either the specificities of the most active mixtures are not very clear or the signal-to-noise ratio of the assay is less than threefold. In these cases, dose–response determinations may not be possible. Another strategy that can be useful is to compare the activity of a given mixture relative to the average mixture activity at that diversity position. The data analysis of complex mixtures is no different from the data obtained using individual compounds. One simply follows the activity that is significantly affecting the assay. In other words, distinguishing between active and inactive samples is independent of the complexity of these samples.

Acknowledgment

This review was supported by DA05084 to IL.

References

1. Fugere, M., and Day, R. (2005) Cutting back on pro-protein convertases: The latest approaches to pharmacological inhibition *Trends Pharmacol Sci* **26**, 294–301.
2. Coppola, J. M., Bhojani, M. S., Ross, B. D., and Rehemtulla, A. (2008) A small-molecule furin inhibitor inhibits cancer cell motility and invasiveness *Neoplasia* **10**, 363–70.

3. Komiyama, T., Coppola, J. M., Larsen, M. J., van Dort, M. E., Ross, B. D., Day, R. et al. (2009) Inhibition of furin/proprotein convertase-catalyzed surface and intracellular processing by small molecules *J Biol Chem* **284**, 15729–38.
4. Jiao, G. S., Cregar, L., Wang, J., Millis, S. Z., Tang, C., O'Malley, S. et al. (2006) Synthetic small molecule furin inhibitors derived from 2,5-dideoxystreptamine *Proc Natl Acad Sci USA* **103**, 19707–12.
5. Becker, G. L., Sielaff, F., Than, M. E., Lindberg, I., Routhier, S., Day, R. et al. (2010) Potent inhibitors of furin and furin-like proprotein convertases containing decarboxylated P1 arginine mimetics *J Med Chem* **53**, 1067–75.
6. Kowalska, D., Liu, J., Appel, J. R., Ozawa, A., Nefzi, A., Mackin, R. B. et al. (2009) Synthetic small-molecule prohormone convertase 2 inhibitors *Mol Pharmacol* **75**, 617–25.
7. Houghten, R. A., Pinilla, C., Appel, J. R., Blondelle, S. E., Dooley, C. T., Eichler, J. et al. (1999) Mixture-based synthetic combinatorial libraries *J Med Chem* **42**, 3743–78.
8. Pinilla, C., Appel, J. R., Borras, E., and Houghten, R. A. (2003) Advances in the use of synthetic combinatorial chemistry: Mixture-based libraries *Nat Med* **9**, 118–22.
9. Nefzi, A., Ostresh, J. M., Yu, Y., and Houghten, R. A. (2004) Combinatorial chemistry: Libraries from libraries, the art of the diversity-oriented transformation of resin-bound peptides and chiral polyamides to low molecular weight acyclic and heterocyclic compounds *J Org Chem* **69**, 3603–9.
10. Houghten, R. A., Pinilla, C., Giulianotti, M. A., Appel, J. R., Dooley, C. T., Nefzi, A. et al. (2008) Strategies for the use of mixture-based synthetic combinatorial libraries: Scaffold ranking, direct testing in vivo, and enhanced deconvolution by computational methods *J Comb Chem* **10**, 3–19.
11. Pinilla, C., Appel, J. R., Blanc, P., and Houghten, R. A. (1992) Rapid identification of high affinity peptide ligands using positional scanning synthetic peptide combinatorial libraries *Biotechniques* **13**, 901–5.
12. Apletalina, E., Appel, J., Lamango, N. S., Houghten, R. A., and Lindberg, I. (1998) Identification of inhibitors of prohormone convertases 1 and 2 using a peptide combinatorial library *J Biol Chem* **273**, 26589–95.
13. Houghten, R. A., Pinilla, C., Blondelle, S. E., Appel, J. R., Dooley, C. T., and Cuervo, J. H. (1991) Generation and use of synthetic peptide combinatorial libraries for basic research and drug discovery *Nature* **354**, 84–6.
14. Zhou, Y., and Lindberg, I. (1993) Purification and characterization of the prohormone convertase PC1(PC3) *J Biol Chem* **268**, 5615–23.
15. Lamango, N. S., Zhu, X., and Lindberg, I. (1996) Purification and enzymatic characterization of recombinant prohormone convertase 2: Stabilization of activity by 21 kDa 7B2 *Arch Biochem Biophys* **330**, 238–50.
16. Fricker, L. D., McKinzie, A. A., Sun, J., Curran, E., Qian, Y., Yan, L. et al. (2000) Identification and characterization of proSAAS, a granin-like neuroendocrine peptide precursor that inhibits prohormone processing *J Neurosci* **20**, 639–48.
17. Hemmer, B., Gran, B., Zhao, Y., Marques, A., Pascal, J., Tzou, A. et al. (1999) Identification of candidate T-cell epitopes and molecular mimics in chronic Lyme disease *Nat Med* **5**, 1375–82.
18. Zhao, Y., Gran, B., Pinilla, C., Markovic-Plese, S., Hemmer, B., Tzou, A. et al. (2001) Combinatorial peptide libraries and biometric score matrices permit the quantitative analysis of specific and degenerate interactions between clonotypic TCR and MHC peptide ligands *J Immunol* **167**, 2130–41.

Chapter 8

Analyses of PCSK9 Post-translational Modifications Using Time-of-Flight Mass Spectrometry

Thilina Dewpura and Janice Mayne

Abstract

Post-translational modification(s) can affect a protein's function – changing its half-life/stability, its protein–protein interactions, biological activity and/or sub-cellular localization. Following translation, a protein can be modified in several ways, including (i) disulfide bridge formation, (ii) chemical conversion of its constituent amino acids (for instance, glutamine can undergo deamidation to glutamic acid), (iii) sulfation, phosphorylation, de/acetylation, and glycosylation (to name a few), (iv) addition of other proteins as occurs during sumoylation and ubiquitination, and (v) proteolytic cleavage(s). There are several techniques available to identify and monitor post-translational modifications of proteins and peptides including mass spectrometry, two-dimensional sodium dodecyl sulfate polyacrylamide electrophoresis (2D-SDS-PAGE), radiolabeling, and immunoblotting. Ciphergen's surface-enhanced laser desorption/ionization time-of-flight mass spectrometer (SELDI-TOF-MS) has been used successfully for protein/peptide profiling in disease states and for the detection of protein/peptide biomarkers (1–4). In this chapter, the secreted proprotein convertase subtilisin/kexin 9 (PCSK9), which we study in our lab, is used to demonstrate coupling of immunoprecipitation with Ciphergen's time-of-flight mass spectrometer and its ProteinChip software to detect and analyze the common post-translational modifications of phosphorylation and glycosylation. The following topics are covered (1): preparation of cell extracts/samples/spent media (2), processing of samples by immunoprecipitation including optimization of conditions and (3) data acquisition by mass spectrometry and its subsequent analyses.

Key words: Proprotein convertase, post-translational modification, ProteinChip, mass spectrometry, immunoprecipitation, phosphorylation, glycosylation.

1. Introduction

The family of mammalian proprotein convertase subtilisin/kexin like (PCSK) serine proteases consists of nine members (PCSK1–9) (5–21). Collectively, the PCSKs are involved in the proteolytic

M. Mbikay, N.G. Seidah (eds.), *Proprotein Convertases*, Methods in Molecular Biology 768,
DOI 10.1007/978-1-61779-204-5_8,

maturation of various substrates into biologically active molecules including proneuropeptides, prohormones, proreceptors, growth factors, cell surface proteins, and serum proteins (22–27). All PCSKs undergo various post-translational modifications: for example, all are N-linked glycoproteins (12, 28, 29). Each PCSK is synthesized as a larger, inactive precursor that undergoes several cleavage events in the endoplasmic reticulum (ER) and during transit through the secretory pathway (30, 31). They are first cleaved within their N-terminal domain by an ER-resident signal peptidase to remove their signal peptide sequence – a domain that directed the preproPCSK into the secretory pathway. Next they undergo autocatalytic cleavage of their N-terminal propeptide domain while still in the ER – a domain that is responsible for correct folding of the nascent protein and that acts as an intramolecular inhibitor (30, 32). Other PCSKs, such as PCSK1, -8, and -9, are sulfated by Golgi-resident sulfotransferases (33–35). PCSK3 – a type 1 transmembrane convertase – undergoes serine phosphorylation of its cytoplasmic tail which affects its sub-cellular partitioning between the plasma membrane, Golgi apparatus, and endosomal compartments (36). We have reported that PCSK9 – a secreted convertase – is phosphorylated within its propeptide domain and near its C terminus (37).

PCSK9 is a 12-exon gene found on chromosome 1 (1p32) that encodes a 692 amino acid preproprotein (13). PCSK9 is primarily expressed in the liver and intestine, and is secreted into circulation (12). Overexpression and knockout studies of PCSK9 in mice have demonstrated that it mediates the post-translational degradation of low-density lipoprotein receptors (LDLRs) – receptors that remove low-density lipoprotein particles from circulation and thus affect cholesterol homeostasis (28, 38–40). PCSK9 transcriptional regulation is well studied, while that of its post-translational regulation is less-so (39–45). However, independent of its transcriptional regulation, PCSK9 post-translational modifications may increase or decrease the half-life of PCSK9 or its availability for LDLR binding, therefore modulating the role of PCSK9 in cholesterol homeostasis. For instance, PCSK9 is post-translationally cleaved by PCSK3 (also known as furin) into an inactive form (of ~53 kDa) which cannot mediate LDLR degradation (34).

PCSK9 is processed from a ~74-kDa proprotein to an ~60-kDa form as it transits through the secretory pathway (12, 28). After autocatalytic cleavage of its propeptide (which remains associated with PCSK9) and while still in the ER, PCSK9 is N-linked glycosylated (at Asn533) in its C-terminal Cys/His-rich domain, and this sugar moiety is further modified in the Golgi apparatus (12, 28). Enzymatic digestion assays specific for the complexity of the carbohydrate addition allow for differentiating between an immature, ER form of PCSK9 (endoglycosidase H sensitive) and Golgi-matured form of PCSK9 (PNGase F sensitive) when studying different PCSK9 variants (28). As PCSK9 transits through the Golgi, its propeptide is sulfated at Tyr38 (34) and PCSK9 is phosphorylated at two sites: Ser47 in its propeptide and Ser688 in its C-terminal Cys/His-rich domain (37). Herein, we overexpress human PCSK9 in a cell culture model as an example of how to identify and study post-translational modifications, in particular phosphorylation and glycosylation, using mass spectrometry-based techniques.

2. Materials

2.1. Cell Culture

1. Human hepatoma cell line HepG2 (ATCC/Cedarlane, Burlington, Canada).
2. Human hepatoma cell line HuH7 (gift from Dr N.G. Seidah; IRCM, Montreal, Canada).
3. cDNA of human PCSK9 with C-terminal V5 tag (gift from Dr N.G. Seidah; IRCM, Montreal, Canada).
4. Plasmid pIRES2-EGFP (enhanced green fluorescent protein; Invitrogen, Burlington, Canada).
5. Dulbecco's modified Eagle's medium (DMEM; GIBCO/Invitrogen).
6. Hank's balanced salt solution (HBSS; GIBCO/Invitrogen).
7. Fetal bovine serum (FBS; GIBCO/Invitrogen).
8. OPTI-MEM media (GIBCO/Invitrogen).
9. Gentamicin (Invitrogen).
10. Lipofectamine 2000 (Invitrogen).
11. Complete protease inhibitor cocktail (Roche, Laval, Canada).
12. 200 μM Sodium orthovanadate (a phosphatase inhibitor; Sigma-Aldrich, Oakville, Canada).
13. Solution of trypsin (0.25%) in Hank's balanced salt solution (GIBCO/Invitrogen).
14. Cell scrappers (Fisher Scientific, Ottawa, Canada).
15. Tunicamycin (Sigma-Aldrich).
16. Corning six-well plates (Fisher Scientific).
17. Cell lysis buffer; 1× radioimmunoprecipitation assay (RIPA) buffer; 50 mM Tris, pH 7.6, 150 mM NaCl,

1% (v/v) NP-40, 0.5% (w/v) deoxycholate, 0.1% (w/v) sodium dodecyl sulfate (SDS) (components; Sigma-Aldrich).

2.2. Immunoprecipitation

1. Mouse monoclonal anti-V5 antibody (Invitrogen).
2. Anti-V5 agarose (Sigma-Aldrich).
3. Protein A agarose (Sigma-Aldrich).
4. 10× Tris-buffered saline–Tween 20 (10× TBS-Tw; 500 mM Tris, pH 7.6, 1 M NaCl + 1% Tween 20) (components; Sigma-Aldrich).
5. 20 mM Tris–HCl (Sigma-Aldrich).
6. Shrimp alkaline phosphatase (SAP; Fermentas, Burlington, Canada).
7. Endoglycosidase H (Sigma-Aldrich).
8. 50 mM Sodium phosphate (pH 5.5; stock 200 mM; Sigma-Aldrich).
9. 0.1% Sodium dodecyl sulfate (SDS; Sigma-Aldrich).
10. 50 mM β-Mercaptoethanol (stock 200 mM; Sigma-Aldrich).

2.3. Elution and Concentration

1. 0.1 M Glycine (pH 2.8; Sigma-Aldrich).
2. 1 M Tris–HCl (pH 9.0; Sigma-Aldrich).
3. Amicon Ultra YM60 Microcon (Millipore Corp., Temecula, CA, USA).
4. 0.1% (v/v) Trifluoroacetic acid (TFA; Sigma-Aldrich).

2.4. Mass Spectrometry

1. Gold ProteinChip Array (Ciphergen Biosystems, Inc., Fremont, CA, USA).
2. Sinapinic acid (SPA); 3,5-dimethoxy-4-hydroxycinnamic acid (Sigma-Aldrich).
3. 50% (v/v) Acetonitrile (ACN; Fisher Scientific).
4. 0.5% (v/v) Trifluoroacetic acid (TFA; Sigma-Aldrich).

2.5. Calibration

All-in-One Protein standards (Ciphergen Biosystems, Inc./Bio-Rad, Mississauga, Canada).

2.6. Software

Ciphergen's ProteinChip 3.1 software (Ciphergen Biosystems, Inc.).

2.7. Equipment

1. Tissue culture incubator (Sanyo, Woodbridge, Canada).
2. Laminar flow hood (Microzone, Ottawa, Canada).
3. Gilson pipetman (Mandel Scientific Company, Guelph, Canada).

4. Vortex (Eppendorf, Mississauga, Canada).
5. Centrifuge (Eppendorf).
6. Thermomixer (Eppendorf).
7. Ciphergen Protein Biology System II (SELDI-TOF-MS; Ciphergen Biosystems, Inc.).

3. Methods

3.1. Cell Culture, Transfection, and Lysis

1. Grow HuH7 and HepG2 cells at 37°C and 5% CO_2 in DMEM + 10% (v/v) FBS + 28 μg/ml gentamicin. To subculture cells, aspirate media from the flask of adherent cells and rinse cells with 1× PBS, pH 7.6. Add 1.5 ml of 0.25% trypsin in HBSS to Corning T75 flask and incubate for 7–10 min at 37°C. Add 5 ml of fresh media to neutralize trypsin, transfer cells to a 15-ml centrifuge tube, and pellet at 850×*g* for 5 min. Remove the supernatant and resuspend cells in fresh media. Count cells using a cytometer and reseed a total of 1.5×10^6 cells in six-well plates to be 80–90% confluent for day of transfection (*see* **Notes 1–5**).
2. On the day of transfection, replace cell growth medium for 500 μl DMEM without FBS and antibiotic, and incubate for 30 min at 37°C and 5% CO_2.
3. Dilute a plasmid expression vector for human PCSK9 (hPCSK9-V5; 4 μg) in 250 μl OPTI-MEM media and mix by inversion.
4. Dilute the Lipofectamine 2000 reagent into 250 μl OPTI-MEM media in a 1:1 ratio (μl/μg) with DNA, mix by inversion, and incubate for 5 min at room temperature (*see* **Notes 6** and 7).
5. Combine DNA/OPTI-MEM with Lipofectamine 2000/OPTI-MEM, mix by inversion, and incubate for 20 min at room temperature to allow liposome complex formation (*see* **Note 8**).
6. Exchange cell medium with fresh 500 μl DMEM without FBS and antibiotic. Add 500 μl of the DNA/Lipofectamine 2000 complexes to cells and incubate at 37°C and 5% CO_2.
7. After 6 h, supplement transfected cells with 1 ml of DMEM + 10% FBS without antibiotic and incubate at 37°C and 5% CO_2 (*see* **Notes 9** and **10**).
8. Forty-eight hours following transfection, collect spent medium from mock-transfected and transfected cells in the presence of a general protease inhibitor cocktail supplemented with the phosphatase inhibitor sodium

orthovanadate (200 μM). Centrifuge at 13,000×g for 3 min to remove suspended cells and debris. Store media at –80°C until analysis (*see* **Notes 11** and **12**).

9. Wash cells twice with 1× PBS, pH 7.6, and lyse them in 150 μl of ice-cold 1× RIPA buffer supplemented with protease inhibitors. Scrap cells in RIPA buffer from plates and transfer them into Eppendorf tubes (*see* **Note 13**). Rotate total cell lysates at 4°C for 10 min. Centrifuge at 4°C and 13,000×g for 10 min and collected supernatants. Store samples at –80°C until analysis.

3.2. Immunoprecipitation

3.2.1. Pre-clearing

1. To a final volume of 1 ml, add 100 μl of 10× TBS-Tw and 500 μl of spent media or 150 μl of total cell lysates collected from mock-transfected and transfected cells.
2. Add 30 μl of protein A or protein G agarose and incubate at 4°C with rotation for 2 h to pre-clear (*see* **Notes 14** and **15**).
3. Centrifuge at 2,000×g for 2 min and transfer pre-cleared supernatant to clean 1.5-ml Eppendorf tube.

3.2.2. Immunoprecipitation

1. To pre-cleared supernatant, add excess anti-V5 Ab and 30 μl of protein A agarose or anti-V5 IgG agarose (40 μl of 50% slurry) since overexpressed PCSK9 carried a C-terminal V5 tag (*see* **Notes 14–16**).
2. Incubate immunoprecipitates overnight at 4°C with rotation.
3. Centrifuge at 2,000×g for 2 min and discard the supernatant.
4. Wash immunoprecipitates by adding ~1 ml of 1× TBS-Tw (*see* **Note 17**).
5. Incubate immunoprecipitates for 10 min at 4°C with rotation.
6. Centrifuge at 2,000×g for 2 min and discard the supernatant. Repeat Steps 4–6 three times.
7. Wash immunoprecipitates by adding ~1 ml of 20 mM Tris–HCl, pH 7.6. Mix by inversion.
8. Centrifuge at 2,000×g for 2 min and discard the supernatant. Repeat Steps 7–8 once more.
9. Centrifuge at 2,000×g for 2 min and discard the supernatant. Immunoprecipitates can be stored at –80°C for long-term storage.
10. Elute immunoprecipitated proteins from agarose beads (proceed to **Section 3.4**) for analyses by mass spectrometry (proceed to **Section 3.5**). For detection of

phosphorylation of immunoprecipitated proteins, proceed to **Section 3.3.1**. For detection of glycosylation of immunoprecipitated proteins by various strategies, proceed to **Section 3.3.2**.

3.3. Sample Immunoprecipitation, Preparation, and Treatment

The presence of a phosphoryl group(s) on an immunoprecipitated protein can be readily detected by comparing mass spectral profiles following treatment in the absence and the presence of shrimp alkaline phosphatase (37). An 80-Da decrease in the molecular mass of a protein (characteristic of protein dephosphorylation) is observable and is advantageous as an alternative to the use of radioactive isotopes. In addition, decreases in the molecular mass of a protein by multiples of 80 Da following phosphatase treatment is indicative of multiple sites of phosphorylation for that particular species.

Ciphergen's time-of-flight mass spectrometer can detect species in molecular mass ranging from 150 to 300 kDa (46, 47). However, as the molecular mass of a species increases, the resolution by time-of-flight mass spectrometry decreases. For example, the propeptide of PCSK9 which is ~14 kDa has half-peak-width resolution of ~100 Da, whereas PCSK9 itself is ~63 kDa in size and defined by a broader peak of lower resolution (half-peak-width resolution of ~2500 Da; **Fig. 8.1**). As a result, if a protein has a single phosphorylation site and the resolution of the species is lower, it would be difficult to detect this 80 Da size shift. If however there were multiple phosphorylation sites, the larger shift in size upon complete dephosphorylation would be observable at lower resolution. This technique is therefore ideal for analyzing lower molecular mass, well-resolved proteins/peptides (where a size shift of 80 Da can be easily detected), but for less-resolved larger proteins, an alternative approach is required (i.e., radiolabeling).

3.3.1. Dephosphorylation

1. Prepare immunoprecipitates for enzymatic dephosphorylation (*see* **Section 3.2**). Prior to phosphatase treatment, divide immunoprecipitates in half so that one half receives phosphatase, while the other half does not.
2. To immunoprecipitates, add 10× reaction buffer (provided) diluted to 1× with dH_2O so that the final volume of the 1× reaction buffer sufficiently covers agarose beads.
3. Add 10 units of shrimp alkaline phosphatase to immunoprecipitates in 1× reaction buffer. For mock treatment, incubate immunoprecipitates in provided 1× reaction buffer alone. Incubate immunoprecipitates for 30 min at 37°C with agitation.
4. Repeat wash steps 4–9 in **Section 3.2**. Elute immunoprecipitated proteins from agarose beads (*see* **Section 3.4**) for

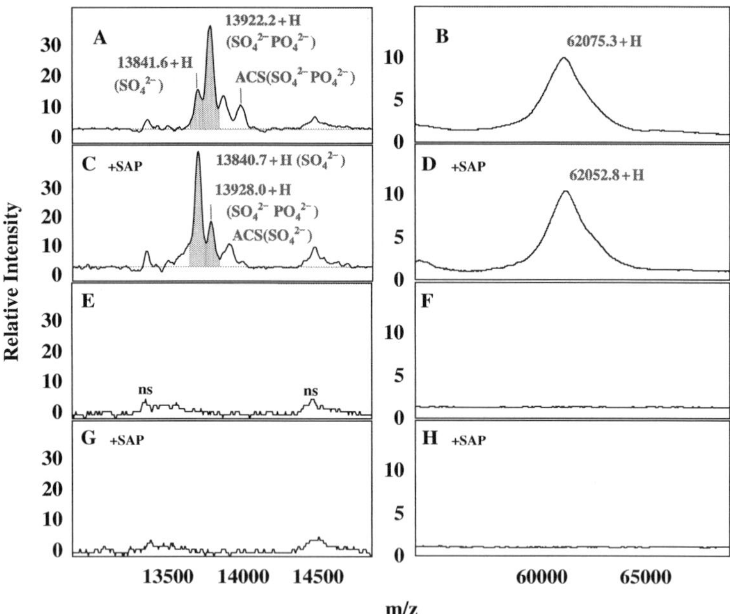

Fig. 8.1. Mass spectral plots of immunoprecipitated PCSK9-V5, its propeptide, and controls incubated in the absence and the presence of shrimp alkaline phosphatase from the media of transiently transfected HuH7 cells. Shown are time-of-flight mass spectrometry analyses of the molecular masses of the secreted proteins/peptides immunoprecipitated with anti-V5 antibody (*see* **Section 3.2**) from the media of HuH7 cells overexpressing either PCSK9-V5 tagged (**a–d**) or the control empty vector (**e–h**). (**a**) and (**b**) show the molecular forms of the propeptide and PCSK9, respectively, in the absence of treatment; (**c**) and (**d**) show the molecular forms of the propeptide and PCSK9, respectively, following dephosphorylation (*see* **Section 3.3.1**). (**e–h**) show the corresponding spectra of immunoprecipitates from HuH7 cells transfected with the control vector. External instrument calibration was performed (*see* **Section 3.5.2.**). The molecular mass (Da) is represented as the mass/charge ratio (*m/z*) for each peak on the *X*-axis and its relative intensity plotted on the *Y*-axis. The sulfated (SO_4^{2-}) and sulfated + phosphorylated ($SO_4^{2-}PO_4^{2-}$) peaks for the major molecular forms of the propeptide are indicated and the area under the peak (shown in *pink*) delineated as per **Section 3.5.5**. ACS, alternative signal peptidase cleavage site for PCSK9 propeptide identified in (37); ns, non-specific.

analyses by mass spectrometry (*see* **Section 3.5**). A mass spectral profile of PCSK9 and its propeptide immunoprecipitated from HuH7 cells treated with or without shrimp alkaline phosphatase is presented in **Fig. 8.1**.

Figure 8.1 shows that while there is no change in the molecular mass of the 62075.3-Da peak representing PCSK9-V5 following phosphatase treatment (panel B versus D) due to resolution limitations (see explanation above), there is a shift from the major sulfated + phosphorylated molecular form of the propeptide at 13922.2 Da (panel A; calculated molecular mass 13915.5 Da) to the unphosphorylated, sulfated form at 13840.7 Da (panel C; calculated molecular mass 13835.5 Da) following phosphatase treatment. For further discussion on mass spectral analyses of phosphorylation, *see* **Section 3.5.5.1**.

3.3.2. Deglycosylation

Protein glycosylation, either N or O linked, is a common post-translational modification that ranges in complexity and size from hundreds to thousands of daltons. Several strategies can be employed for detecting glycoproteins by mass spectrometry, including treatment of cells in culture with tunicamycin (which inhibits an enzyme involved in the first stage of N-linked glycosylation) or enzymatic deglycosylation (28). For instance, endoglycosidase H will cleave N-linked mannose-rich oligosaccharides, while PNGase F will deglycosylate all N-linked glycoproteins.

Glycoproteins generally produce a broad peak when analyzed by time-of-flight mass spectrometry. The post-translational modification of a protein by glycosylation can be shown by a size shift to a lower molecular mass following deglycosylation treatment (**Fig. 8.2**). The peak representing the deglycosylated protein is often observed as a sharper peak.

3.3.2.1. Tunicamycin Treatment

1. Grow HepG2 and HuH7 cells at 37°C/5% CO_2 in DMEM/10% FBS/28 μg/ml gentamicin. Reseed cells in six-well plates to be 70–80% confluent for day of transfection (*see* **Section 3.1**).
2. A day prior to tunicamycin treatment, exchange cell media for serum-free DMEM. Incubate cells at 37°C/5% CO_2 overnight (*see* **Note 10**).
3. On the day of treatment, exchange media for fresh serum-free DMEM and incubate cells for 18–24 h either in the presence or the absence of 5 μg/ml tunicamycin.
4. Collect spent media and total cell lysates from cells in the presence of a complete protease inhibitor cocktail (*see* **Section 3.1**).
5. Prepare spent media and/or total cell lysates for immunoprecipitation (*see* **Section 3.2**) (*see* **Note 17**). Immunoprecipitated proteins can now be eluted from agarose beads (**Section 3.4**) for analyses by mass spectrometry (**Section 3.5**).

A mass spectral profile of PCSK9 and its propeptide immunoprecipitated from the media of overexpressing HepG2 and HuH7 cells treated with or without tunicamycin is presented in **Fig. 8.2**. The data show that the secreted propeptide is not glycosylated and hence its molecular mass does not change upon tunicamycin treatment, while the molecular mass of secreted PCSK9-V5 decreases by 2047 Da following deglycosylation in HepG2 cells and by 2173.1 Da in HuH7 cells. The larger molecular mass of secreted PCSK9-V5 from HuH7 cells suggests the presence of

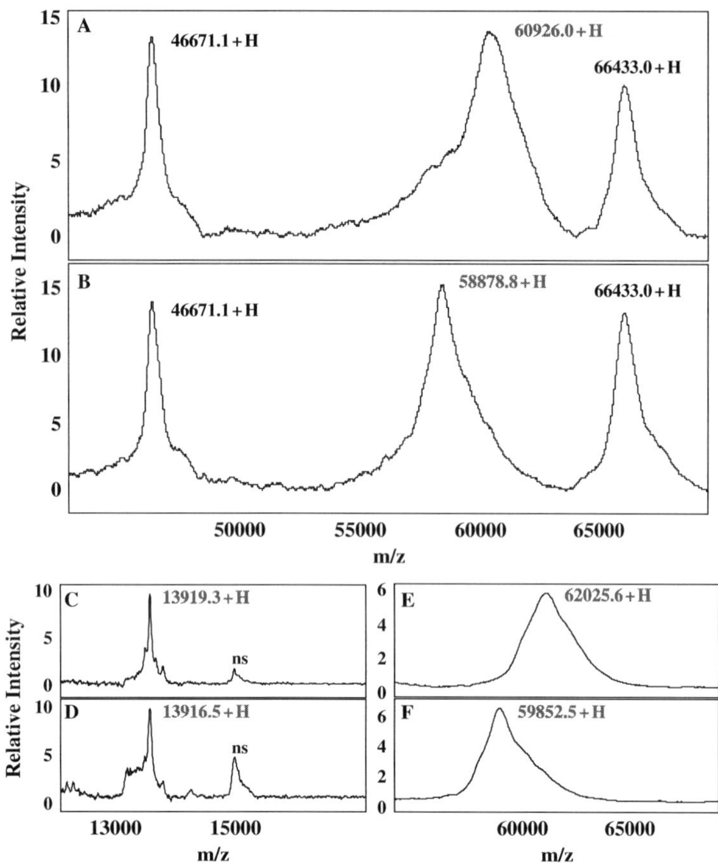

Fig. 8.2. Mass spectral plot of immunoprecipitated PCSK9 and its propeptide from HepG2 and HuH7 cells incubated in the absence and the presence of tunicamycin. Shown are time-of-flight mass spectrometric analyses of the molecular masses of PCSK9-V5 immunoprecipitated with anti-V5 antibody from the media of HepG2 cells overexpressing PCSK9-V5 tagged and incubated in the absence (**a**) and the presence of tunicamycin (**b**) (*see* **Section 3.3.2**). Spectra from (**a**) and (**b**) were internally calibrated (*see* **Section 3.5.4**). (**c–f**) represent time-of-flight mass spectrometric analyses of the molecular masses of the PCSK9-V5 (**e** and **f**) and its propeptide (**c** and **d**) immunoprecipitated with anti-V5 antibody from the media of overexpressing HuH7 cells and incubated in the absence (**c** and **e**) and the presence of tunicamycin (**d** and **f**) (*see* **Section 3.3.2.**) and calibrated externally (*see* **Section 3.5.4.**). The molecular mass (Da) is represented as the mass/charge ratio (*m/z*) for each peak on the *X*-axis and its relative intensity plotted on the *Y*-axis. ns, non-specific.

yet, unidentified PCSK9 modifications in this cell type. For further discussion of analyses of glycosylation, *see* **Section 3.5.5.2**.

3.3.2.2. Enzymatic Protein Deglycosylation Treatments: Endoglycosidase H

1. Immunoprecipitates for endoglycosidase H treatment are prepared (*see* **Section 3.2**).
2. For deglycosylation under native conditions and for 40 μl reaction volume, add 10 μl of 200 mM sodium phosphate buffer (final concentration 50 mM) and dH_2O to bring

reaction volume up to 40 μl without endoglycosidase H (*see* **Note 18**).

3. Incubate immunoprecipitates in the presence and the absence of 1 μl of endoglycosidase H overnight at 37°C in a thermomixer with moderate agitation (~800 rpm). Repeat wash steps 4–9 in **Section 3.2**.
4. Proceed to antigen elution (*see* **Section 3.4.1**) and analyze by mass spectrometry (**Section 3.5**).

For other enzymatic deglycosylation strategies, several companies provide kits containing a mixture of enzymes (such as PNGase F, O-glycosidase, and neuraminidase) to remove both N- and O-linked glycan moieties from proteins.

3.4. Elution and Concentration

3.4.1. Antigen Elution

1. Immunoprecipitates for time-of-flight mass spectral analysis of immunocaptured PCSK9 and its propeptide were prepared as described in **Section 3.2** above.
2. To elute antibody antigen complex from agarose beads, add 150 μl of 0.1 M glycine (pH 2.8) to immunoprecipitates (*see* **Note 19**).
3. Incubate at room temperature for 10 min, with shaking (~700 rpm).
4. Centrifuge at 2,000×*g* for 2 min and transfer the supernatant to clean 1.5-ml Eppendorf tube. Repeat from Step 1 with another 150 μl of 0.1 M glycine. Transfer the supernatant to same tube as before.
5. Add 30 μl of 1 M Tris–HCl (pH 9.0) to neutralize the supernatant.
6. Concentrate supernatants 20× with an Amicon Ultra YM60 Microcon (Millipore Corp.) by centrifugation at 4,000×*g* (*see* **Note 20**).
7. To equilibrate sample concentrates, add 200 μl of 0.1% trifluoroacetic acid (TFA) to sample retained in inner chamber of Microcon.
8. Concentrate by centrifugation at 4,000×*g*. Repeat Step 7 five times to exchange sample buffer for 0.1% TFA alone. Samples are ready for mass spectrometry analyses (**Section 3.5**).

3.5. Mass Spectrometry

3.5.1. Sample Loading

1. Following antigen elution, 2.5 μl of sample is applied to a spot on the eight-spot Gold ProteinChip Array and allowed to air-dry.
2. 1 μl of saturated SPA in 50% (v/v) CAN/0.5% TFA is added to each sample-containing spot and air-dried to crystallize sample in an energy-absorbing matrix for positive ion mass spectrometry (*see* **Note 21**).

3. Open Ciphergen's ProteinChip 3.1 software and allow program to connect to Ciphergen Protein Biology System II mass spectrometer.
4. Load Gold ProteinChip Array into Ciphergen Protein Biology System II mass spectrometer.

3.5.2. External Calibration

It is important to calibrate the mass spectrometer before reading your samples to ensure reliable and reproducible mass-to-charge (m/z) results. Ciphergen provides two ranges of standards for external calibration: The All-in-One Peptide standards (molecular mass range ~1–7 kDa) and the All-in-One Protein standards (molecular mass range ~7–150 kDa). Choose the standard for calibration that overlaps with your proteins and/or peptides of interest:

1. Prepare the standard mixtures and apply to a spot on the eight-spot Gold ProteinChip Array as directed by the protocol provided by Ciphergen Biosystems, Inc.
2. Load Gold ProteinChip Array into Ciphergen Protein Biology System II.
3. Follow steps as directed by the protocol provided by Ciphergen Biosystems, Inc. for reading and calibrating the mass spectrometer using the appropriate molecular mass range standards.

We have achieved a mass accuracy of <0.17% for protein and polypeptides from ~12000 to ~67000 Da using the Ciphergen Protein Biology System II mass spectrometry as illustrated in **Fig. 8.3**. We perform external calibration of the instrument before each time-of-flight mass spectral experiment.

3.5.3. Data Acquisition

Prior to data collection/acquisition, a protocol defining instrument parameters such as mass range to collect, sensor sensitivity, laser intensity, and laser position on ProteinChip Array sample spots needs to be specified. The ProteinChip software 3.1 operation manual provides a description of each of these parameters and how to manipulate them.

PCSK9 (unlike its family members) forms a tight 1:1 complex with its propeptide after its autocatalytic cleavage in the ER, throughout the secretory pathway and in circulation (12, 28). We use this feature to analyze both the propeptide and PCSK9 which carries the C-terminal V5 tag together in our analyses of immunoprecipitates. To carry out mass spectrometry data acquisition and analyses of the propeptide:PCSK9 heterodimer, each sample-containing spot is read with two separate protocols to optimize parameters for the lower molecular mass propeptide (of apparent molecular mass of ~14 kDa) and the higher molecular mass PCSK9 (of apparent molecular mass of ~63 kDa) (28, 37). For propeptide data acquisition, the mass optimization window

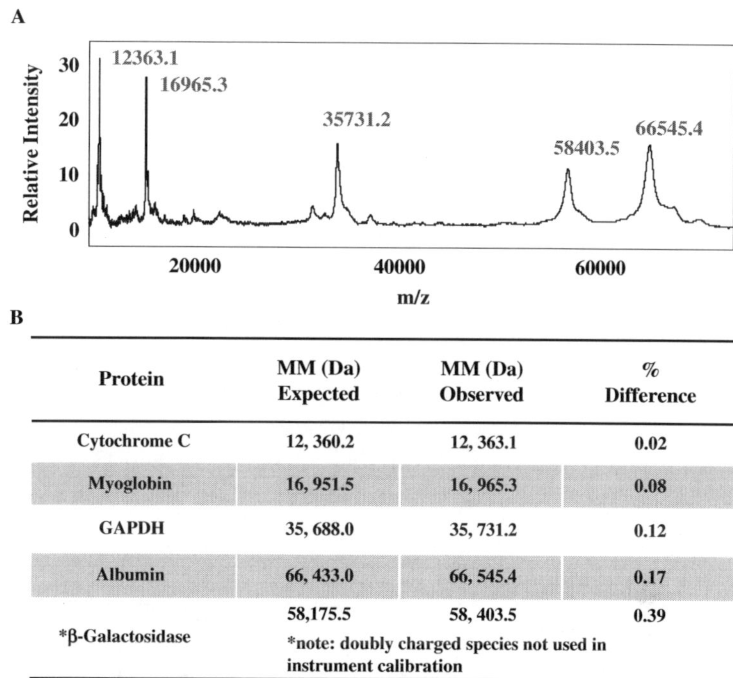

Protein	MM (Da) Expected	MM (Da) Observed	% Difference
Cytochrome C	12, 360.2	12, 363.1	0.02
Myoglobin	16, 951.5	16, 965.3	0.08
GAPDH	35, 688.0	35, 731.2	0.12
Albumin	66, 433.0	66, 545.4	0.17
*β-Galactosidase	58,175.5	58, 403.5	0.39
	*note: doubly charged species not used in instrument calibration		

Fig. 8.3. Mass spectral plot and expected versus observed molecular mass table of All-in-One ProteinChip standards for external calibration of mass spectrometer. (**a**) Ciphergen's All-in-One Protein standards were reconstituted and applied to a spot on an eight-spot Gold ProteinChip Array following product insert instructions and using sinapinic acid as the energy-absorbing matrix. Protein standards were read, analyzed, and assigned according to product insert with the external calibration carried out as described in Ciphergen's ProteinChip manual. The molecular mass (Da) is represented as the mass/charge ratio (*m/z*) for each peak on the *X*-axis and its relative intensity plotted on the *Y*-axis. (**b**) Molecular mass table of expected versus observed molecular masses (Da) of components of All-in-One Protein standard used for instrument calibration. Percent difference (%) between expected versus observed molecular masses is shown. External instrument calibration was performed before each experiment to increase mass accuracy.

range was set from 7.5 to 20 kDa, laser power was set at 175, and detector sensitivity at 8. For PCSK9 data acquisition, the mass optimization window range was set from 40 to 80 kDa, laser power was set at 210, and detector sensitivity at 8.

3.5.4. Data Normalization

1. Concurrently run samples in the absence and the presence of a specific treatment (e.g., dephosphorylation) on the same Gold ProteinChip Array and with the same data acquisition/analyses conditions to allow for spot-to-spot or sample-to-sample comparisons.
2. Generate each mass spectrum from an average of 100 shots/spot using an automated data collection protocol

within the ProteinChip software (version 3.1) with masses externally calibrated with All-in-One Protein standards (*see* **Section 3.5.2.** In **Fig. 8.1**, spectra were externally calibrated).

3. Use the "total ion current" normalization within the ProteinChip software (version 3.1) to normalize for the impacts of variables such as sample amount and application, matrix application, and mass spectrometer laser/detector performance. This normalization is based on the assumption that such technical parameters are largely responsible for greatest differences between samples. To normalize against total ion current, first select all spectra in an experiment to be normalized to one another:
   (a) Select the Normalization icon (in the Experimental Toolbar) to initiate the normalization spectra screen.
   (b) Select total ion current and the mass-to-charge range over which the total ion current is averaged.
   (c) Click on baseline subtraction and then apply to normalize selected spectra in experiment.
4. Alternatively, one can add an internal known peptide or protein equally into their samples and use the ProteinChip software to normalize against the intensity and mass of that "internal" calibrant.

For panels a and b in **Fig. 8.2**, the spectra were internally calibrated. All-in-One Protein standards were spiked onto the ProteinChip Array with immunoprecipitates. Shown are internal calibrants: yeast enolase at 46671.1 and bovine albumin at 66433.0 Da. Not shown is carbonic anhydrase at 29023 Da. All were used for the three-point internal calibration.

3.5.5. Data and Software Analyses

The preprocessing of data which includes instrument calibration and spectrum normalization (which includes baseline correction) is an important step in Ciphergen's time-of-flight mass spectrometry analyses, eliminating or decreasing differential effects due to external factors.

Data are further processed by assigning peaks using the peak detection option in the ProteinChip software (version 3.1). If internal calibrants are added (as in **Fig. 8.2**) or if there is a known common peptide or protein between samples, it can be assigned and then used for peak and mass spectrum alignment.

We use a freeware program (called PAWS) that uses the amino acid sequence of a protein to calculate its theoretical molecular mass. This software (and others like it) allows the user to modify any known amino acids to further define a known protein or peptide theoretical/calculated mass. This information can be combined with mass spectral data to predict the presence of post-translation modifications as exemplified below and illustrated in **Figs. 8.1** and **8.2**.

3.5.5.1. Phosphorylation Analyses

For phosphorylation analyses, all data were normalized for total ion current between 2500 and 80000 Da using Ciphergen's ProteinChip 3.1 software. The software was used to assign molecular mass to peaks appearing in the spectrum based upon their mass-to-charge ratio for time-of-flight mass spectrometry.

The predicted molecular mass of PCSK9 and its propeptide based upon amino acid prediction (PAWS) is 58856.4 (with disulfides) and 13835.5 Da (pyroQ at amino acid 31 and sulfation at Tyr38), respectively. The size observed by mass spectrometry is 62075.3 and 13922.2 Da, respectively (**Fig. 8.1**). We previously used a combination of mass spectrometry and shrimp alkaline phosphatase treatment to determine that the propeptide of PCSK9 is phosphorylated (37).

In **Fig. 8.1**, we illustrate this technique using immunoprecipitates treated with shrimp alkaline phosphatase from the media of HuH7 cells transiently overexpressing PCSK9-V5 and its propeptide. The resulting shift of 80 Da in the propeptide of PCSK9 following phosphatase treatment demonstrates that it is phosphorylated at one residue. There is no major shift in the peak for PCSK9, and as previously mentioned the resolution of the mass spectrometry would not be sufficient to detect few phosphorylation events. However, the large size discrepancy between the calculated (58856.4 Da) and observed (62075.3 Da) molecular masses for secreted PCSK9 indicates the presence of other post-translational modifications (e.g., glycosylation).

To compare the ratio of unphosphorylated to phosphorylated proteins/peptides the peak areas can be calculated using the indirect method (with a bracket height of 0.4 and width expansion factor of 2) contained within the Analysis Protocol Properties Icon (under the Area Tab) in Ciphergen's ProteinChip 3.1 software. The data is exported from the ProteinChip software to Microsoft Excel using the export peak information tab under the File menu. Dewpura et al. (37) use this technique to compare phosphorylation levels of the propeptide of PCSK9 between cell lines and in the presence of naturally occurring variants found in the propeptide.

In **Fig. 8.1**, the sulfated (SO_4^{2-}) and sulfated + phosphorylated ($SO_4^{2-}PO_4^{2-}$) peaks for the major molecular forms of the propeptide are indicated, and the "area under the peak" detected – using the above parameters – is illustrated in pink. We sum the areas under the peak for both forms to give total area for the major molecular forms of propeptide and then divide each one by the total area to give proportion of SO_4^{2-} propeptide and/or $SO_4^{2-}PO_4^{2-}$ propeptide. Approximately 70% of the propeptide of PCSK9 was secreted from HuH7 in its

$SO_4^{2-}PO_4^{2-}$ form, while following phosphatase treatment, 20% remained in its $SO_4^{2-}PO_4^{2-}$ form and 80% was now in a dephosphorylated SO_4^{2-} form. For complete dephosphorylation, the incubation time with phosphatase or units of phosphatase activity can be increased. These analyses should be conducted on at least three independent experiments for statistical analyses.

3.5.5.2. Glycosylation Analyses

For glycosylation analyses, all data were normalized for total ion current between 2500 and 80000 Da using Ciphergen's ProteinChip 3.1 software. The software was used to assign molecular masses to peaks appearing in the spectrum.

The predicted molecular mass of PCSK9-V5 tagged based upon amino acid prediction (PAWS) is 58856.4 Da (with disulfides). The observed molecular mass of PCSK9-V5 immunoprecipitated from the media of HepG2 cells is 60926.0 Da which is larger than its calculated mass. We have previously shown PCSK9 to be an N-linked glycoprotein (28). In **Fig. 8.2** we show that tunicamycin treatment of HepG2 cells transiently overexpressing PCSK9-V5 resulted in a shift in the peak for PCSK9-V5 to 58878.8 Da. The difference in mass of secreted PCSK9-V5 in the absence and the presence of tunicamycin is due to its N-linked glycan moiety. The size differential determined by mass spectrometry is much more accurate than that determined by SDS-PAGE analyses.

3.6. Statistical Analyses

1. Experiments are repeated at least three times for each sample and three times using independent samples.
2. Data, such as protein and peptide molecular masses represented as peak mass-to-charge ratios, peak intensity, and area under the peak can be exported from ProteinChip software to Microsoft Excel and copied into GRAPHPAD PRISM 5.0 statistical software (or a statistical software package of choice).
3. Express analyzed data as mean ± standard error (SE). Student's two-tailed t-test was used for statistical analysis of differences between PCSK9 propeptide phosphorylation in several cell lines (37) with significance set as $p < 0.05$.

3.7. Conclusions

Mass spectrometry is a powerful means to identify and define post-translational modifications in a protein or a peptide. Although not illustrated within this chapter, it can be combined with trypsinization and peptide analyses to identify and define the tryptically generated peptide and even the residue that is phosphorylated within that peptide. To confirm, these results can then be combined with site-directed mutagenesis and mass spectrometry analyses. These methodologies were used in 37 to identify the site of phosphorylation in the propeptide of PCSK9 as Ser47. Beyond detecting and analyzing post-translational modifications, this methodology can also be adapted to define substrate/convertase relationships (48, 49). We have used this technology to identify and define differences in tissue-specific substrate cleavage patterns of several proprotein convertases while comparing wild-type and knockout mice models (48, 49).

4. Notes

1. Cell lines of interest can be used however, be aware that the post-translational modifications do differ between cell lines (37).
2. Specific growth medium used to culture cells will depend on the cell line.
3. Other transfection reagents (for example, Effectene from QIAGEN, Missassauga, Canada) may be more appropriate/effective for your cell line of interest.
4. Level of confluency of cells recommended for day of transfection depends upon the specific protocol of the transfection reagent used.
5. Depending on the transfection reagent used, be aware that growth media containing antibiotics may affect transfection efficiency.
6. For Lipofectamine 2000 transfection protocol, OPTI-MEM media is optimal for transfection.
7. Ratio of DNA to transfection reagent is dependent upon transfection reagent and cell line used.
8. When mixing DNA with Lipofectamine 2000, do *not* vortex, as this will disrupt liposome formation.
9. It is not necessary to change media after transfection using Lipofectamine 2000. Approximately 6 h following transfection, simply add media supplemented with growth serum. Be aware as this may vary depending on the transfection protocol used.
10. For immunoprecipitation and analysis of secreted proteins/peptides by mass spectrometry, incubating cells in serum-free media can reduce non-specific binding of com mon, high-abundant serum proteins such as albumin. In the case of cells expressing PCSK9-V5, we have not found that serum-free media affects its phosphorylation, or its glycosylation.

11. The length of post-transfection incubation for protein expression depends on the cell type: We have found that 24 h is sufficient for PCSK9 expression and secretion from Chinese hamster ovary (CHO) cells, 24–48 h for human embryonic kidney (HEK 293) cells, and 8–72 h for HuH7 and HepG2 cells.
12. If interested in labile modification like phosphorylation, ensure your protease inhibitor cocktail contains phosphatase inhibitors.
13. Other cell lysis protocols are available.
14. Depending upon the species in which your antibody of interest is raised, protein A agarose or protein G agarose may be more appropriate. Check binding efficiencies.
15. If available, preimmune sera can be used to pre-clear.
16. Use antibody specific to protein or tag on protein of interest.
17. While immunoprecipitations of media are washed with 1× TBS-Tw, immunoprecipitations of total cell lysates are washed in 1× RIPA followed by washes with 1× TBS-Tw, since cell lysates are more complex.
18. The samples can also be deglycosylated, either before or after immunoprecipitation, under denaturing conditions which can increase the rate of deglycosylation of the glycoprotein of interest. Denaturing protocols for deglycosylation for endoglycosidase H treatment are also provided by Sigma-Aldrich – briefly β-mercaptoethanol is added to a final concentration of 50 mM, SDS to a final concentration 0.1% (w/v), samples heated to 99°C for 5 min, and then cooled before endo H incubation.
19. If using an antibody for immunoprecipitations that is directed against a peptide tag on your protein of interest, such as the V5 tag on PCSK9, an alternative elution method is to compete bound antigen from the antibody:agarose bead complex with excess peptide (e.g., V5 peptide; Invitrogen).
20. The molecular mass cutoff (MWCO) for the Centricon/Microcon depends on the protein or the peptide you are studying: It must have a lower MWCO than the molecular mass of your protein/peptide to allow buffer exchange and protein/peptide concentration while ensuring protein/peptide retention.
21. Other matrices may be used as an energy-absorbing matrix for time-of-flight mass spectrometry. One consideration is the size of the protein or the peptide of interest: for

instance, α-cyano-4-hydroxycinnamic acid (CHCA) is better suited for studying small peptides (of <10 kDa) by mass spectrometry.

Acknowledgements

The work was supported by grants from Canadian Institute of Health Research, Heart and Stroke Foundation of Canada and the Strauss Foundation.

References

1. Zinkin, N. T., Grall, F., Bhaskar, K., Otu, H. H., Spentzos, D., Kalmowitz, B., Wells, M., Guerrero, M., Asara, J. M., Libermann, T. A., and Afdhal, N. H. (2008) Serum proteomics and biomarkers in hepatocellular carcinoma and chronic liver disease *Clin Cancer Res* **14**, 470–7.
2. Boggs, S. E. (2004) Protein profiling in respiratory disease: Techniques and impact *Expert Rev Proteomics* **1**, 29–36.
3. Liu, A. Y., Zhang, H., Sorensen, C. M., and Diamond, D. L. (2005) Analysis of prostate cancer by proteomics using tissue specimens *J Urol* **173**, 73–8.
4. Rubin, R. B. and Merchant, M. (2000) A rapid protein profiling system that speeds study of cancer and other diseases *Am Clin Lab* **19**, 28–9.
5. Roebroek, A. J., Schalken, J. A., Bussemakers, M. J., van Heerikhuizen, H., Onnekink, C., Debruyne, F. M., Bloemers, H. P., and Van de Ven, W. J. (1986) Characterization of human c-fes/fps reveals a new transcription unit (fur) in the immediately upstream region of the proto-oncogene *Mol Biol Rep* **11**, 117–25.
6. Seidah, N. G., Gaspar, L., Mion, P., Marcinkiewicz, M., Mbikay, M., and Chretien, M. (1990) cDNA sequence of two distinct pituitary proteins homologous to Kex2 and furin gene products: Tissue-specific mRNAs encoding candidates for prohormone processing proteinases *DNA Cell Biol* **9**, 415–24.
7. Nakagawa, T., Hosaka, M., Torii, S., Watanabe, T., Murakami, K., and Nakayama, K. (1993) Identification and functional expression of a new member of the mammalian Kex2-like processing endoprotease family: Its striking structural similarity to PACE4 *J Biochem* **113**, 132–5.
8. Smeekens, S. P., and Steiner, D. F. (1990) Identification of a human insulinoma cDNA encoding a novel mammalian protein structurally related to the yeast dibasic processing protease Kex2 *J Biol Chem* **265**, 2997–3000.
9. Nakayama, K., Kim, W. S., Torii, S., Hosaka, M., Nakagawa, T., Ikemizu, J., Baba, T., and Murakami, K. (1992) Identification of the fourth member of the mammalian endoprotease family homologous to the yeast Kex2 protease. Its testis-specific expression *J Biol Chem* **267**, 5897–900.
10. Kiefer, M. C., Tucker, J. E., Joh, R., Landsberg, K. E., Saltman, D., and Barr, P. J. (1991) Identification of a second human subtilisin-like protease gene in the fes/fps region of chromosome 15 *DNA Cell Biol* **10**, 757–69.
11. Munzer, J. S., Basak, A., Zhong, M., Mamarbachi, A., Hamelin, J., Savaria, D., Lazure, C., Hendy, G. N., Benjannet, S., Chretien, M., and Seidah, N. G. (1997) In vitro characterization of the novel proprotein convertase PC7 *J Biol Chem* **272**, 19672–81.
12. Seidah, N. G., Benjannet, S., Wickham, L., Marcinkiewicz, J., Jasmin, S. B., Stifani, S., Basak, A., Prat, A., and Chretien, M. (2003) The secretory proprotein convertase neural apoptosis-regulated convertase 1 (NARC-1): Liver regeneration and neuronal differentiation *Proc Natl Acad Sci USA* **100**, 928–33.
13. Abifadel, M., Varret, M., Rabes, J. P., Allard, D., Ouguerram, K., Devillers, M., Cruaud, C., Benjannet, S., Wickham, L., Erlich, D., Derre, A., Villeger, L., Farnier, M., Beucler,

I., Bruckert, E., Chambaz, J., Chanu, B., Lecerf, J. M., Luc, G., Moulin, P., Weissenbach, J., Prat, A., Krempf, M., Junien, C., Seidah, N. G., and Boileau, C. (2003) Mutations in PCSK9 cause autosomal dominant hypercholesterolemia *Nat Genet* **34**, 154–6.

14. Constam, D. B., Calfon, M., and Robertson, E. J. (1996) SPC4, SPC6, and the novel protease SPC7 are coexpressed with bone morphogenetic proteins at distinct sites during embryogenesis *J Cell Biol* **134**, 181–91.
15. Seidah, N. G., Hamelin, J., Gaspar, A. M., Day, R., and Chretien, M. (1992) The cDNA sequence of the human pro-hormone and pro-protein convertase PC1 *DNA Cell Biol* **11**, 283–9.
16. Seidah, N. G., Hamelin, J., Mamarbachi, M., Dong, W., Tardos, H., Mbikay, M., Chretien, M., and Day, R. (1996) cDNA structure, tissue distribution, and chromosomal localization of rat PC7, a novel mammalian proprotein convertase closest to yeast kexin-like proteinases *Proc Natl Acad Sci USA* **93**, 3388–93.
17. Smeekens, S. P., Avruch, A. S., LaMendola, J., Chan, S. J., and Steiner, D. F. (1991) Identification of a cDNA encoding a second putative prohormone convertase related to PC2 in AtT20 cells and islets of Langerhans *Proc Natl Acad Sci USA* **88**, 340–4.
18. Espenshade, P. J., Cheng, D., Goldstein, J. L., and Brown, M. S. (1999) Autocatalytic processing of site-1 protease removes propeptide and permits cleavage of sterol regulatory element-binding proteins *J Biol Chem* **274**, 22795–804.
19. Lusson, J., Vieau, D., Hamelin, J., Day, R., Chretien, M., and Seidah, N. G. (1993) cDNA structure of the mouse and rat subtilisin/kexin-like PC5: A candidate proprotein convertase expressed in endocrine and nonendocrine cells *Proc Natl Acad Sci USA* **90**, 6691–5.
20. Seidah, N. G., Mowla, S. J., Hamelin, J., Mamarbachi, A. M., Benjannet, S., Toure, B. B., Basak, A., Munzer, J. S., Marcinkiewicz, J., Zhong, M., Barale, J. C., Lazure, C., Murphy, R. A., Chretien, M., and Marcinkiewicz, M. (1999) Mammalian subtilisin/kexin isozyme SKI-1: A widely expressed proprotein convertase with a unique cleavage specificity and cellular localization *Proc Natl Acad Sci USA* **96**, 1321–6.
21. Shennan, K. I., Smeekens, S. P., Steiner, D. F., and Docherty, K. (1991) Characterization of PC2, a mammalian Kex2 homologue, following expression of the cDNA in microinjected *Xenopus* oocytes *FEBS Lett* **284**, 277–80.
22. Bergeron, E., Basak, A., Decroly, E., and Seidah, N. G. (2003) Processing of alpha4 integrin by the proprotein convertases: Histidine at position P6 regulates cleavage *Biochem J* **373**, 475–84.
23. Bravo, D. A., Gleason, J. B., Sanchez, R. I., Roth, R. A., and Fuller, R. S. (1994) Accurate and efficient cleavage of the human insulin proreceptor by the human proprotein-processing protease furin. Characterization and kinetic parameters using the purified, secreted soluble protease expressed by a recombinant baculovirus *J Biol Chem* **269**, 25830–7.
24. Duguay, S. J., Milewski, W. M., Young, B. D., Nakayama, K., and Steiner, D. F. (1997) Processing of wild-type and mutant proinsulin-like growth factor-IA by subtilisin-related proprotein convertases *J Biol Chem* **272**, 6663–70.
25. Rholam, M., and Fahy, C. (2009) Processing of peptide and hormone precursors at the dibasic cleavage sites *Cell Mol Life Sci* **66**, 2075–91.
26. Rouille, Y., Westermark, G., Martin, S. K., and Steiner, D. F. (1994) Proglucagon is processed to glucagon by prohormone convertase PC2 in alpha TC1-6 cells *Proc Natl Acad Sci USA* **91**, 3242–6.
27. Valore, E. V. and Ganz, T. (2008) Posttranslational processing of hepcidin in human hepatocytes is mediated by the prohormone convertase furin *Blood Cells Mol Dis* **40**, 132–8.
28. Benjannet, S., Rhainds, D., Essalmani, R., Mayne, J., Wickham, L., Jin, W., Asselin, M. C., Hamelin, J., Varret, M., Allard, D., Trillard, M., Abifadel, M., Tebon, A., Attie, A. D., Rader, D. J., Boileau, C., Brissette, L., Chretien, M., Prat, A., and Seidah, N. G. (2004) NARC-1/PCSK9 and its natural mutants: Zymogen cleavage and effects on the low density lipoprotein (LDL) receptor and LDL cholesterol *J Biol Chem* **279**, 48865–75.
29. Seidah, N. G., and Chretien, M. (1999) Proprotein and prohormone convertases: A family of subtilases generating diverse bioactive polypeptides *Brain Res* **848**, 45–62.
30. Seidah, N. G., Chretien, M., and Day, R. (1994) The family of subtilisin/kexin like pro-protein and pro-hormone convertases: Divergent or shared functions *Biochimie* **76**, 197–209.
31. Seidah, N. G., and Prat, A. (2007) The proprotein convertases are potential targets in the treatment of dyslipidemia *J Mol Med* **85**, 685–96.

32. Muller, L., and Lindberg, I. (1999) The cell biology of the prohormone convertases PC1 and PC2 *Prog Nucleic Acid Res Mol Biol* **63**, 69–108.
33. Boudreault, A., Gauthier, D., Rondeau, N., Savaria, D., Seidah, N. G., Chretien, M., and Lazure, C. (1998) Molecular characterization, enzymatic analysis, and purification of murine proprotein convertase-1/3 (PC1/PC3) secreted from recombinant baculovirus-infected insect cells *Protein Expr Purif* **14**, 353–66.
34. Benjannet, S., Rhainds, D., Hamelin, J., Nassoury, N., and Seidah, N. G. (2006) The proprotein convertase (PC) PCSK9 is inactivated by furin and/or PC5/6A: Functional consequences of natural mutations and post-translational modifications *J Biol Chem* **281**, 30561–72.
35. Elagoz, A., Benjannet, S., Mammarbassi, A., Wickham, L., and Seidah, N. G. (2002) Biosynthesis and cellular trafficking of the convertase SKI-1/S1P: Ectodomain shedding requires SKI-1 activity *J Biol Chem* **277**, 11265–75.
36. Jones, B. G., Thomas, L., Molloy, S. S., Thulin, C. D., Fry, M. D., Walsh, K. A., and Thomas, G. (1995) Intracellular trafficking of furin is modulated by the phosphorylation state of a casein kinase II site in its cytoplasmic tail *Embo J* **14**, 5869–83.
37. Dewpura, T., Raymond, A., Hamelin, J., Seidah, N. G., Mbikay, M., Chretien, M., and Mayne, J. (2008) PCSK9 is phosphorylated by a Golgi casein kinase-like kinase *ex vivo* and circulates as a phosphoprotein in humans *FEBS J* **275**, 3480–93.
38. Maxwell, K. N., and Breslow, J. L. (2004) Adenoviral-mediated expression of Pcsk9 in mice results in a low-density lipoprotein receptor knockout phenotype *Proc Natl Acad Sci USA* **101**, 7100–5.
39. Park, S. W., Moon, Y. A., and Horton, J. D. (2004) Post-transcriptional regulation of low density lipoprotein receptor protein by proprotein convertase subtilisin/kexin type 9a in mouse liver *J Biol Chem* **279**, 50630–8.
40. Rashid, S., Curtis, D. E., Garuti, R., Anderson, N. N., Bashmakov, Y., Ho, Y. K., Hammer, R. E., Moon, Y. A., and Horton, J. D. (2005) Decreased plasma cholesterol and hypersensitivity to statins in mice lacking Pcsk9 *Proc Natl Acad Sci USA* **102**, 5374–9.
41. Costet, P., Cariou, B., Lambert, G., Lalanne, F., Lardeux, B., Jarnoux, A. L., Grefhorst, A., Staels, B., and Krempf, M. (2006) Hepatic PCSK9 expression is regulated by nutritional status via insulin and sterol regulatory element-binding protein 1c *J Biol Chem* **281**, 6211–18.
42. Dubuc, G., Chamberland, A., Wassef, H., Davignon, J., Seidah, N. G., Bernier, L., and Prat, A. (2004) Statins upregulate PCSK9, the gene encoding the proprotein convertase neural apoptosis-regulated convertase-1 implicated in familial hypercholesterolemia *Arterioscler Thromb Vasc Biol* **24**, 1454–9.
43. Langhi, C., Le May, C., Kourimate, S., Caron, S., Staels, B., Krempf, M., Costet, P., and Cariou, B. (2008) Activation of the farnesoid X receptor represses PCSK9 expression in human hepatocytes *FEBS Lett* **582**, 949–55.
44. Li, H., Dong, B., Park, S. W., Lee, H. S., Chen, W., and Liu, J. (2009) HNF1{alpha} plays a critical role in PCSK9 gene transcription and regulation by a natural hypocholesterolemic compound berberine *J Biol Chem* **284**(42), 2885–95.
45. Dong, B., Wu, M., Li, H., Kraemer, F. B., Adeli, K., Seidah, N. G., Park, S. W., and Liu, J. (2010) Strong induction of PCSK9 gene expression through HNF1alpha and SREBP2: mechanism for the resistance to LDL-cholesterol lowering effect of statins in dyslipidemic hamsters *J Lipid Res* **51**, 1486–95.
46. Poon, T. C. (2007) Opportunities and limitations of SELDI-TOF-MS in biomedical research: Practical advices *Expert Rev Proteomics* **4**, 51–65.
47. Wright, G. L., Jr. (2002) SELDI ProteinChip MS: A platform for biomarker discovery and cancer diagnosis *Expert Rev Mol Diagn* **2**, 549–63.
48. Mbikay, M., Croissandeau, G., Sirois, F., Anini, Y., Mayne, J., Seidah, N. G., and Chretien, M. (2007) A targeted deletion/insertion in the mouse Pcsk1 locus is associated with homozygous embryo preimplantation lethality, mutant allele preferential transmission and heterozygous female susceptibility to dietary fat *Dev Biol* **306**, 584–98.
49. Anini, Y., Mayne, J., Gagnon, J., Sherbafi, J., Chen, A., Kaefer, N., Chretien, M., and Mbikay, M. (2010) Genetic deficiency for proprotein convertase subtilisin/kexin type 2 in mice is associated with decreased adiposity and protection from dietary fat-induced body weight gain *Int J Obes (Lond)* **34**(11), 1599–607.

Section III

Molecular Biology and Genetics

Chapter 9

The Molecular Biology of Furin-Like Proprotein Convertases in Vascular Remodelling

Philipp Stawowy and Kai Kappert

Abstract

Vascular smooth muscle cell (VSMC) proliferation and migration represent key features in atherosclerosis and restenosis. The proprotein convertases (PCs) furin and PC5 are highly expressed in human atheroma and are putatively involved in vascular lesion formation via the activation of precursor proteins, essential for cell proliferation and migration. In vitro assays have identified these PCs to govern cell functions via endoproteolytic cleavage of key substrates, including pro-integrins and pro-matrix metalloproteinases. In vivo gene expression studies of furin/PC5 and their substrates demonstrate their coordinated regulation in animal models of vascular remodelling and in human atherosclerotic lesions. Here we describe in vitro and in vivo models to investigate the function of furin/PC5 in VSMCs and in vascular lesion formation. In conjunction with the development of novel PC inhibitors, this should facilitate the development of new strategies targeting PCs in cardiovascular disease.

Key words: Vascular smooth muscle cells, proprotein convertases, furin, atherosclerosis, laser capture microdissection.

1. Introduction

Changes in risk factor prevalence and medical breakthroughs, including thrombolysis, surgery, and percutaneous coronary interventions (PCIs), as well as pharmacological approaches, such as enzyme inhibitors and statins, have contributed to a recent decline in cardiovascular mortality (1). Despite these achievements, however, coronary heart disease is anticipated to remain the number one killer in the Western world over the next decades (2).

M. Mbikay, N.G. Seidah (eds.), *Proprotein Convertases*, Methods in Molecular Biology 768,
DOI 10.1007/978-1-61779-204-5_9,

Formation and progression of atherosclerotic lesions involves the complex interaction of growth factors, cytokines and chemokines with various vascular and non-vascular cell types, including endothelial cells, mononuclear inflammatory cells (MNCs), thrombocytes and vascular smooth muscle cells (VSMCs) (3). All of these cells have been found to express pro-protein convertases (PCs) in vitro (4–7). Furthermore, PCs such as furin and PC5 are highly expressed in human atherosclerotic lesions (**Fig. 9.1**) (8). Thus, based on their capacity to activate pro-peptides of growth factors, chemokines, adhesion molecules or proteolytic enzymes via endoproteolytic cleavage at dibasic amino acid residues, furin and PC5 may well be suitable targets in coronary heart disease (8, 9).

Typically regarded as a chronic inflammatory disease, recent studies suggest that VSMCs are significant contributors to the initiation and early progression of atherosclerosis as well (10). Furthermore, VSMC proliferation and migration into the neointima are the key aspects in clinical conditions such as restenosis and vein bypass graft failure (11). However, because direct cellular data are difficult to obtain from human studies, our understanding of VSMC responses to vascular injury is in particular derived from in vivo animal models and in vitro assays.

Accordingly, studies have demonstrated that furin and PC5 are upregulated in VSMCs following balloon injury in rodents

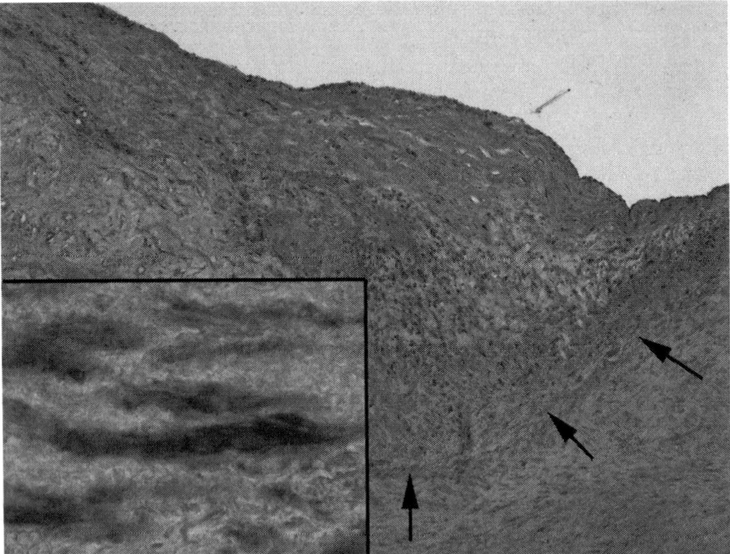

Fig. 9.1. Immunohistochemical staining of furin in an advanced human atherosclerotic lesion (femoral artery). *Arrows* denote the internal elastic lamina. Furin is found in vascular smooth muscle cells and mononuclear inflammatory cells in the lesion. The *inset* demonstrates cytoplasmic expression of furin in spindle-shaped myofibroblasts. Nuclei are stained with haematoxylin.

in vivo (5, 12). Gene expression studies demonstrated the co-regulation and colocalization of these PCs with their substrates αvβ3 integrin and membrane-bound matrix metalloproteinase-1 (MT1-MMP), both in animal models and in advanced human vascular lesions (8, 12–14). Complementary, in vitro assays identified the activation of pro-integrins and pro-MMPs by furin/PC5 in VSMCs, thereby controlling essential cell functions for vascular remodelling, including adhesion, migration and invasion (6, 13, 15, 16).

This chapter describes basic methods to study PCs in VSMC functions, namely primary rat VSMC migration, and in vivo models, in combination with in situ gene expression profiling. We believe that the complementary use of the described methods is well suited for assaying/ monitoring the critical role of PCs in the vasculature and should pave the way for identifying and establishing PCs as novel drug targets in vascular disease.

2. Materials

2.1. Migration In Vitro Assay

1. Dilute gelatin in sterile PBS (with calcium/magnesium) to a concentration of 0.2%.
2. Dilute bovine serum albumin (fraction V, for instance, from Sigma Aldrich, St. Louis, Missouri, USA) in sterile PBS (with calcium/magnesium) to a concentration of 1%.

2.2. Animals and Materials

1. Male Sprague–Dawley rats, weighing 350–430 g at day of intervention.
2. Topical anti-septic/ bactericide agent.
3. Absolute alcohol.
4. Anaesthetics (e.g. isoflurane).
5. Pre-warmed sterile PBS/H_2O.
6. Sterile scissors, needles, syringes, restrainers; gauze, cotton-tipped applicators ("Q-tips"), surgical sutures.
7. Rodent operating table/ surgery platform with heating source.
8. Fogarty balloon embolectomy catheters, 2 French (Edwards Lifesciences Corp., Irvine, CA, USA).
9. Operation lamps and microscope.

2.3. Immunohistochemistry

1. Gelatin solution (1 l): 5 g gelatin, 0.3 g chromium potassium sulphate and 0.2 g Na azide.

2. Antibody diluent: 10 ml Goat serum (for primary antibodies derived from goat, use a different species), 10 ml PBS (10×), 0.3 ml Triton X-100; add H_2O to a final volume of 100 ml. Prepare aliquots of 10 ml and store at –20°C.
3. Mounting medium: For permanent storage, medium is provided by commercial sources (e.g. Fisher Scientific Permount). Temporal covering of stained slides can be done with glycerol solution (90 ml glycerol, 10 ml of 1× PBS).
4. 10× Phosphate buffered saline (PBS): 80 g NaCl, 2 g KCl, 14.2 g $Na_2HPO_4{\bullet}2H_2O$, 2 g KH_2PO_4 and 1,000 ml dist. H_2O. Throughout the protocol, 1× PBS is used. Check pH to be 7.2.
5. Citrate buffer (0.01 M, pH 6.0): 2.1 g Citric acid monohydrate ($C_6H_8O_7{\bullet}H_2O$), 900 ml dist. H_2O, then adjust pH with 2 M NaOH, add dist. H_2O to 1 l.
6. Tris buffer for alkaline phosphatase kit: 100 mM Tris, pH 8.2–8.5 (for blue colour; other colour may require a different pH, consult manufacturer's manual). The pH needs to be monitored from time to time.

2.4. Laser Capture Microdissection

1. Cryomolds.
2. Liquid nitrogen.
3. Commercial Tissue-Tek O.C.T. compound (for instance, Sakura Finetek, Zoeterwoude, The Netherlands).
4. SuperFrost Plus charged glass slides (Menzel Gläser, Braunschweig, Germany).
5. RNasin (for instance, Promega UK, Southampton, UK).
6. RNase Zap (for instance, Sigma, St. Louis, MO, USA).
7. RNase-free water (for instance, Ambion, Austin, TX, USA).
8. Zincfix buffer: 5 g $ZnCl_2$, 6 g $ZnAc_2{\bullet}2H_2O$, 0.1 g $CaAc_2$ in 1 l of 0.1 M Tris–HCl, pH 7.4. Make fresh if required. Buffer is stable at room temperature for 1 week.
9. 50-ml Tubes (for instance, Falcon).
10. Ethanol (70, 95, 100%).
11. Xylene.
12. Arcturus laser capture microdissection device (Arcturus Bioscience, Mountain View, USA; from MDS Analytical Technologies).
13. Arcturus PicoPure^{TM} RNA Isolation Kit (Arcturus Bioscience).
14. Agilent 2100 Bioanalyzer (Agilent Biotechnologies, Palo Alto, CA, USA).

15. Oligo-dT primer or random primer, HPRT primer, and target gene primer.
16. SuperScript II kit (Invitrogen GmbH, Karlsruhe, Germany).
17. Power SYBR® Green PCR Master Mix (Applied Biosystems, Inc., Foster City, CA, USA).
18. Cycler: for instance, ABI PRISM 7500HT RT-PCR cycler (Applied Biosystems) or the Mx3000P™ QPCR System (Stratagene, Agilent Technologies, La Jolla, CA).

3. Methods

3.1. Carotid Artery Balloon Injury Model

The rat carotid balloon injury model, originally described by Clowes and colleagues (17), represents a well-characterized and commonly used in vivo model for studying the arterial response on an experimentally mechanically induced injury. Determined parameters in this vascular remodelling process include changes in the morphological architecture of the vessel, as well as biochemical, molecular and cellular modifications in the composition of the vessel wall. This in vivo model is characterized by a rapid tissue remodelling process, including development of a newly formed neointima, which is driven by proliferation and migration of VSMCs, and a phenotypic switch of VSMCs from a contractile to a synthetic phenotype. In vivo assays on the arterial response to vessel injury is of crucial relevance in the understanding of the pathological development and progression of vascular disorders. Even though the rat carotid artery balloon injury model comprises features of a restenotic remodelling process in humans, terming it "neointima formation assay" is – strictly speaking – more accurate. Nonetheless, this tissue remodelling assay allows investigation of several crucial pathophysiological processes and serves as a valuable and helpful tool for verifying results from in vitro assays ("proof of concept"). In the following description, an introduction of the rat carotid balloon injury model is given. For additional information, the reader may also consult the recent paper by Tulis (18):

1. Anaesthetize animals by inhalative isoflurane administration throughout the operation.
2. Place the animal on the operation table.
3. Remove the ventral hair of the rat and sterilize the skin. Make a straight incision below chin until the sternum.
4. Gently separate the skin from the underlying tissue.
5. Use a skin retractor to keep the skin out of the way.

6. Separate muscular tissues.
7. Dissect alongside the left carotid artery distally towards the head. Expose the left common carotid artery bifurcation into its internal and external branches. Expose the common carotid artery in its entire length of the skin incision down to the sternum.
8. Moisten all exposed tissues during the surgery.
9. After clamping and ligation of several (smaller) arteries (e.g. the ascending pharyngeal, occipital and/or superior thyroid arteries) (*see* details by David A. Tulis (18)), arteriotomize the external carotid artery and gently insert a 2-F embolectomy catheter.
10. Gently advance the uninflated catheter to the aortic arch.
11. Inflate the catheter with a pre-determined volume and pressure.
12. Retract the catheter under constant rotation to the arteriotomy site (this induces endothelial denudation = injury). Deflate the catheter. Push the catheter forward to the aortic arch, inflate, and repeat injury process four additional times.
13. Place a suture around the external carotid artery, completely retract catheter and tie a knot at the arteriotomy site. Perform surgical wound closure.
14. Perform general post-operative care.
15. After defined times *post operationem* (for instance, 0, 2, 7 and 14 days), sacrifice the rats under anaesthesia, extract the common carotid arteries (injured and contralateral control non-injured sites) and fragment them (*see* **Note 1**).

3.2. Immunohistochemistry

Immunohistochemistry allows the detection of PCs and their substrates in vivo, in either human specimens **(Fig. 9.1)** or animal models **(Fig. 9.2)**. Even though not very suitable for quantitative analyses, it is an invaluable tool to investigate protein expression of targets and, thus, provides important support for the biological significance of in vitro findings (e.g. cell culture or enzyme assays).

Colocalization studies are best done combining a polyclonal antibody for one target and a monoclonal antibody for the other. Antibodies are then detected by different methods/enzymes, e.g. one staining is done with HRP-SA and the other with alkaline phosphatase (AP). It is best to use the polyclonal antibody with HRP-SA first.

The following protocol is for paraffin-embedded samples:

1. Deparaffinize sample sections (4 μm thickness) on gelatin precoated slides in xylene (3 × 15 min).

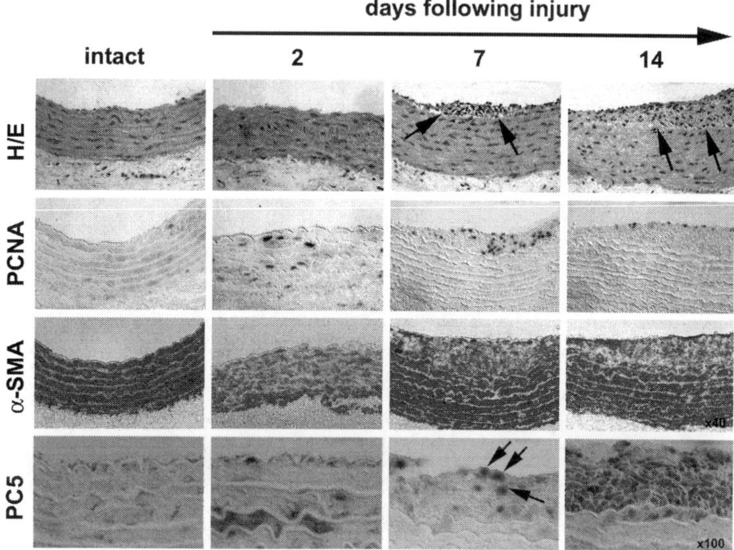

Fig. 9.2. Comparative analysis of proliferating cell nuclear antigen (PCNA), α-smooth muscle actin (α-SMA), and PC5 expression following vascular injury of the rodent artery. The *upper row* depicts haematoxylin/eosin (H/E) staining. Following balloon injury, from day 7 on, formation of a neointima is found (*arrows* denote the internal elastic lamina). Whereas little PCNA staining is found in intact vessels, cell proliferation is evident on day 2 following balloon injury. On days 7 and 14, proliferating, PCNA-positive cells are almost exclusively found in the neointima. Neointima formation is accompanied by a decrease in α-SMA content. This indicates a phenotypic shift, from the "contractile" resting cell to a "proliferating" secretory one. VSMC proliferation and phenotypic differentiation are characterized by increases in proprotein convertase PC5 expression. PC5 co-localizes with PCNA in neointima VSMCs (*bottom row*, *arrows* denote PCNA-positive nuclei, PC5 displays cytoplasmic staining).

2. Rehydrate sections in a graded series of ethanol 100–75% (15 min each), followed by PBS/0.1% Triton X-100 (2 × 15 min).
3. Immerse sections in a 3% H_2O_2/methanol solution to quench endogenous peroxidases.
4. Wash sections in PBS/0.1% Triton X-100.
5. Rinse sections in distilled H_2O and soak them then in 0.01 M citrate buffer, pH 6.
6. Microwave sections at "high" for 2 × 2 min (refill buffer in-between). Let the sections cool down in the citrate buffer (*see* **Note 2**).
7. Block background staining by incubation with non-immune serum (e.g. 1:20 in PBS) for 30 min (*see* **Note 3**).
8. Wash the sections in PBS/0.1% Triton X-100 (3 × 5 min).
9. Incubate the sections with the primary antibody in antibody diluent, typically overnight at 4°C in a humidified chamber box (*see* **Note 4**).

10. Wash the sections in PBS/0.1% Triton X-100 (3 × 15 min).
11. Retrieve the primary antibody and incubate the sections with a biotinylated secondary antibody to react with the primary antibody (15 min at room temperature) (*see* **Note 5**).
12. Wash the slides in PBS/0.1% Triton X-100 (3 × 5 min).
13. Link horseradish peroxidase-conjugated streptavidin (HRP-SA) to the secondary antibody (10 min incubation).
14. Wash the slides with PBS/0.1% Triton X-100 (3 × 5 min).
15. Visualize antibody/enzyme complex with a chromogen, such as 3,3′-diaminobenzidine (DAB) which deposits a specific brown staining in the presence of horseradish peroxidase (HPR). Monitor staining by microscopy (*see* **Note 6**).
16. Stop reactions by washing in PBS (3 × 5 min).
17. Counter-stain samples with haematoxylin and cover permanently; or use them for colocalization studies (do not counter-stain at this step in colocalization studies; *see* Steps 18–25 below).

Colocalization

18. Following the steps for the first (polyclonal) antibody outlined above (Steps 1–17), wash sections in 100 mM Tris, pH 8.2 (2 × 10 min).
19. Incubate with the next primary (monoclonal) antibody overnight at 4°C (*see* **Note 7**).
20. Wash sections in 100 mM Tris, pH 8.2.
21. Incubate sections with an anti-mouse AP-conjugated antibody at 37°C for 1 h.
22. Wash sections in Tris buffer.
23. Use an AP kit according to the instructions of the manufacturer (e.g. Vector Blue Alkaline Phosphatase Kit; Vector Laboratories). Monitor the reaction by microscopy.
24. Stop reaction by washing in PBS.
25. Counter-stain the sections with haematoxylin.
26. Briefly dip slides in 70% ethanol and cover them with glycerol (temporarily) or commercially available mounting medium permanently (*see* **Note 8**).

3.3. Vascular Smooth Muscle Cell Migration Assay

The transwell migration assay is suitable for evaluating chemokine/growth factor-directed migration of various cell types, including VSMCs. Depending on the scientific question, filters may be coated with different matrix types (e.g. collagens, fibronectin, vitronectin or laminin) and cells are being pretreated with inhibitors or stimulates. In a first approach of generally assaying chemotactic responses of VSMCs, the use of gelatin (heat-denatured collagen) has been well proven:

1. Insert 8-μm micropore transwell chambers in a 24-well plate and add 200 and 700 μl of gelatin to the inner and outer chambers, respectively. Incubate either for 24 h at 4°C or for 2 h at 37°C.
2. Gently remove gelatin (e.g. with a pipette tip connected to a water pump). Block unspecific binding sites in the chamber by adding 200 and 700 μl of 1% BSA to the inner and outer chambers, respectively. Incubate for 1 h at 37°C.
3. During blocking time, trypsinize rat primary pre-confluent VSMCs (80–90% confluent, serum starved at 0.4% FCS for 16 h) by the use of 1× trypsin for 2–10 min.
4. Gently detach cells during trypsinization by mechanical tapping of the incubation plate.
5. After detachment of the majority of cells, stop trypsinization by adding 10% FCS containing DMEM.
6. Collect media (e.g. in a 15-ml Falcon tube) and spin cells down for 5 min in a pre-warmed centrifuge (37°C).
7. Discard the supernatant and wash cells with 0.4% FCS containing DMEM. Spin cells down for 5 min in a centrifuge at 37°C.
8. Resuspend cells in ~ 1.0–2.0 ml of 0.4% FCS containing DMEM and count cells (for instance, in a Neubauer chamber) (*see* **Note 9**).
9. Dilute cells in 0.4% FCS containing DMEM to a concentration of 250 VSMCs/μl.
10. Add 700 μl of 0.4% FCS containing DMEM to the outer chamber and then add 200 μl of cell suspension (= 50,000 cells) to the inner chamber.
11. Incubate under standard culture conditions (37°C, 5% CO_2, 95% humidity) for 1 h. This will allow the VSMCs to fully attach to the matrix without interfering with this adhesion/attachment process due to any chemotactic compound in the outer chamber.
12. Discard the media of the outer chamber by gentle suction and add chemotactic media (e.g. 10% FCS containing DMEM or 0.4% FCS containing DMEM with 10 ng/ml PDGF-BB) to the outer chamber.
13. Incubate chambers under culture conditions (37°C, 5% CO_2, 95% humidity) for 3 or 4 h (*see* **Note 10**).

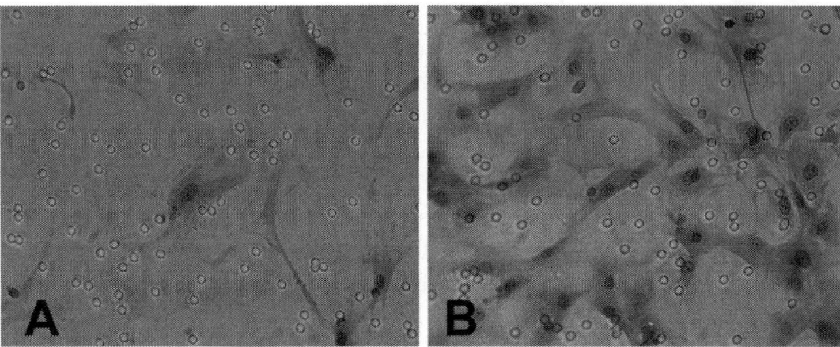

Fig. 9.3. Transwell vascular smooth muscle cell (VSMC) migration. VSMCs were allowed to adhere for 1 h and then subjected to PDGF-BB (10 ng/ml)-directed migration through gelatin-coated filters (8 μm pores) for 4 h. (**a**) Control cells are shown in which the lower chamber contains serum-depleted cell culture medium (0.4% FCS). (**b**) Migrated cells below the filter are demonstrated. Cells are stained with Mayer's reagent.

14. Gently discard the media in both the inner and outer chambers (not too many wells at the same time to avoid drying out). Then add 200 μl of 1× PBS to the inner chamber and immediately discard the fluid again. This will allow apoptotic cells and cell debris to be removed.
15. Add 200 μl of 1× PBS to the inner chamber and mechanically clean the inner chamber (= upper surface of the migration filter) by using a cotton Q-tip. Discard PBS and repeat washing with PBS twice.
16. Discard PBS and add 200 and 700 μl of ice-cold 100% methanol (to be kept at –20°C) to the inner and outer chambers, respectively, to fix cells. Incubate at –20°C for at least 30 min and up to 4 days. Wash inner and outer chambers three times with sterile water (200 μl inner and 700 μl outer chambers, respectively).
17. Stain cells by adding 500 μl haematoxylin or Mayer's reagent for 30 min.
18. Wash inner and outer chambers three times with sterile water (200 μl inner and 700 μl outer chambers, respectively) until no further blue colour is detected in discarded fluid (*see* **Note 11**).
19. Detect migrated cells in migration pores and count at least four different high-power fields per filter (**Fig. 9.3**).

3.4. Laser Capture Microdissection of Vascular Tissue

Computer-based laser capture microdissection (LCM) of tissues has enabled isolation and collection of morphologically defined cell populations. The Arcturus system provides a viable platform for microdissection of a variety of different cell types and tissues. The procedure is rather simple and, if RNase-free conditions are accurately followed, robust. In general, the LCM method

Laser Capture Microdissection

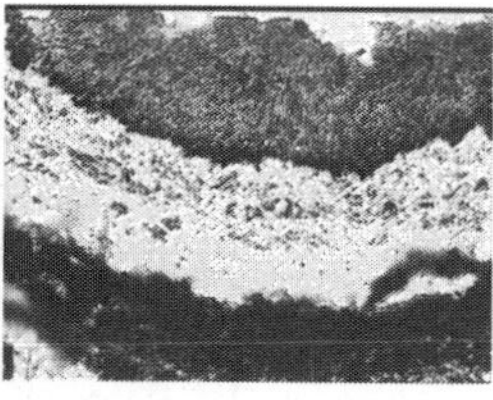
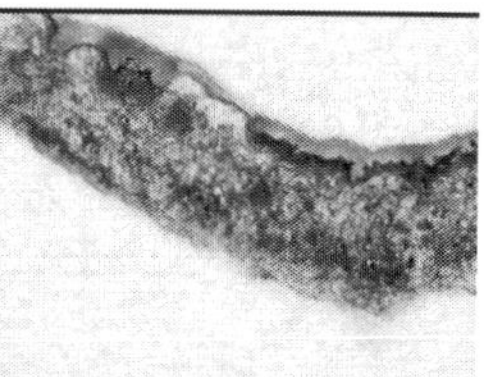

Fig. 9.4. Layer-specific laser capture microdissection (LCM) of a rat restenotic lesion. The panel depicts the specific isolation of the tunica media (*right*) through LCM (*left* and *middle*). Microdissection was performed on 10-μm-thick cryosections derived from snap-frozen vessel tissue in cryomedium (Tissue-Tek® O.C.T. Compound). (Reproduced from (20) with permission from The Federation of American Societies for Experimental Biology.)

is a process to capture biomolecules (such as RNA), for later quantitative amplification-based analyses. An in-depth outline to LCM methods is beyond the scope of this chapter, but here we provide details on how high-quality RNA is obtained from vessel tissues for in situ PC gene expression profiling of restenotic vascular lesions (**Fig. 9.4**). For further details, the reader may also consult other recent articles on LCM (19).

After sacrificing balloon-injured Sprague–Dawley rats, the vessel of interest (e.g. the injured common carotid artery with or without neointima formation) is surgically removed, and repetitively flushed with sterile PBS for removal of remaining excessive intravascular blood.

3.4.1. Preparation of Vascular Tissues

1. Place vessel in cryomedium in cryomolds (smallest size as possible, this will simplify later sectioning). The vessel should be placed in a vertical manner for appropriate later cutting (cryo-sectioning) of the tissue.
2. Place cryomolds in liquid nitrogen for approximately 20 s, until no further gas is seen to be released from the tissue.
3. After this snap-frozen in liquid nitrogen, store the frozen tissue at –70°C until further processing.
4. Cut frozen tissue into 10-μm-thick cryosections and mount them on SuperFrost Plus charged glass slides (*see* **Note 12**).
5. Immediately transfer slides to a –70°C freezer and store them at –70°C until further use.

3.4.2. Staining of Vascular Tissues for Laser Capture Microdissection

1. Add 50 μl haematoxylin mixed with 1 μl RNasin (5,000 U/l) on the tissue section. This will facilitate morphological differentiation. Incubate for 60 s.
2. Incubate slide in a cuvette (or a 50-ml Falcon tube) filled with Zincfix buffer for 60 s.

3. Incubate slide in a cuvette (or a 50-ml Falcon tube) in 70% ethanol for 60 s.
4. Incubate slide in a cuvette (or a 50-ml Falcon tube) in 95% ethanol for 60 s.
5. Incubate slide in a cuvette (or a 50-ml Falcon tube) in 100% ethanol for 60 s.
6. Incubate slide in a cuvette (or a 50-ml Falcon tube) in xylene for 60 s, followed by air-drying under a hood.

3.4.3. Laser Capture Microdissection

1. Microdissect vessel areas of interest following the manufacturer's protocol. Several similar vascular areas (for instance, neointima area) can be microdissected to the same Arcturus CapSure Cap; this will allow simplified pooling of the material.
2. Pool material to enable isolation of a sufficient number of cells of interest. In our experience, a cell number of >1,000 cells is necessary for valid in situ gene expression profiling using high-quality RNA. An example of tunica media isolation from rat restenotic carotid lesions is depicted in **Fig. 9.4**.

3.4.4. RNA Isolation and Complementary DNA Synthesis from Microdissected Tissue

1. Isolate RNA using the Arcturus PicoPure™ RNA Isolation Kit. Elute the RNA in the final isolation step in 11.5 μl and transfer 1.5 μl for further analyses of RNA quality (for instance, use the Agilent 2100 Bioanalyzer and an RNA 6000 Nano Assay Kit) (*see* **Note 13**).
2. For complementary DNA (cDNA) synthesis, use the SuperScript II reverse transcriptase kit (Invitrogen, as recommended by the manufacturer). Input of the whole RNA amount (10 μl) is recommended. Either use oligo-dT primers (5′-AAA CGA CGG CCA GTG AAT TGT AAT ACG ACT CAC TAT AGG CGC TTT TTT TTT TTT TTT-3′) or random primers (hexamers) at 1 μg/μl for amplifications in the 20 μl reaction volume.

3.4.5. Quantitative Reverse Transcriptase Polymerase Chain Reaction (RT-PCR) (See ***Note 14****)*

1. Add 1 μl (5%) of cDNA to a 20 μl reaction volume for a 40 (-50)-cycle RT-PCR assay using the following primers: CTC ATG GAC TGA TAT GGA CAG GAC (forward; 100 nM), and GCA GGT CAG CAA AGA ACT TAT AGC C (reverse; 100 nM, gene = hypoxanthine-guanine phosphoribosyltransferase, HPRT).
2. Perform the reaction (in duplicate at the least) using, for instance, either the ABI PRISM 7500HT RT-PCR Cycler or the Stratagene Mx3000P™ QPCR System under conditions recommended by the manufacturer.
3. Quantify HPRT levels using the ΔΔCT method (20), followed by adjustment for cDNA amount.

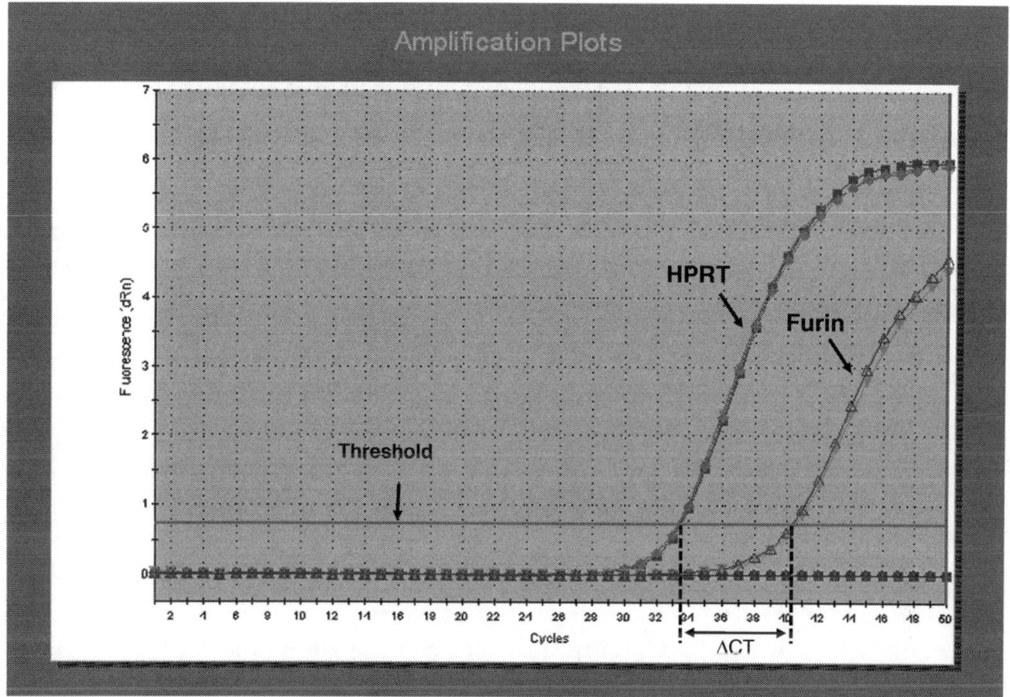

Fig. 9.5. Quantitative real-time RT-PCR of cDNA derived from microdissected vascular tissue. Amplification curves represent single wells in a 96-well format, with each microdissected material being amplified in triplicate per primer pair. The example shows microdissected tunica media 8 days after balloon injury. Amplification curves show the housekeeping gene *HPRT* and furin. According to the $\Delta\Delta$CT method (20), comparisons and ratios to other microdissected material can be performed. As negative control (H_2O) a sample of water should be processed the exact same way as the tissue, resulting in no detectable amplificates. Prior to gene expression analyses, a detailed primer optimization including various dilutions of cDNA (for instance, 1, 1:4, 1:16, 1:64), a minus RT (–RT) sample, a water control and different primer concentrations (range 50–200 nM) is highly recommended. The cDNA dilutions are required to show amplification curves with CT differences of 2.0 and dissociation curves at anticipated temperature. Water and –RT samples should not show amplification curves.

4. Perform RT-PCR using HPRT primers and target gene primers, such as – for instance – rat furin CCA GCA CAT CTC AAC GCT AA (forward; 100 nM) and GGC CAC TGT TGT CCA GTT CT (reverse; 100 nM).

An example of real-time RT-PCR amplification of microdissected material is depicted in **Fig. 9.5**.

4. Notes

1. Fragmentation of the arteries into several (e.g. 3–5) parts will allow processing to desired applications, such as for further total tissue protein isolation, total tissue RNA isolation, laser capture microdissection, and immunohistochemistry.

2. Formalin/aldehyde fixation of tissue may lead to protein cross-links. Microwave treatment in citrate buffer breaks them, thereby unmasking epitopes in formalin-fixed, paraffin-embedded sections.
3. "Ready-to-use" blocking solutions are also provided with the ABC kit.
4. Volumes should be between 50 and 100 μl (depending on the sample size) to prevent drying. The concentration and time of the primary antibody needs to be determined by the investigator. Take care that sections do not dry throughout the complete assay, because this will result in non-specific staining. Drying can be further prevented by coverage of samples with parafilm.
5. Antibody retrieval is typically conducted with an "ABC kit" from commercial sources according to the manufacturer's manual (e.g. Histostain-Plus Kit; Zymed Laboratories).
6. Development of staining should be monitored by microscopy and may vary according to the expression of the target. Usually, an incubation time beyond 15–30 min is not recommended (either because of "over-staining" and that no signal will develop beyond this time).
7. Depending on the tissue used, endogenous AP may be blocked by 0.1 M levamisole (in Tris buffer) incubation prior to the use of the AP kit.
8. Avoid xylene-based clearing agents/mounting media when using AP as enzyme.
9. In case of a large proportion of dead cells and cell debris, an appropriate (e.g. 1:20) dilution of cells with trypan blue is recommended.
10. The appropriate chemotaxis time needs to be determined in the individual laboratory and under individual conditions. In our experience, 4 h incubation time is best to induce migration of VSMCs, and cells will "stuck" in the filter pores after this time. After this time, a large proportion of cells have indeed *trans*-migrated and are found at the lower surface of the porous migration filter.
11. Transferring the migration filters to a new 24-well plate is recommended. Add sterile water to inner and outer chambers to avoid drying out.
12. Repetitive mounting of several sections to one slide is recommended for decreasing number of processed slides, but keeping the slide at a temperature well below 0°C during cryo-sectioning is highly recommend for generating high-quality RNA.

13. We highly recommend further processing of RNA to subsequent DNA synthesis and gene expression profiling only if *good* RNA quality is documented. This accounts in particular when RNA samples in a nanogram range are processed.
14. For validation of the RNA isolation and cDNA synthesis procedures, and for testing of cDNA suitability for RT-PCR-based analyses, primers for the housekeeping gene rat hypoxanthine–guanine phosphoribosyltransferase (HPRT; Ensembl gene ID: ENSRNOG00000031367), previously established as valid housekeeper in rat tissue (21), are used in SYBR Green® real-time RT-PCR.

References

1. Ford, E. S., Ajani, U. A., Croft, J. B., Critchley, J. A., Labarthe, D. R., Kottke, T. E., Giles, W. H., and Capewell, S. (2007) Explaining the decrease in US deaths from coronary disease, 1980–2000 *N Engl J Med* **356**, 2388–98.
2. Lloyd-Jones, D., Adams, R., Carnethon, M., De Simone, G., Ferguson, T. B., Flegal, K., Ford, E., Furie, K., Go, A., Greenlund, K., Haase, N., Hailpern, S., Ho, M., Howard, V., Kissela, B., Kittner, S., Lackland, D., Lisabeth, L., Marelli, A., McDermott, M., Meigs, J., Mozaffarian, D., Nichol, G., O'Donnell, C., Roger, V., Rosamond, W., Sacco, R., Sorlie, P., Stafford, R., Steinberger, J., Thom, T., Wasserthiel-Smoller, S., Wong, N., Wylie-Rosett, J., and Hong, Y. (2009) Heart disease and stroke statistics—2009 update: A report from the American heart association statistics committee and stroke statistics subcommittee *Circulation* **119**, e21–181.
3. Libby, P., and Theroux, P. (2005) Pathophysiology of coronary artery disease *Circulation* **111**, 3481–8.
4. Leblond, J., Laprise, M. H., Gaudreau, S., Grondin, F., Kisiel, W., and Dubois, C. M. (2006) The serpin proteinase inhibitor 8: An endogenous furin inhibitor released from human platelets *Thromb Haemost* **95**, 243–52.
5. Stawowy, P., Blaschke, F., Kilimnik, A., Goetze, S., Kallisch, H., Chretien, M., Marcinkiewicz, M., Fleck, E., and Graf, K. (2002) Proprotein convertase PC5 regulation by PDGF-BB involves PI3-kinase/p70(s6)-kinase activation in vascular smooth muscle cells *Hypertension* **39**, 399–404.
6. Stawowy, P., Meyborg, H., Stibenz, D., Borges Pereira Stawowy, N., Roser, M., Thanabalasingam, U., Veinot, J. P., Chretien, M., Seidah, N. G., Fleck, E., and Graf, K. (2005) Furin-like proprotein convertases are central regulators of the membrane type matrix metalloproteinase-pro-matrix metalloproteinase-2 proteolytic cascade in atherosclerosis *Circulation* **111**, 2820–7.
7. Varshavsky, A., Kessler, O., Abramovitch, S., Kigel, B., Zaffryar, S., Akiri, G., and Neufeld, G. (2008) Semaphorin-3B is an angiogenesis inhibitor that is inactivated by furin-like proprotein convertases *Cancer Res* **68**, 6922–31.
8. Stawowy, P., Kallisch, H., Borges Pereira Stawowy, N., Stibenz, D., Veinot, J. P., Grafe, M., Seidah, N. G., Chretien, M., Fleck, E., and Graf, K. (2005) Immunohistochemical localization of subtilisin/kexin-like proprotein convertases in human atherosclerosis *Virchows Arch* **446**, 351–9.
9. Chretien, M., Seidah, N. G., Basak, A., and Mbikay, M. (2008) Proprotein convertases as therapeutic targets *Expert Opin Ther Targets* **12**, 1289–300.
10. Doran, A. C., Meller, N., and McNamara, C. A. (2008) Role of smooth muscle cells in the initiation and early progression of atherosclerosis *Arterioscler Thromb Vasc Biol* **28**, 812–19.
11. Dzau, V. J., Braun-Dullaeus, R. C., and Sedding, D. G. (2002) Vascular proliferation and atherosclerosis: New perspectives and therapeutic strategies *Nat Med* **8**, 1249–56.
12. Stawowy, P., Graf, K., Goetze, S., Roser, M., Chretien, M., Seidah, N. G., Fleck, E., and Marcinkiewicz, M. (2003) Coordinated regulation and colocalization of alphav integrin and its activating enzyme proprotein convertase PC5 in vivo *Histochem Cell Biol* **119**, 239–45.
13. Stawowy, P., Kallisch, H., Veinot, J. P., Kilimnik, A., Prichett, W., Goetze, S., Seidah,

N. G., Chretien, M., Fleck, E., and Graf, K. (2004) Endoproteolytic activation of alpha(v) integrin by proprotein convertase PC5 is required for vascular smooth muscle cell adhesion to vitronectin and integrin-dependent signaling *Circulation* **109**, 770–6.

14. Stawowy, P., and Fleck, E. (2005) Proprotein convertases furin and PC5: Targeting atherosclerosis and restenosis at multiple levels *J Mol Med* **83**, 865–75.
15. Kappert, K., Furundzija, V., Fritzsche, J., Margeta, C., Kruger, J., Meyborg, H., Fleck, E., and Stawowy, P. (2010) Integrin cleavage regulates bidirectional signalling in vascular smooth muscle cells *Thromb Haemost* **103**, 556–63.
16. Kappert, K., Meyborg, H., Baumann, B., Furundzija, V., Kaufmann, J., Graf, K., Stibenz, D., Fleck, E., and Stawowy, P. (2009) Integrin cleavage facilitates cell surface-associated proteolysis required for vascular smooth muscle cell invasion *Int J Biochem Cell Biol* **41**, 1511–17.
17. Clowes, A. W., Reidy, M. A., and Clowes, M. M. (1983) Mechanisms of stenosis after arterial injury *Lab Invest* **49**, 208–15.
18. Tulis, D. A. (2007) Rat carotid artery balloon injury model *Methods Mol Med* **139**, 1–30.
19. Graeme, I. M., and Curran, S. (2005) *Laser Capture Microdissection: Methods and Protocols.* Clifton, NJ: Humana Press. 319 pp.
20. Schefe, J. H., Lehmann, K. E., Buschmann, I. R., Unger, T., and Funke-Kaiser, H. (2006) Quantitative real-time RT-PCR data analysis: Current concepts and the novel "gene expression's CT difference" formula *J Mol Med* **84**, 901–10.
21. Kappert, K., Paulsson, J., Sparwel, J., Leppanen, O., Hellberg, C., Ostman, A., and Micke, P. (2007) Dynamic changes in the expression of DEP-1 and other PDGF receptor-antagonizing PTPs during onset and termination of neointima formation *FASEB J* **21**, 523–34.

Chapter 10

Identification of the Myosin Heavy Polypeptide 9 as a Downstream Effector of the Proprotein Convertases in the Human Colon Carcinoma HT-29 Cells

Nathalie Scamuffa, Peter Metrakos, Fabien Calvo, and Abdel-Majid Khatib

Abstract

In addition to the large spectrum of the protein precursors processed and activated by the proprotein convertases (PCs) that are crucial for the maintenance of the malignant phenotype of colon cancer cells such as matrix metalloproteases, adhesion molecules, growth factors, and growth factor receptors, the PCs also regulate the expression and the activity of other proteins that are not PC substrates and involved in the acquisition of the metastatic and tumorigenic potential of these tumor cells. The identification in colon cancer cells of such proteins is thereby crucial for the understanding of the cascade of molecular events regulated by the PCs leading to tumorigenesis and metastasis and thus may constitute potential candidates for new colon cancer-specific targets and/or biomarkers. Using the human colon cancer cells HT-29 and ProteinChip arrays analysis that apply the surface-enhanced laser desorption ionization time-of-flight mass spectrometry (SELDI-TOF-MS), we identified the myosin heavy polypeptide 9 as new downstream effector of PCs in these cells. This protein was reported to be involved in the processes of malignant epithelial transformation and its role in colon cancer is unknown.

Key words: Proteinchip array SELDI-TOF MS, proprotein convertases, PCs, HT-29 cells, α1-PDX.

1. Introduction

A large majority of secretory proteins are synthesized as inactive precursors that are converted to their bioactive forms by one or more of the seven known dibasic mammalian subtilisin/kexin-like proprotein convertases (PCs) family (1–6). Precursors are usually cleaved at the general motif K/R-X-K/R↓. While the convertases PC1 and PC2 are found within secretory granules and

M. Mbikay, N.G. Seidah (eds.), *Proprotein Convertases*, Methods in Molecular Biology 768,
DOI 10.1007/978-1-61779-204-5_10, © Springer Science+Business Media, LLC 2011

process precursors therein, the convertases Furin, PC5-B, and PC7 are the only members of the mammalian PCs with trans-membrane domains and together with PC5-A and PACE4 are the main enzymes that process precursors sorted to the constitutive secretory pathway (1–8). Among the PCs substrates, matrix metalloproteases (MMPs), growth factors, growth factor receptors, and adhesion molecules have been identified to have significant roles during the development of the advanced malignant phenotype and the acquisition of the metastatic potential of various tumor cells including colon cancer (9–13). These include the metalloproteinases stromelysin-3 (str-3), membrane-type MMPs (MT-MMPs), the adamalysin metalloproteinases (ADAMs), and the adamalysin metalloproteinases with thrombospondin motifs (ADAM-TS) of which the expression and the activity have been correlated with increased local aggressiveness, metastasis, and poor clinical outcome all were found to contain the amino acid motif recognized by the PCs (1–13).

The most promising protein-based-specific inhibitors of the PCs is α1-antitrypsin variant known as α1-antitrypsin Portland or α1-PDX that was generated by site-directed mutagenesis to contain the PC site K/R-X-K/R↓ (1, 4, 8, 12). α1-PDX was first showed to be a potent inhibitor of Furin-mediated cleavage of HIV gp160, but subsequently demonstrated to also inhibit all PCs involved in processing within the constitutive secretory pathway (1, 4, 8, 12). Previously we demonstrated that PC inhibition by this inhibitor dramatically affects the malignant phenotype of the human colon carcinoma cell line HT-29, which was associated with a blockade in the processing and the function of IGF-1 receptor and MT1-MMP (5, 12, 13). The objective of this study is to identify in these cells other proteins (substrates or downstream mediators) regulated by the PCs and involved in their malignant phenotypes using the proteinChip technology (14).

The proteinChip system is based on established characteristics of the surface-enhanced laser desorption/ionization time-of-flight mass spectrometry (SLDI-TOF-MS) (14). It associates the capture of protein on specific surface with mass spectrometry. The arrays of proteinChip system present various affinity matrices similar to those of conventional chromatography media including reverse phase, metal affinity, and reverse phase. This technology was found suitable for proteins expression profiling, sequencing, purification, characterization, and identification. Usually protein analysis by proteinChip system consists of application of very small amount (1 μl) of crude sample such as fluids, cells extracts, or media, followed by buffer washes to remove the non-specific bounds. This step facilitates the processes of desorption and ionization of retained proteins in the ProteinChip reader and their mass spectrometry analysis. Retained proteins are eluted from the proteinChip array by laser desorption/ionization. Thereby the ionized proteins are detected and their mass accurately determined by time-of-flight mass spectrometry (14).

2. Materials

2.1. Cell Culture

1. Control and stably α1-PDX-transfected HT-29 (HT-29/PDX) human colon adenocarcinoma cells. Their characteristics were described previously (13).
2. Culture medium: Dulbecco modified Eagle medium supplemented with 10% fetal bovine serum (Invitrogen), 100 U/ml penicillin (Invitrogen), 100 mg/ml streptomycin (Invitrogen), and 200 μg/ml G418 (Invitrogen).
3. Cell pseudomonas exotoxin A (13): 1 μg/ml (Sigma). This toxin mediates cells death only after its cleavage by the PCs (15) and is used for the selection of cells that expressed the PC inhibitor α1-PDX (15).

2.2. Protein Identification

1. ProteinChip system/SELDI-TOF-MS (Ciphergen).
2. 4–20% acrylamide gradient gel (Invitrogen). This gel provides good separation over a wide range of protein sizes.
3. Coomassie blue R-250.
4. Trifluoroacetic acid (TFA).
5. HPLC-grade acetonitrile.
6. 100 ml of buffer A: 50% methanol and 10% acetic acid.
7. 100 ml of buffer B: 100 mM ammonium bicarbonate, pH 8.
8. 100 ml of buffer C: 50% acetonitrile and 0.1 M ammonium bicarbonate.
9. 12 ml buffer D: 100% acetonitrile.
10. 25 μg modified trypsin (Roche Molecular Biochemicals, Indianapolis, IN).
11. The reversed phase H4 proteinChip array (Ciphergen Biosystems).
12. All-in-One peptide standard (lyophilized) (Ciphergen Biosystems): This standard mix consists of five peptides ranging in molecular weight from 1,084.24 to 7,033.61 Da. To use, just reconstitute in buffer. This mix was optimized for use with NP2 and H4 proteinChip arrays. The use of other Chips arrays may lead to loss of one or more peaks due to retention by the array surface.

13. Saturated CHCA solution: Need is 20% CHCA (alpha-cyano-4-hydroxycinnamic acid): Dilution is made as follows: fivefold with 50% acetonitrile and 0.25% TFA (*see* **Note 1**).
14. Trypsin solution: The bovine pancreatic trypsin is provided as a lyophilized enzyme. The later is stable when stored under this form at 4°C. It loses its activity when exposed to repeated freeze-thaw procedures. For trypsin solution preparation, the lyophilized trypsin is dissolved in 125 μl of 10 mM HCl to make 10x trypsin solution (0.2 μg/μl). 5–10 μl aliquots of the 10× solution can be stored at –20°C or –80°C for up to 6 months. The active 1x trypsin is obtained by adding 45 μl of 25 mM ammonium bicarbonate to 5 μl of 10× trypsin solution.

3. Methods

3.1. Gel Separation and Digestion of Proteins

1. Lyse HT-29 and HT-29/PDX tumor cells in phosphate-buffered saline (PBS) containing 2% Nonidet P-40.
2. Subject lysates to SDS-PAGE gel electrophoresis (4–20% acrylamide gradient gels).
3. SDS-PAGE gel was stained with Coomassie blue R-250 (0.1% W/V) in 40% methanol and 10% acetic acid (*see* **Note 2**).
4. Localize on the gel the bands of which the expression was modified in HT-29/PDX cells. For example, in **Fig. 10.1**, the band that corresponds to a protein of about 228 kD was found to be downregulated in HT-29 cells following the expression of the PC inhibitor α1-PDX (HT-29/PDX).
5. Excise this band from the gel and transfer to 0.5 ml microcentrifuge tube. Also excise a portion from the side edge of the gel to serve as a negative control.
6. Prior to protein analysis and identification, destain the gels to remove the SDS by incubating in Solution A for 1 h and in Solution B for 10 min. Repeat these incubations two times by replacing the solution with a fresh one each time.
7. Incubate the gels in 0.4 ml of Solution C with agitation for 1 h.
8. Remove the buffer, add 50 μl of Solution D to gels pieces, and incubate with agitation for 15 min.
9. Remove the buffer and dry gel pieces before protein digestion by trypsin.

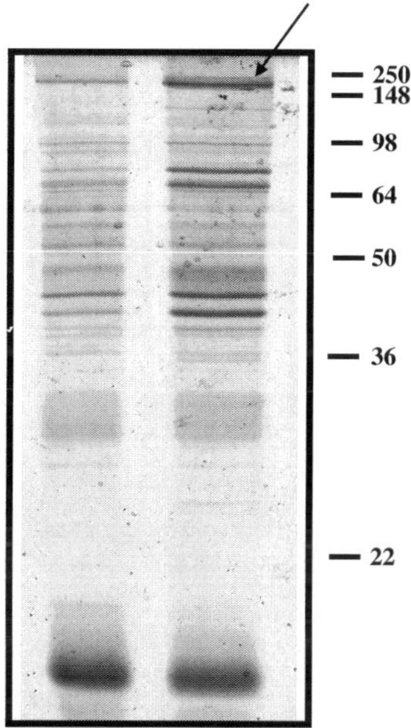

Fig. 10.1. SDS-PAGE gel of HT-29 and HT-29/PDX. Indicated is the protein that was excised before digestion with trypsin.

3.2. Protein Digestion by Trypsin

1. Add 10 μl of 1x trypsin to the dried gel pieces and incubate at room temperature during 15 min to rehydrate the gel.
2. Allow digestion to take place by incubating the gel pieces at 37°C for 16 h.
3. Centrifuge the tube content for 1 min at high speed.
4. Add 20 μl of 25 mM ammonium bicarbonate (pH 8) and allow the mixture to equilibrate for 1 h at room temperature.
5. Apply the trypsin digest (1–2 μl) to a spot on an H4 proteinChip array. Air-dry the latter for 10 min (*see* **Note 3**).
6. Add 0.5–1 μl of 10% saturated CHCA solution to each H4 proteinChip spot and allow them to dry prior to analysis or addition of MW standards (*see* **Section 3.4** for internal calibration paragraph).

3.3. Spot Analysis by Proteinchip Reader

1. Generate TOF mass spectra using the Ciphergen Protein Biology system II (PBS II).
2. Analyze the spectra with the Ciphergen software. When searching for peptides with size smaller than 5,000 MW, the time lag focus could be set at 2,000, the sensitivity at 4–6, and the intensity at 120. The laser was fired and the intensity was increased until we obtained 10–50% of major peaks in the 1–5 kDa range.

3. After collection of data on all spots, calibrate them all internally (*see* **Section 3.4**).
4. Establish the peptide maps of the trypsin-digested protein and of control spot (**Fig. 10.2**).
5. For background subtraction, label only peaks in the protein sample map that are not detected in the control map.
6. Collect the molecular weights of peptides derived only from the target protein and analyze them with ProFound software (*see* **Note 4**).

In our study, the *Z* score correspond to 2.33 (percentile: 99.0) and identify the myosin heavy polypeptide 9 as the protein affected by the expression of α1-PDX in HT-29 cells (HT-29/PDX) (**Figs. 10.3** and **10.4**). Note that MW 228.08 corresponds to the size of the excised band in **Fig. 10.1**. This protein was reported to be involved in the processes of malignant epithelial transformation. Its role in tumorigenesis and metastasis is poorly investigated (16).

3.4. Internal Calibration

The internal calibration is necessary for obtaining the best accurate mass of the peaks. This calibration is performed during analysis of unknown proteins. Add molecular mass markers directly to the unknown samples.

1. To determine the optimal calibrant concentration, apply a small amount of the trypsin digest to each of the spots on the H4 Chip and allow to dry.
2. In parallel, perform a series of three dilutions of the peptide calibrants in water and mix an aliquot of each dilution with equal volume of 10% saturated CHCA.
3. Add a small amount of the lowest concentration peptide calibrant solution to one of the sample spots.

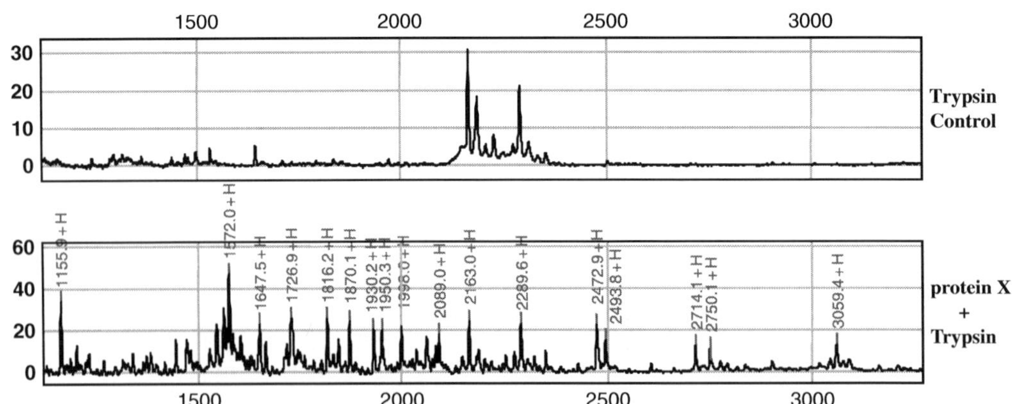

Fig. 10.2. Peptide map of the in-gel tryptic digestion of the 228 kD protein excised from gel in **Fig. 10.1**. Also peaks obtained by trypsin autolysis are indicated and used as a background signal.

ProFound - Search Result Summary

Version 4.105
The Rockefeller University Edition

Protein Candidates for search BDF3B8D1-0360-2FE4D59E [88967 sequences searched]

Rank	Probability	Est'd Z	Protein Information and Sequence Analyse Tools (T)	%	pI	kDa	®
+1	1.0e+000	2.33	T gi|12667788|ref|NP_002464.1| (NM_002473) myosin, heavy polypeptide 9, non-muscle [Homo sapiens]	9	5.5	228.08	®
+2	5.0e–008	-	T gi|16904208|gb|AAL30811.1|AF435925_1 (AF435925) very large G protein-coupled receptor 1b [Homo sapiens]	3	4.5	695.07	®
+3	1.4e–008	-	T gi|14530760|emb|CAC42394.1| (AL096841) dJ575L21.1 (KIAA0684 protein) [Homo sapiens]	10	5.8	123.44	®

The following is a list for Z score and its corresponding percentile in an estimated random match population:

Z	percentile
1.282	90.0
1.645	95.0
2.326	99.0
3.090	99.9

Fig. 10.3. ProFound database search using peptide masses determined by the Ciphergen Protein Biology system II (PBS II). *Z* score indicates 2.33 meaning that there is up to 99 chances that the protein identified is the myosin heavy polypeptide 9.

Measured Mass (M)	Avg/ Mono	Computed Mass	Error (ppm)	Residues Start	To	Missed Cut	Peptide sequence
1571.980	A	1571.797	116	374	387	0	VSHLLGINVTDFTR
1726.900	A	1726.993	-54	210	225	0	QLLQANPILEAFGNAK
1816.190	A	1815.955	129	1816	1830	1	IAQLEEQLDNETKER
1870.060	A	1870.050	6	1755	1770	0	ANLQIDQINTDLNLER
1950.270	A	1950.136	69	1418	1433	0	LQQELDDLLVDLDHQR
1998.020	A	1998.224	-102	1754	1770	1	KANLQIDQINTDLNLER
2089.020	A	2089.160	-67	1698	1716	1	QAQQERDELADEIANSSGK
2472.860	A	2472.602	104	1731	1751	0	IAQLEEELEEEQGNTELINDR
2493.780	A	2493.624	63	1302	1322	0	DFSALESQLQDTQELLQEENR
2714.130	A	2713.936	72	1731	1753	1	IAQLEEELEEEQGNTELINDRLK
2750.130	A	2749.928	73	1302	1324	1	DFSALESQLQDTQELLQEENRQK
3059.420	A	3059.442	-7	1053	1080	1	LEGDSTDLSDQIAELQAQIAELKMQLAK

Unmatched Average Masses:

1155.900 1647.490 1930.170 2162.990 2289.600 3329.810 3922.910

Search again using unmatched masses:

Search for | single protein only | in | All taxa

Fig. 10.4. Results of profound databases searching, using peptide masses determined by Ciphergen software (PBS II reader). ProFound Software compares the experimental obtained masses and those of the databases that help for the establishment of the *Z* score.

4. Also add higher concentrations of the calibrant to the remaining spots. Keep one sample spot free of calibrant for comparison.
5. Collect data using the ProteinChip reader and select the spectrum that shows the best sample/calibrant ratio (*see* **Note 5**).

4. Notes

1. Reagents 6–16 are components of the Ciphergen Peptide Mapping kit.
2. The Coomassie blue R-250 is the preferred stain. Avoid using any staining method that uses oxidants or cross-linking reagents.
3. Additional microlitre can be added on the same spot to increase to signal response. In parallel, a spot with solution obtained from the digestion of protein-free gel pieces was also prepared in order to be used for the establishment of the baseline and localization of the background signal obtained from trypsin autolysis (*see* **Fig. 10.2**).
4. This software (http://129.85.19.192/profound_bin/WebProFound.exe) provides a list of *Z* scores and corresponding percentiles in an estimated random match population. It ranks the protein sequences in the database according to their probability of producing the peptide map. For protein identification, this site can also be used: http://www.expasy.ch/tools/#proteome.
5. In most cases, the primary known peak that corresponds to 2163.3 Da is visible in the digested sample and can be used as one of the internal control.

Acknowledgments

This work was supported by grants to AM K from INSERM and INCA.

References

1. Lahlil, R., Calvo, F., and Khatib, A. M. (2009) The potential anti-tumorigenic and anti-metastatic side of the proprotein convertases inhibitors *Recent Pat Anticancer Drug Discov* **4**, 83–91.
2. Scamuffa, N., Basak, A., Lalou, C., Wargnier, A., Marcinkiewicz, J., Siegfried, G., Chrétien, M., Calvo, F., Seidah, N. G., and Khatib, A. M. (2008) Regulation of prohepcidin processing and activity by the subtilisin-like proprotein convertases Furin, PC5, PACE4 and PC7 *Gut* **57**, 1573–82.
3. Creemers, J. W., and Khatib, A. M. (2008) Knock-out mouse models of proprotein convertases: Unique functions or redundancy? *Front Biosci* **13**, 4960–71.
4. Bontemps, Y., Lapierre, M., Siegfried, G., Calvo, F., and Khatib, A. M. (2008) Inhibitory feature of the proprotein convertases prosegments *Med Chem* **4**, 116–20.
5. Scamuffa, N., Siegfried, G., Bontemps, Y., Ma, L., Basak, A., Cherel, G., Calvo, F., Seidah, N. G., and Khatib, A. M. (2008) Selective inhibition of proprotein convertases

represses the metastatic potential of human colorectal tumor cells *J Clin Invest* **118**, 352–63.

6. Lapierre, M., Siegfried, G., Scamuffa, N., Bontemps, Y., Calvo, F., Seidah, N. G., and Khatib, A. M. (2007) Opposing function of the proprotein convertases furin and PACE4 on breast cancer cells' malignant phenotypes: Role of tissue inhibitors of metalloproteinase-1 *Cancer Res* **67**, 9030–4.
7. Tzimas, G. N., Chevet, E., Jenna, S., Nguyên, D. T., Khatib, A. M., Marcus, V., Zhang, Y., Chrétien, M., Seidah, N., and Metrakos, P. (2005) Abnormal expression and processing of the proprotein convertases PC1 and PC2 in human colorectal liver metastases *BMC Cancer* **5**, 149.
8. Khatib, A. M., Bassi, D., Siegfried, G., Klein-Szanto, A. J., and Ouafik, L. (2005) Endo/exo-proteolysis in neoplastic progression and metastasis *J Mol Med* **83**, 856–64.
9. Siegfried, G., Basak, A., Prichett-Pejic, W., Scamuffa, N., Ma, L., Benjannet, S., Veinot, J. P., Calvo, F., Seidah, N., and Khatib, A. M. (2005) Regulation of the stepwise proteolytic cleavage and secretion of PDGF-B by the proprotein convertases *Oncogene* **24**, 6925–35.
10. Siegfried, G., Basak, A., Cromlish, J. A., Benjannet, S., Marcinkiewicz, J., Chrétien, M., Seidah, N. G., and Khatib, A. M. (2003) The secretory proprotein convertases furin, PC5, and PC7 activate VEGF-C to induce tumorigenesis *J Clin Invest* **111**, 1723–32.
11. Siegfried, G., Khatib, A. M., Benjannet, S., Chrétien, M., and Seidah, N. G. (2003) The proteolytic processing of pro-platelet-derived growth factor-A at RRKR(86) by members of the proprotein convertase family is functionally correlated to platelet-derived growth factor-A-induced functions and tumorigenicity *Cancer Res* **63**, 1458–63.
12. Khatib, A. M., Siegfried, G., Chrétien, M., Metrakos, P., and Seidah, N. G. (2002) Proprotein convertases in tumor progression and malignancy: Novel targets in cancer therapy *Am J Pathol* **160**, 1921–35.
13. Khatib, A. M., Siegfried, G., Prat, A., Luis, J., Chrétien, M., Metrakos, P., and Seidah, N. G. (2001) Inhibition of proprotein convertases is associated with loss of growth and tumorigenicity of HT-29 human colon carcinoma cells: Importance of insulin-like growth factor-1 (IGF-1) receptor processing in IGF-1-mediated functions *J Biol Chem* **276**, 30686–93.
14. Merchant, M., and Weinberger, S. R. (2000) Recent advancements in surface-enhanced laser desorption/ionization-time of flight-mass spectrometry *Electrophoresis* **21**, 1164–77.
15. Gu, M., Gordon, V. M., Fitzgerald, D. J., and Leppla, S. H. (1996) Furin regulates both the activation of Pseudomonas exotoxin A and the quantity of the toxin receptor expressed on target cells *Infect Immun* **64**, 524–7.
16. Chiavegato, A., Bochaton-Piallat, M. L., D'Amore, E., Sartore, S., and Gabbiani, G. (1995) Expression of myosin heavy chain isoforms in mammary epithelial cells and in myofibroblasts from different fibrotic settings during neoplasia *Virchows Arch* **426**, 77–86.

Chapter 11

Regulation of 7B2 mRNA Translation: Dissecting the Role of Its 5′-Untranslated Region

Haidy Tadros, Gunther Schmidt, Francine Sirois, and Majambu Mbikay

Abstract

7B2 is a chaperone for the prohormone/proneuropeptide convertase PC2. Its mRNA is readily detectable in most neuronal and endocrine cells; the protein, in contrast, is often found at relatively low levels, suggesting that translation of the corresponding mRNA may be repressed. Because the 5′ untranslated region (5′-UTR) of this mRNA is relatively long and burdened with multiple AUGs, it has been speculated that it contributes to this repression. In this report, the influence of this region was assessed using in vitro and ex vivo approaches. The results showed that, in a cell-free system, full-length 7B2 mRNA was a poor template for translation. Its translatability dramatically improved when its 5′-UTR was truncated or when it was replaced with the 5′-UTR of carboxypeptidase E mRNA. These observations were confirmed in transfected mouse insulinoma MIN6 cells and human embryonic kidney HEK293 cells. Acute exposure of MIN6 cells to high glucose increased endogenous 7B2 biosynthesis without affecting the levels of its mRNA, suggesting that translation repression of this mRNA can be relieved by physiological stimuli.

Key words: 7B2, translational regulation, translation initiation, 5′ untranslated region, pancreatic beta-cell, glucose stimulation.

Abbreviations

aa	Amino acid
CIPIC	Complete protease inhibitor cocktail
CPE	Carboxypeptidase E
DEPC	Diethylpyrocarbonate
ER	Endoplasmic reticulum
IP	Immunoprecipitation
ir7B2	immunoreactive 7B2
MCS	Multiple cloning sites
ML	Metabolic labeling
nt	Nucleotide

M. Mbikay, N.G. Seidah (eds.), *Proprotein Convertases*, Methods in Molecular Biology 768,
DOI 10.1007/978-1-61779-204-5_11,

ORF	Open reading frame
PBS	Phosphate-buffered saline
PC	Proprotein convertase
PL	Pulse labeling
qRT-PCR	quantitative reverse transcriptase-polymerase chain reaction
RIPA	Radioimmunoprecipitation assay
sqPI	semi-quantitative phosphorImaging
TGN	*Trans* golgi network
uAUG	upstream AUG
UTR	Untranslated region

1. Introduction

7B2 is a resident protein of neuroendocrine secretory granules. In mouse, it is biosynthesized in the endoplasmic reticulum (ER) as a secretory proprotein of 186 amino acids (aa). This precursor is cleaved in the *trans* Golgi network (TGN) by furin after the basic motif $RRKRR^{151-155}$; this motif is then removed by carboxypeptidase E (CPE), generating the N-terminal $7B2^{1-150}$ protein and the C-terminal $7B2^{156-186}$ peptide (1). In secretory granules, $7B2^{156-186}$ is further processed by proprotein convertase 2 (PC2) and CPE to produce $7B2^{156-171}$ and $7B2^{174-186}$ peptides (2). Pro7B2 facilitates the exit of newly biosynthesized proPC2 from the ER and its subsequent maturation (3). Its C-terminal $7B2^{156-186}$ is a potent inhibitor of the active enzyme (4, 5). Ablation of the 7B2 gene in mouse results in PC2 deficiency and impaired processing of several precursors to hormones and neuropeptides (6, 7).

7B2 transcripts are widely expressed in neuronal and endocrine cells (8, 9). Yet, except for the anterior pituitary, most neuroendocrine tissues contain relatively low amounts of immunoreactive 7B2 (ir7B2) (10, 11), suggesting that translation of the mRNA may be repressed in these tissues. We and others have hypothesized that the length of the 5′-UTR of 7B2 mRNA, its nucleotide composition, and the presence of upstream AUGs (uAUGs) within it (**Fig. 11.1a**) could make this mRNA a poor template for translation (12, 13). To determine whether the 5′-UTR plays any role in the translation of 7B2 mRNA, we evaluated in vitro and ex vivo the translatability of this mRNA when the UTR is truncated or replaced. We also examined whether exogenous stimulation of an endocrine cells could enhance endogenous 7B2 biosynthesis.

We produced in vitro capped and polyadenylated transcripts of rat 7B2 cDNA templates in which the 5′-UTR was either full length [342 nucleotides (nts)] or variably truncated. The translatability of these mRNAs was compared in an in vitro translation

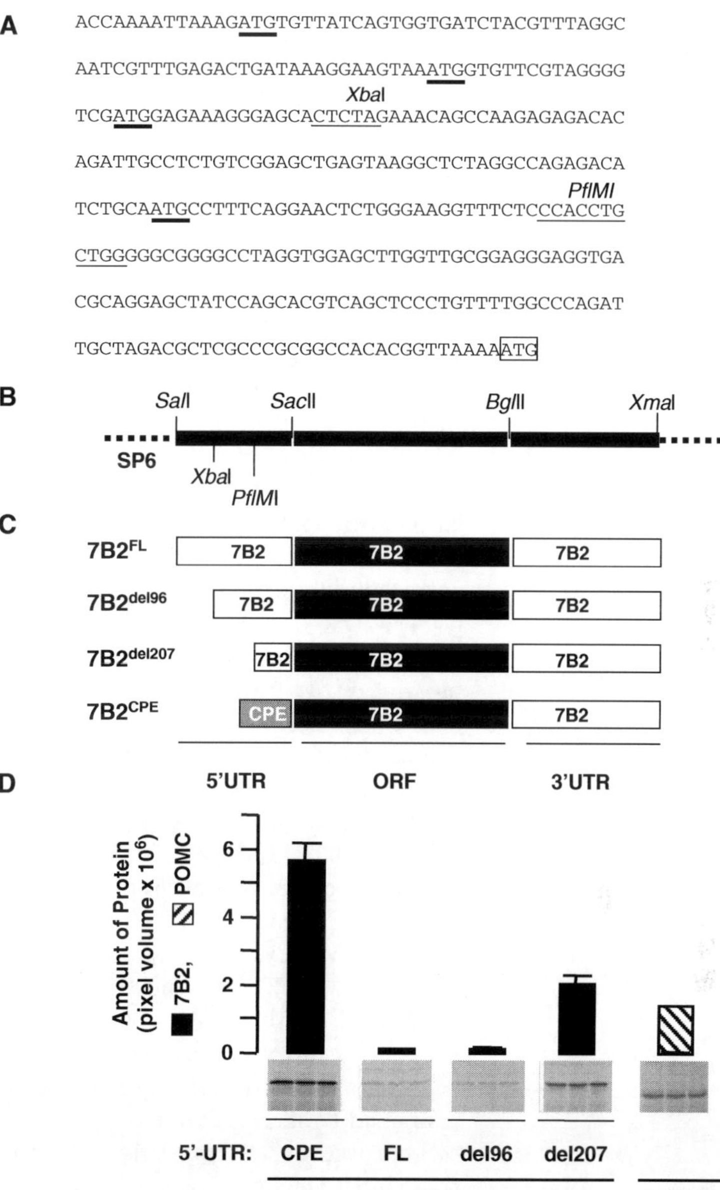

Fig. 11.1. 7B2 cDNA constructs. (**a**) Sequence of the 5′-UTR of rat 7B2 cDNA. Upstream ATGs are *underlined* in *bold*; the initiator ATG is *boxed*. Internal restriction enzyme cleavage sites used for truncation are *underlined* and specified. (**b**) Diagram of the inserted rat 7B2 cDNA. The three modules representing the 5′-UTR, the ORF, and 3′-UTR of rat 7B2 cDNA are flanked by *dashed lines* representing plasmid sequences. The cDNA is preceded by a promoter sequence recognized by the SP6 RNA polymerase. (**c**) Diagrams of the expected 7B2 mRNAs differing by their 5′UTR. (**d**) Relative expression of 7B2 at 0.2 μg mRNA/25 μl reaction for 90 min. Very little 7B2 was translated from either 7B2FL or $7B2^{del96}$ mRNA. By comparison, ~20-fold and ~50-fold more 7B2 was translated from $7B2^{del207}$ and $7B2^{CPE}$ mRNA, respectively. $POMC^{del102}$ mRNA was as efficiently translated as $7B2^{del207}$ mRNA (after correction for the fact that 7B2 has 5 methionine residues in its sequence and $POMC^{del102}$ has 3). These results indicate that full-length 7B2 mRNA translation is restrained by its 5′-UTR.

system based on rabbit reticulocyte lysate, by evaluating [^{35}S]-methionine incorporation into newly biosynthesized 7B2. The radioactive translation products were analyzed by electrophoresis through a polyacrylamide gel containing sodium dodecyl sulfate (SDS-PAGE) followed by semi-quantitative phosphorImaging (sqPI).

Since, during cellular biosynthesis, mRNAs compete for limited amounts of translation initiation factors and these factors are preferentially used by some mRNA templates at the expense of others (14), we examined how the presence of an equal amount of a competing proopiomelanocortin (POMC) mRNA would affect the translation of the various 7B2 mRNA isoforms in vitro, quantifying the levels of the radioactive translation products by SDS-PAGE/sqPI as above.

Within cells, however, 7B2 mRNA must compete for translation with a multitude of mRNAs. We therefore examined the ability of the various mRNAs to be translated into 7B2 when the corresponding expression vectors are transfected in mouse insulinoma MIN6 or human embryonic kidney HEK293 cells. Biosynthesis was captured by metabolic pulse labeling (PL) of de novo biosynthesized proteins with [^{35}S]-methionine, immunoprecipitation (IP) of 7B2 molecular forms, their fractionation by SDS-PAGE, and their quantification by sqPI.

Increased mRNA translation represents a most immediate mechanism of producing more protein in response to regulatory stimuli. Acute exposure of MIN6 cells to high glucose is known to stimulate the translation of several mRNAs (e.g., mRNAs for insulin, PC1, and PC2), but not of others (e.g., CPE mRNA) (15, 16). We examined whether such an exposure would also induce greater 7B2 biosynthesis, without affecting the levels of its mRNA. Biosynthesis was assessed by PL/IP/SDS-PAGE/qPI as above. 7B2 mRNA levels were measured by quantitative RT-PCR.

Overall, the results suggest that 7B2 mRNA translation is restrained by its 5′-UTR and can be stimulated under conditions of improved overall translation. Such regulation might influence the cellular level of PC2 activity.

2. Materials (*See* Note 1)

2.1. In Vitro Transcription and Translation

1. In vitro transcription plasmid vectors: pSP64pA-7B2^{FL}, pSP64pA-7B2^{del96}, pSP64pA-7B2^{del207}, pSP64pA-7B2^{CPE}, and pSP64pA-POMCtr. They were produced by inserting the open reading frame (ORF) of rat 7B2 cDNA or truncated mouse proopiomelanocortin (POMCt) cDNA into the multiple cloning site (MCS) of the pSP64(polyA) plasmid

(Promega, Ottawa, ON) (*see* **Note 2**). The 7B2 ORF was preceded by a 5′-UTR derived from either rat 7B2 or rat carboxypeptidase (CPE) cDNA. The 7B2 5′-UTR was either full length (FL) or truncated (del) of 96 or 207 nucleotides from the 5′ end. These inserts are diagrammatically represented in **Fig. 11.1c**. The POMCt cDNA ORF carried an in-frame internal deletion of 102 nucleotides. This truncated ORF specified a translation product 34 aa shorter than full-length pre-POMC and which, unlike the full POMC, migrated differentially from prepro7B2 on SDS-PAGE.

2. The mMessage mMachine™ kit (Ambion, Austin, TX). The kit contains optimized 10x reaction buffer, a 2x cap analog m7G(5′)pppG(5′)G and ribonucleotides mix, and a 10x SP6 RNA polymerase mix. Store at –80°C.
3. Retic Lysate IVT™ kit, methionine-free (Ambion, Austin, TX). The kit contains nuclease-free water, mRNA-depleted rabbit reticulocyte lysate (optimized for an ATP regenerating system as well as for hemin, yeast tRNA, and calf liver tRNA contents), and 20x methionine-free translation mix. Store at –80°C.
4. DNAase I (Fermentas AB, Scarborough, ON).
5. L-[^{35}S]-methionine, translation, and cell-labeling grade (Amersham Canada, Mississauga, ON, and now renamed GE Healthcare Canada), concentration: 10 mCi/ml; specific activity: 10 mCi/mmol.

2.2. Cell Culture

1. Eukaryotic expression plasmid vectors: pcDNA3-7B2^{FL}, pcDNA3-7B^{del96}, and pcDNA3-7B2^{del207}. To produce these vectors, the cDNA inserts of the pSP64pA-7B2 vectors (*see* **Section 2.1**) were excised and introduced into the MCS of the pcDNA3 plasmid (Stratagene, La Jolla, CA) (*see* **Note 3**).
2. Dulbecco's modified Eagle Medium (DMEM) (Invitrogen, Burlington, ON).
3. Fetal bovine serum (FBS) (Invitrogen).
4. Heat-inactivated FBS (Invitrogen).
5. Mouse insulinoma MIN6 cells (a gift from Dr. Donald F. Steiner, University of Chicago). They were grown at 37°C in DMEM 10% heat-inactivated FBS, 1 mM Na-pyruvate, 2 mM L-glutamine, and 28 μM β-mercaptoethanol, in a 5% CO_2–95% air atmosphere. As endocrine cells, they endogenously express 7B2.
6. Human embryonic kidney HEK293 cells: (American Tissue Type Collection, Manassas, VA). They were grown at 37°C in DMEM medium containing 5% FBS, in a 5% CO_2–95%

air atmosphere. They are non-endocrine and do not endogenously express 7B2.

2.3. Transfection and Metabolic Labeling

1. Lipofectamine 2000 (Life Technologies, Burlington, ON).
2. RPMI (Roswell Park Memorial Institute) 1640 medium, methionine-free: (ICN Biomedicals, Nepean, ON).
3. Complete protease inhibitor cocktail (CPIC, Boehringer Mannheim, Laval, QC). Prepare a 25x stock solution by dissolving 1 tablet/2 ml distilled and deionized water (ddH_20). Store at –20°C in 0.1 ml aliquots.
4. RIPA (radioimmunoprecipitation assay) buffer: 50 mM Tris–HCl, 1% Nonidet P-40, 0.5% sodium-deoxycholate, 150 mM NaCl, 1% SDS.
5. RIPA-CPIC: RIPA supplemented with 0.04 volumes of 25x CPIC.
6. Phosphate-buffered saline (PBS): 137 mM NaCl, 2.7 mM KCl, 10 mM Na_2HPO_4, 2 mM KH_2PO_4, pH of 7.4.

2.4. Immunoprecipitation

1. Labquake® Rotisserie shaker (Fisher Scientific, Ottawa, ON).
2. Rabbit anti-$7B2^{23-39}$ polyclonal antibody (a gift from Dr. Nabil G. Seidah, Clinical Research Institute of Montreal). This antibody recognizes a highly conserved sequence in the N-terminal region of 7B2.
3. Protein A-agarose, a 50% (w/v) suspension (Sigma, St. Louis, MO).
4. PBS-E: PBS containing 1 mM EDTA.
5. NTE buffer: 1 M NaCl, 10 mM Tris–HCl, and 1 mM EDTA, pH 8.

2.5. SDS-PAGE

1. Acrylamide/bis-acrylamide solution (30.8%T) (neurotoxin, working with gloves!): 30% acrylamide and 0.8% bis-acrylamide. Store at 4°C.
2. 10% SDS.
3. 10% ammonium persulfate.
4. *N, N, N, N′*-tetramethylethylenediamine (TEMED, Bio-Rad).
5. SDS electrophoresis buffer: 25 mM Tris, 192 mM glycine, 0.1% SDS.
6. Gel fixation solution: 50% methanol–10% acetic acid–40% water.
7. Amplify fluor solution (Amersham).

8. Kodak Biomax XAR film (Perkin Elmer Canada, Woodbridge, ON).
9. Autoradiographic cassettes with a phosphor screen (Fisher Scientific, Ottawa, ON).
10. Typhoon PhosphorImager (Molecular Dynamics, Sunnyvale, CA).

2.6. Quantitative RT-PCR

1. RNeasy total RNA extraction kit (Qiagen, Mississauga, ON).
2. dN_6 random primer: (Invitrogen).
3. Superscript II RNase H–Reverse Transcriptase (Invitrogen).
4. 2X QuantiTect SYBR Green PCR Master Mix (Qiagen) (containing optimized concentrations of $MgCl_2$, KCl, dNTP, SYBR Green dye, and Taq DNA polymerase).
5. Oligodeoxynucleotide primers: For PCR amplification of cDNA fragments of mouse (m) 7B2 and ribosomal protein L30: m7B2-F(orward): 5′-GGG GGA TTT TTT TGA TGT GGA-3′; m7B2-R(everse): 5′-CCC AAA CAG CAT AAC CCA AAA-3′; mL30-F: 5′-TGC GCA CAA GCC ATC TAC TC-3′; mL30-R: 5′-CTG GTG AAG CCC AAG ATC GT-3′. They were ordered from Invitrogen.
6. The LightCycler Instrument (Roche, Laval, QC, cat no. 2011-468).
7. Standards: Mouse 7B2 or L30 DNA amplicons preamplified by PCR from MIN6 cell total cDNA. They were purified by agarose gel electrophoresis, eluted from gel, dissolved in 10 mM Tris–HCl/1 mM EDTA, and quantified by spectrophotometry at 260 nm.

3. Methods

3.1. In Vitro Transcription (See Note 4)

1. Linearize the pSP64pA expression plasmids by cleavage at 3′-end of the polyA tail. Purify the cleaved plasmids by phenol–chloroform extraction and ethanol precipitation. Dissolve the DNA in DPEC-treated water and determine their concentrations by spectrophotometry at 260 nm.
2. Mix 1 μg of each linear plasmid with 2 μl of 10x reaction buffer, 10 μl of 2x NTP/cap, and 2 μl of SP6 RNA polymerase from the mMessage mMachine^{TM} kit in a final volume of 20 μl.
3. Incubate mixture at 37°C for 2 h to allow transcription.

4. Add 1 μl of DNAse I and incubate at 37°C for 10 min to degrade the plasmid.
5. Purify the transcripts by phenol–chloroform extraction and isopropanol precipitation.
6. Dissolve the transcripts into DPEC-treated water and determine their concentrations by spectrophotometry at 260 nm. Adjust concentration to 1 mg/ml (*see* **Note 5**).

3.2. In Vitro *Translation* (See Note 6)

1. Mix the transcripts (0.25–4 ng, singly or in combination) with the recommended amounts of methionine-free master mix of the Retic Lysate IVTTM kit and 20 μCi of [^{35}S]-methionine in a 25 μl final volume.
2. Incubate at 30°C for up to 90 min to allow translation.
3. Add 2.5 μl of 1 μg/ml RNase A and resume incubation for 10 min to degrade the transcripts, stopping translation.
4. Analyze translation by SDS-PAGE and sqPI (**Figs. 11.1d** and **11.2**).

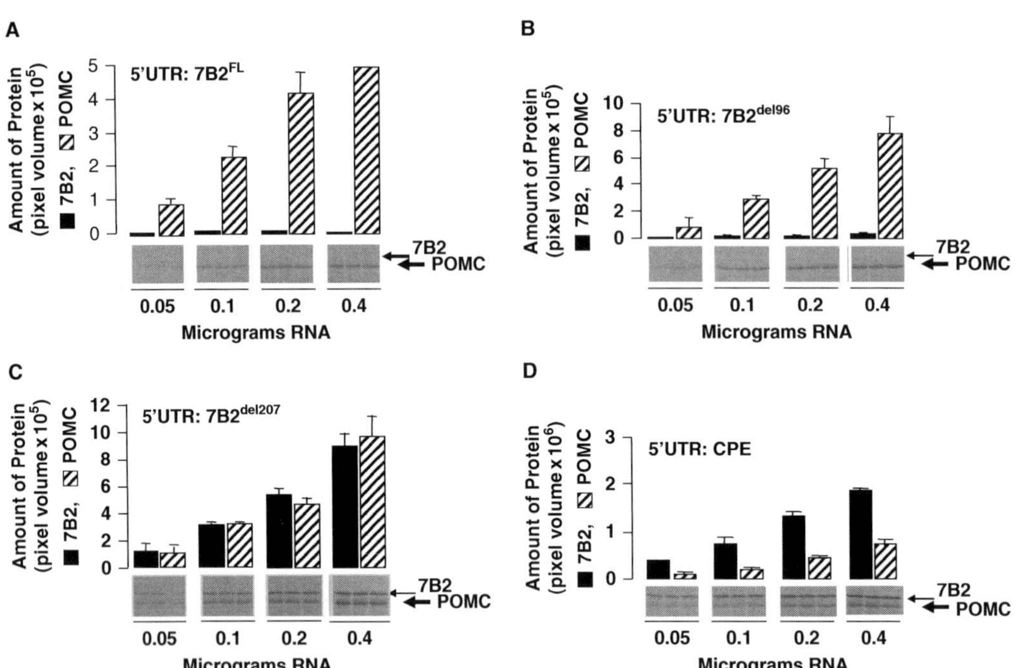

Fig. 11.2. Competition between POMC^{del102} mRNA and 7B2^{FL} (**a**), 7B2^{del96} (**b**), 7B2^{del207} (**c**), or 7B2^{CPE} (**d**) mRNA. 7B2 and POMC mRNAs were added in equal amounts for a total mRNA input ranging from 0.05 to 0.4 μg per reaction mix. Whereas POMC^{del102} accumulated with increasing concentration of mRNAs, expression of pro7B2 from 7B2^{FL} or 7B2^{del96} mRNA remained very low and unchanged (**Fig. 11.2a, b**). 7B2^{del207} mRNA isoform was able to better compete with POMC mRNA as the amounts of their respective products increased in parallel with the concentrations of total mRNAs (**Fig. 11.2c**). 7B2CPE mRNA also proved to be a good template since its translation was unaffected by the presence of POMC mRNA in the reaction mix (**Fig. 11.2d**). Collectively, these results suggest that native 7B2 mRNA may not efficiently compete for translation factors because of its 5′-UTR.

3.3. Transfection and Metabolic Pulse Labeling

1. Seed into 6-well plates with 7.5 × 10^5 HEK393 cells/well or 4 × 10^5 MIN6 cells/well in 3 ml of the proper culture medium (*see* **Section 2.2**).
2. Grow the cell monolayers to 65–75% confluence.
3. Transfect triplicate cell monolayers, each with 1 μg of pcDNA3-7B2 vector DNA/well, using Lipofectamine 2000 liposomes as recommended by the manufacturer.
4. Incubate the cells at 37°C for 48 h.
5. Rinse the cell monolayers twice with 5 ml of PBS/well.
6. Overlay the cell monolayers with 1 ml/well of methionine-free RPMI medium and incubate at 37°C for 30 min to deplete endogenous methionine.
7. Remove the medium and substitute 1 ml/well of RPMI medium containing 200 μCi of [^{35}S]-methionine/well; resume incubation for 10–30 min.
8. Remove and discard the radioactive medium. Add 400 μl of RIPA-PIC lysing buffer/well. Scrape off the cells off the dishes and transfer cell lysates into Eppendorf tubes.
9. Vigorously vortex the lysates for 30 s and place the tubes on ice for 30 min.
10. Centrifuged the lysates at 10,000×*g* for 10 min at 4°C
11. Transfer the post-mitochondrial supernatants into fresh Eppendorf tubes and store at 80°C until analysis.

3.4. Glucose Stimulation of MIN6 Cells

1. Seed MIN6 cells in 6-well dish and grow into monolayers (*see* **Section 3.3**, steps 1–2).
2. Substitute fresh medium containing 2.4 mM glucose and incubate at 37°C for 16 h.
3. Substitute fresh medium containing either 2.4 mM (controls, 6 wells) or 16.7 mM (stimulatory, 6 wells) glucose and incubate for 30 min.
4. Perform pulse labeling of de novo biosynthesized proteins (*see* **Section 3.3**, steps 5–11) on 3 wells of controls and 3 wells of stimulated cells.
5. Extract total RNA from cells in each of the remaining wells (*see* **Section 3.7**).

3.5. Immuno-precipitation

1. To each 200 μl aliquot of supernatant of lysates from pulse-labeled cells (*see* **Section 3.3**, step 11) in an Eppendorf tube, add 50 μl of 100 mM L-methionine, 250 μl of PBS, and 2 μl of rabbit anti-7B2$^{23-39}$ antibody (*see* **Note 7**).

2. Clamp the tubes to a Rotisserie shaker and mix the content at 4°C for 16 h to allow antigen–antibody binding.
3. Add to each tube 15 μl of a Protein A-agarose suspension and incubate further at 4°C for 16 h with mixing to allow attachment of the immune complexes to the resin.
4. Sediment the resin by centrifugation at 3,000×g and 4°C for 5 min.
5. Rinse the pellets three times with ice-cold RIPA buffer, twice with ice-cold NTE, and twice with PBS-E.
6. Analyze the precipitates by SDS-PAGE and sqPI (**Figs. 11.3 and 11.4b**).

3.6. SDS-PAGE and Semi-quantitative PhosphorImaging

1. Prepare 1 mm thick, 19 × 23 cm, 12%T polyacrylamide gels in the Hoefer DALT Gel Caster. Vertical slab SDS gels are cast in a Multiple Gel Caster. For assembly of the gel cassette, refer to manufacturer's user manual (Hoefer DALT System User Manual, Amersham Biosciences). To cast 25 slab gels, mix the following solutions: 600 ml of acrylamide/bis-acrylamide solution (30.8%T), 375 ml of 1.5 M Tris–Cl (pH 8.8), 15 ml of 10% SDS, and 15 ml of 10% ammonium persulfate. Add water to a final volume of 1,500 ml. Before gel casting, add 215 μl of TEMED.

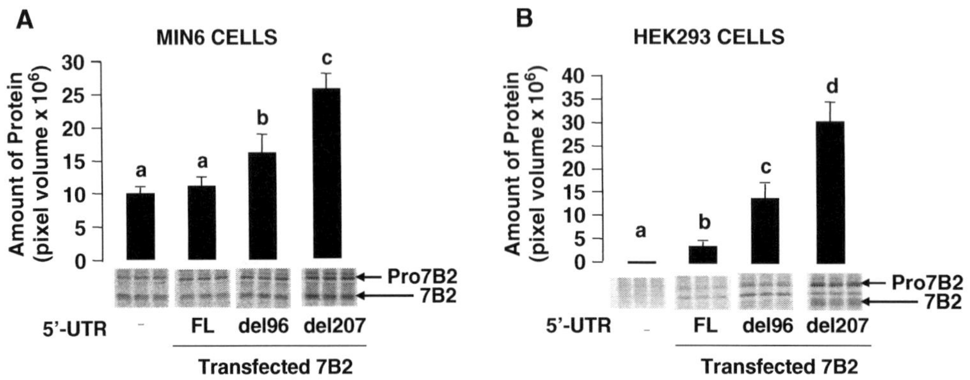

Fig. 11.3. Expression of 7B2 mRNA variants in MIN6 (**a**) and HEK293 (**b**) cells. Cells were transfected with pcDNA3-7B2 expression vectors; 48 h later, they were metabolically labeled with [^{35}S]-methionine for 30 min; ir7B2 proteins were immunoprecipitated from cell lysates, resolved and quantified. Difference in the *letter* above the *bars* indicates significant difference. Note that, while both cell types contained the expected 29 kDa pro7B$^{21-186}$ and the 24 kDa mature 7B2$^{1-150}$, HEK293 cells showed two additional forms above and below the 7B2$^{1-150}$ form. In MIN6 cells, there was no significant increase of 7B2 level above the endogenous content when the 7B2^{FL} mRNA was transduced. Transduction of 7B2^{del96} and 7B2^{del207} mRNAs significantly increased intracellular 7B2 to 1.6- and 2.6-fold above the endogenous content, respectively (**Fig. 11.3a**). In HEK293 cells, transduced 7B2^{del96} and 7B2^{del207} mRNA induced, respectively, 3-fold and 6.6-fold more 7B2 than 7B2^{FL} mRNA (**Fig. 11.3b**). These results reinforce those from the in vitro translation studies indicating that native 7B2 mRNA is poorly translated. They also revealed that removal of the first 96 nts of the UTR, which has no effect in vitro (*see* **Figs. 11.1d and 11.2**), could significantly improve intracellular translation of the transduced rat 7B2 mRNA.

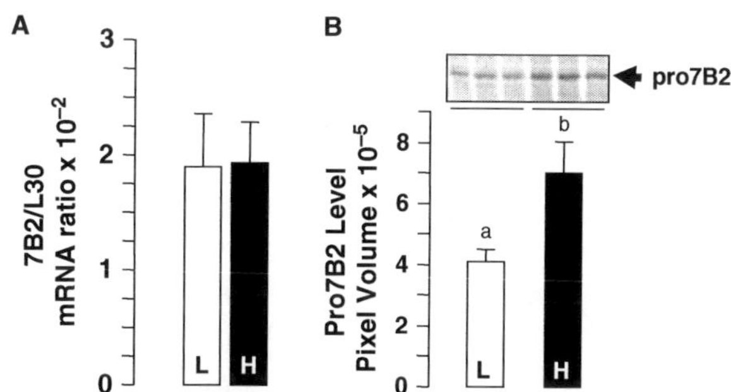

Fig. 11.4. Glucose stimulates 7B2 expression in MIN6 cells. Cells were cultured overnight in low (L, 2.4 mM) glucose DMEM medium. Fresh medium containing either low or high (H, 16.7 mM) glucose and incubation was resumed for 30 min. (**a**) Total RNA was extracted and analyzed for quantitative RT-PCR for 7B2 and L30 mRNA content (*see* **Section 3.7**). (**b**) Alternatively, cells were metabolically labeled with [^{35}S]-methionine for 10 min. ir7B2 proteins were immunoprecipitated and quantified. Because of the short duration of metabolic labeling, the prominent immunoprecipitated protein was pro7B2$^{1-186}$. Difference in the *letter* above the *bars* indicates significant difference.

2. To 12.5 μl of a dissolved sample in an Eppendorf tube, add an equal volume of 2x Laemmli buffer; to a pelleted sample in an Eppendorf tube, add 25 μl of 1x Laemmli buffer and suspend.
3. Place the tubes in boiling water bath and heat for 5 min.
4. Centrifuge the tubes at 10,000×*g* for 5 min.
5. Load 10–20 μl of the supernatant onto an SDS-12% polyacrylamide gel and fractionate their proteins by electrophoresis.
6. Soak the gel in the Gel Fixation Solution for 30 min and then in Amplify fluor Solution for another 30 min.
7. Dry the gel under vacuum and expose it overnight to a phosphor screen in an exposure cassette (*see* **Note 8**).
8. Reveal and quantify specific bands on a phosphorImager (**Figs. 11.1c, 11.2, 11.3,** and **11.4b**).

3.7. Quantitative RT-PCR

1. Extract total RNA from MIN6 cells using the Qiagen RNeasy total RNA extraction kit as specified by the manufacturer.
2. Reverse transcribe an aliquot of the total RNA to total cDNA using dN_6 random primers and the Superscript II RNase H$^-$ Reverse Transcriptase as previously described (17).
3. Prepare a 20 μl reaction mixture containing aliquot of the total cDNA (or, to derive a concentration curve, of known increasing amounts of pre-amplified cDNA amplicons),

1X LightCycler DNA Master SYBR Green I buffer (as source of $MgCl_2$, dNTP, SYBR Green dye, and Taq polymerase), and 0.25 mM of forward and reverse primers.

4. Amplify 7B2 or L30 cDNA amplicons in a LightCycler in a program including a 2 min pre-heating to 95°C, followed by 40 amplification cycles, each involving a 5 s denaturation at 94°C, a 10 s annealing at 52°C and a 25 s polymerization at 72°C, and a 2 s fluorescence data acquisition at 80°C.
5. Derive 7B2 or L30 mRNA concentrations in experimental samples from fluorescence values at the maximum of the log-linear amplification using the second derivative method (18), by reference to the cognate standard concentration curves.
6. Compute the ratios between 7B2 and L30 mRNA concentrations in the same sample (C_{PC2}/C_{L30}) to determine the relative concentrations of 7B2 mRNA (**Fig. 11.4a**).

4. Notes

1. Basic molecular cloning methods, such as the construction of expression vectors, are not described in this protocol. They were conducted following standard molecular biology techniques as described in popular protocol manuals e.g., (19).
2. The MCS in the pSP4(polyA) vector is located downstream to the SP6 viral promoter and upstream to 31-mer of deoxyadenylate tail (polyA). After insertion of a cDNA ORF of interest, the recombinant vector is linearized with a single cleavage past the polyA stretch. Transcript from the SP6 promoter terminates with a polyA tail which stabilizes it and allows its efficient translation in a eukaryotic translation system.
3. The MCS in the pcDNA3 vector is located downstream to a promoter enhancer from human cytomegalovirus (CMV) and upstream to a bovine growth mRNA sequence containing a polyadenylation signal. The CMV promoter in the expression drives strong transcription of the inserted gene in most mammalian cells transfected with this vector.
4. RNase contaminations from dust, desk surface, fingers, and saliva constitute a constant danger when producing and manipulating RNA. Manipulations are to be conducted with gloved hands in a clean laboratory area with minimal traffic. Special care should be taken to ensure that all glassware,

plasticware, and solutions are RNase free. One decontamination procedure consists of soaking the materials in a 0.1–1% DPEC in water for at least 1 h before autoclaving to evaporate DPEC decomposition product (CO_2 and ethanol). Most commercial kits for RNA manipulation provide sterile DPEC-treated ddH_2O; but one can prepare it in-house as described above.

5. Absorbance values could be misleading, as partially degraded RNA tends to yield greater values. It is therefore highly advised to verify the integrity of the transcripts by electrophoresis in a denaturing agarose gel followed by an ethidium bromide staining and visualization under ultraviolet light. Intact in vitro produced transcript should appear as a relatively sharp band.
6. Commercial in vitro translation cell-free systems have limited mRNA translation capacity. It is important to determine this capacity by running an mRNA concentration and a time course. A study comparing the translatability of mRNAs should be conducted with final mRNA concentrations and for a period of time within the linear range of translation.
7. The addition of non-radioactive methionine serves to minimize non-specific background radioactivity during IP. The amount of antibody to be used for IP is determined empirically. It should be saturating to permit quantitative comparison among experimental samples.
8. Prior to utilization, the phosphor screen should be exposed to phosphorescent light for at least 1 h to erase any remnant signals captured during prior exposure to a radioactive dried gel or a membrane.

Acknowledgments

This work was supported by grants from the Canadian Diabetic Association and the Strauss Foundation.

References

1. Paquet, L., Bergeron, F., Boudreault, A., Seidah, N. G., Chretien, M., Mbikay, M., and Lazure, C. (1994) The neuroendocrine precursor 7B2 is a sulfated protein proteolytically processed by a ubiquitous furin-like convertase *J Biol Chem* **269**, 19279–85.
2. Zhu, X., Rouille, Y., Lamango, N. S., Steiner, D. F., and Lindberg, I. (1996) Internal cleavage of the inhibitory 7B2 carboxyl-terminal peptide by PC2: A potential mechanism for its inactivation *Proc Natl Acad Sci USA* **93**, 4919–24.

3. Muller, L., Zhu, X., and Lindberg, I. (1997) Mechanism of the facilitation of PC2 maturation by 7B2: Involvement in ProPC2 transport and activation but not folding *J Cell Biol* **139**, 625–38.
4. Martens, G. J., Braks, J. A., Eib, D. W., Zhou, Y., and Lindberg, I. (1994) The neuroendocrine polypeptide 7B2 is an endogenous inhibitor of prohormone convertase PC2 *Proc Natl Acad Sci USA* **91**, 5784–7.
5. van Horssen, A. M., van den Hurk, W. H., Bailyes, E. M., Hutton, J. C., Martens, G. J., and Lindberg, I. (1995) Identification of the region within the neuroendocrine polypeptide 7B2 responsible for the inhibition of prohormone convertase PC2 *J Biol Chem* **270**, 14292–6.
6. Westphal, C. H., Muller, L., Zhou, A., Zhu, X., Bonner-Weir, S., Schambelan, M., Steiner, D. F., Lindberg, I., and Leder, P. (1999) The neuroendocrine protein 7B2 is required for peptide hormone processing in vivo and provides a novel mechanism for pituitary Cushing's disease *Cell* **96**, 689–700.
7. Sarac, M. S., Zieske, A. W., and Lindberg, I. (2002) The lethal form of Cushing's in 7B2 null mice is caused by multiple metabolic and hormonal abnormalities *Endocrinology* **143**, 2324–32.
8. Marcinkiewicz, M., Touraine, P., Mbikay, M., and Chrétien, M. (1993) Expression of neuroendocrine secretory protein 7B2 mRNA in the mouse and rat pituitary gland *Neuroendocrinology* **58**, 86–93.
9. Marcinkiewicz, M., Touraine, P., and Chrétien, M. (1994) Pan-neuronal mRNA expression of the secretory polypeptide 7B2 *Neurosci Lett* **177**, 91–4.
10. Iguchi, H., Chan, J. S., Dennis, M., Seidah, N. G., and Chretien, M. (1985) Regional distribution of a novel pituitary protein (7B2) in the rat brain *Brain Res* **338**, 91–6.
11. Iguchi, H., Chan, J. S., Seidah, N. G., and Chrétien, M. (1984) Tissue distribution and molecular forms of a novel pituitary protein in the rat *Neuroendocrinology* **39**, 453–8.
12. Braks, J. A., Broers, C. A., Danger, J. M., and Martens, G. J. (1996) Structural organization of the gene encoding the neuroendocrine chaperone 7B2 *Eur J Biochem* **236**, 60–7.
13. Mbikay, M., Seidah, N. G., and Chretien, M. (2001) Neuroendocrine secretory protein 7B2: Structure, expression and functions *Biochem J* **357**, 329–42.
14. Ray, B. K., Brendler, T. G., Adya, S., Daniels-McQueen, S., Miller, J. K., Hershey, J. W., Grifo, J. A., Merrick, W. C., and Thach, R. E. (1983) Role of mRNA competition in regulating translation: Further characterization of mRNA discriminatory initiation factors *Proc Natl Acad Sci USA* **80**, 663–7.
15. Skelly, R. H., Schuppin, G. T., Ishihara, H., Oka, Y., and Rhodes, C. J. (1996) Glucose-regulated translational control of proinsulin biosynthesis with that of the proinsulin endopeptidases PC2 and PC3 in the insulin-producing MIN6 cell line *Diabetes* **45**, 37–43.
16. Goodge, K. A., and Hutton, J. C. (2000) Translational regulation of proinsulin biosynthesis and proinsulin conversion in the pancreatic beta-cell *Semin Cell Dev Biol* **11**, 235–42.
17. Tadros, H., Chretien, M., and Mbikay, M. (2001) The testicular germ-cell protease PC4 is also expressed in macrophage-like cells of the ovary *J Reprod Immunol* **49**, 133–52.
18. Rasmussen, R. (2001) Quantification on the LightCycler. In *Rapid Cycle Real-Time PCR* (Meuer, S., Wittwer, C., and Nakagawara, K., Eds.), Chap 3, pp. 21–34, Springer, Berlin.
19. Sambrook, J., and Russell, D. W. (2001) *Molecular Cloning: A Laboratory Manual*, CSHL Press, Cold Spring Harbor, NY.

Chapter 12

Analysis of Epigenetic Alterations to Proprotein Convertase Genes in Disease

YangXin Fu and Mark W. Nachtigal

Abstract

Epigenetic alterations produce heritable changes in phenotype or gene expression without changing DNA sequence. Modified levels of gene expression contribute to a variety of human diseases encompassing genetic disorders, pediatric syndromes, autoimmune disease, aging, and cancer. Alterations in proprotein convertase gene expression are associated with numerous disease states; however, the underlying mechanism for changes in PC gene expression remains understudied. Epigenetic changes in gene expression profiles can be accomplished through modification of chromatin, specifically via chemical modification of DNA bases (methylation of cytosine) or associated histone proteins (acetylation or methylation). In general, active chromatin is associated with low DNA methylation status and histone acetylation, whereas silenced gene are typically in inactive regions of chromatin exhibiting DNA hypermethylation and histone deacetylation. This chapter will provide in-depth protocols to analyze epigenetic alterations in proprotein convertase gene expression using the *PCSK6* gene in the context of human ovarian cancer as a model system.

Key words: Epigenetics, DNA methylation, histone modification, bisulfite genomic sequencing, ovarian cancer, proprotein convertase, *PCSK6*, demethylating agents, histone deacetylase inhibitor.

1. Introduction

Epigenetics is defined as heritable changes in gene expression that are not due to changes in DNA sequence (1). Epigenetic modifications, including DNA methylation, histone modifications (e.g., methylation, acetylation, ubiquitination, and phosphorylation), nucleosome remodeling, and posttranscriptional gene regulation by micro-RNAs (miRNA), regulate gene expression and play roles in various biological processes (1–3). Epigenetic aberrations lead

M. Mbikay, N.G. Seidah (eds.), *Proprotein Convertases*, Methods in Molecular Biology 768,
DOI 10.1007/978-1-61779-204-5_12, © Springer Science+Business Media, LLC 2011

to abnormal gene expression and may contribute to various diseases, including cancers (1, 2, 4). DNA methylation, the addition of a methyl moiety to the cytosine-5 position within CpG dinucleotides, is the most studied epigenetic modification. CpG dinucleotides are usually found in short GC-rich DNA sequences, known as "CpG islands," frequently, but not exclusively, in the promoter regions of genes. In normal tissues, they either are unmethylated or have a low frequency of methylated residues. However, in cancer cells, hypermethylation or hypomethylation of CpG islands can occur, resulting in the silencing of tumor suppressor or activation of tumor promoter genes, respectively, thus contributing to cancer development (1, 2). This chapter will focus on gene silencing.

DNA methylation, histone modification, and nucleosomal remodeling are intimately linked and represent the molecular basis for gene silencing by epigenetic alterations (1). Specifically, DNA cytosine methylation recruits methylated DNA binding proteins that in turn recruit histone deacetylases (HDACs), nucleosomal remodeling complexes, and chromatin remodeling complexes to the methylated DNA, leading to the formation of transcriptionally repressive chromatin and permanent silencing of cancer-related genes (5–8). Reactivation of these silenced genes using DNA demethylating agents or HDAC inhibitors has become an attractive therapeutic strategy for diseases including cancer (9).

Proprotein convertases (PCs), a family of serine endoproteases, play a vital role in normal physiology by converting proproteins to biological active molecules or by inactivating substrates in response to biological cues. Aberrant expression of PCs has been shown to play roles in human cancer (10, 11). Ovarian cancer (OvCa) is the leading cause of death related to gynecologic cancers (12). OvCa is comprised of at least five distinct histological subtypes (serous, clear cell, endometrioid, mucinous, and transitional) (13). Approximately 10% of all OvCa is classified as familial, and ~50% of these are correlated with mutations in BRCA1 or BRCA2 breast/ovarian cancer genes. By contrast, the etiology for the remaining 90% of OvCa remains elusive. There is considerable debate regarding the origin of OvCa, but it is believed that the majority of OvCas arise from the cells covering the surface of the ovary, the ovarian surface epithelium (OSE) (14–18). There are numerous candidate molecules under investigation to evaluate their participation in the transformation of OSE cells to OvCa cells, and we and others have evidence to support a role for PCs in human OvCa pathogenesis.

To study the potential role of altered PC expression in ovarian tumorigenesis, we examined the expression of PCs in human OvCa and found that the mRNA for one of the PCs, PACE4 (*PCSK6*), was downregulated in OvCa cells compared with normal OSE cells (19). Further analysis revealed that there was no *PCSK6* gene deletion in OvCa cells, and these cells can support exogenous *PCSK6* promoter activity, suggesting that epigenetic alterations may have been responsible for reduced expression of *PCSK6*. Indeed, our data demonstrated that reduced *PCSK6* expression was due to DNA hypermethylation and histone deacetylation of CpG islands within the *PCSK6* promoter and first exon, indicating that epigenetics plays a role in PC expression in human OvCa cells (19). Therefore, our study provides an example suggesting that PC epigenetics should be examined as a potential mechanism for altering PC expression levels and thus contributing to human disease.

A number of techniques have been developed to examine DNA methylation status and each of them has documented advantages and disadvantages (20). Among these techniques, bisulfite genomic sequencing is the most direct and accurate method to analyze the methylation of specific CpG sites and is a standard technique for DNA methylation analysis (20, 21). Bisulfite sequencing relies on the conversion of the cytosine residues to uracil by sodium bisulfite treatment, which is represented by thymidine in the subsequent PCR step. However, methylated cytosine residues are not reactive to sodium bisulfite and thus will remain unchanged. Therefore, the cytosine residues in the final sequencing results will represent methylated cytosines, rendering bisulfite genomic sequencing a direct approach to analyze methylation status of specific CpG sites or genes (21).

In this chapter, we will provide a detailed protocol to evaluate DNA methylation status of CpG islands using the bisulfite genomic sequencing method (21, 22). We will also present an example of reactivation of the endogenous *PCSK6* gene, which is silenced by epigenetic alterations, using a demethylating agent and/or a HDAC inhibitor singly or in combination in cultured OvCa cells (19).

2. Materials

2.1. Cell Culture

1. Growth medium for primary human OvCa cells: MCDB105/M199 supplemented with 10% fetal bovine serum (Cansera, Rexdale, Ontario, Canada) and 100 units/ml penicillin G and 100 μg/ml streptomycin.
2. Growth medium for established human OvCa cell lines: RPMI 1640 (Invitrogen/Gibco, Burlington, Ontario, Canada) supplemented with 10% fetal bovine serum.

2.2. Genomic DNA Isolation

1. Phosphate-buffered saline (PBS): (1x PBS: 137 mM NaCl, 2.7 mM KCl, 10 mM Na_2HPO4, 1.76 mM KH_2PO4, pH 7.4), stored at room temperature after autoclave.
2. DNA lysis buffer (100 mM Tris, pH 8.0, 5 mM EDTA, 0.2% SDS, 200 mM NaCl) containing 0.1 mg/ml Proteinase K. Proteinase K (20 mg/ml in H_2O, stored at –20°C) is added to the buffer prior to use.
3. Phenol:chloroform:isoamyl alcohol (25:24:1, saturated with 10 mM Tris, pH 8.0, 1 mM EDTA, for molecular biology, Sigma).
4. Chloroform:isoamyl alcohol (24:1, Sigma).
5. 100% ethanol.
6. 70% ethanol.

2.3. Bisulfite Treatment

1. Sodium bisulfite (Sigma).
2. Hydroquinone (Sigma).
3. Bisulfite solution (make freshly prior to use): dissolve 2.025 g of sodium bisulfite in 4 ml of sterile H_2O and adjust the pH to 5.0 with 10 M NaOH (about 150 μl). In a separate tube, dissolve 0.11 g of hydroquinone in 5 ml of sterile H_2O. Add 250 μl of hydroquinone solution to the sodium bisulfite solution resulting in a final concentration of 10 mM. Adjust the volume to 5 ml with sterile H_2O and pass the solution through a 0.45 μm filter membrane.
4. QIAEX II (Qiagen) or Wizard DNA Clean-Up systems (Promega).
5. 3 M NaOH.
6. 3 M sodium acetate (NaOAc, pH 5.2): to make 100 ml of 3 M NaOAc, add 40.81 g of sodium acetate ($CH3COONa{\cdot}3H_2O$, trihydrate) in 80 ml of H_2O, adjust the pH to 5.2 with glacial acetic acid. Store at room temperature after autoclave.

2.4. PCR Amplification

1. PCR cycler.
2. PCR tubes.
3. FastStart Taq DNA Polymerase (Roche) (*see* **Note 1**).
4. 100 mM dNTP set (Invitrogen): to make 10 mM dNTP mix, add 100 μl of each 100 mM dNTP (dATP, dCTP, dGTP, and dCCT) to 600 μl of sterile H_2O. Store at –20°C in 100 μl of aliquots.
5. Primers: primers are dissolved in sterile H_2O to make 100 μM stocks. To prepare 10 μM primers for PCR reaction, add both forward and reverse primers in one tube. For example, add 10 μl of 100 μM forward primers, 10 μl of 100 μM

reverse primer, and 80 μl of sterile H_2O to make 100 μl. Store at –20°C.

2.5. Subcloning of PCR Products

1. Agarose.
2. TAE buffer.
3. Gel purification kit, such as Qiagen gel purification kit.
4. Subcloning kits: TOPO TA Cloning® kit (Invitrogen) or pDrive Cloning Vector (Qiagen). Vectors provided in these kits accept single 3′-deoxyadenosine overhang of PCR products generated at end of PCR amplification and allow for the blue/white screening of positive clones.

2.6. Transformation and Miniprep of Plasmid DNA

1. LB medium.
2. Competent cells (e.g., DH5α).
3. Miniprep of plasmid DNA kits from Qiagen or Sigma.

2.7. Treatment of Cells with a DNA Methylating Agent or HDAC Inhibitor

1. Dimethyl sulfoxide (DMSO, Sigma).
2. 5-aza-2′-deoxycytidine (DAC, Sigma): dissolved in DMSO to make a 50 mM stock and store frozen in aliquots at –20°C. Dilute to appropriate working concentrations in cell culture medium prior to use.
3. Trichostatin A (TSA, Sigma, 5 mM in DMSO (0.2 μm filtered)): Store frozen in aliquots at –20°C. Dilute to appropriate working concentrations in cell culture medium prior to use.

3. Methods

DNA methylation is catalyzed by DNA methyltransferases (DNMTs). Genes aberrantly silenced by DNA methylation can be reactivated by demethylating agents that inhibit DNMTs activity. These demethylating agents, such as 5-aza-2′-deoxycytidine (DAC), are incorporated into DNA during DNA synthesis of dividing cells and form a stable complex between DAC in DNA and DNMTs, thereby trapping the DNMTs on the DNA and resulting in loss of DNA methylation in the nascent strand (23, 24).

Histone acetylation is a hallmark of active genes. The histone acetylation status of a given gene is controlled by the balance between histone acetyltransferases (HATs) and histone deacetylases (HDACs). HDAC inhibitors, such as TSA, inhibit the removal of an acetyl group from lysine residues in histones by HDACs, resulting in hyperacetylation of histone proteins and increased gene expression (25).

In this section, we will provide a detailed protocol of bisulfite genomic sequencing as an approach to examine DNA methylation status in cultured cells (*see* **Note 16,** which discusses an alternative approaches to analyze DNA methylation status). Using this approach, we have demonstrated that the promoter and first exon DNA of *PCSK6* are hypermethylated in OvCa cells compared with normal OSE cells (19). We will also present an example of the induction of *PCSK6* gene expression using DAC, TSA, or both in cultured OvCa cells. Please *see* **Note 17** that mentions alternative approaches to analyze chromatin modifications.

3.1. Cell Culture

1. Primary human OvCa cells and the normal surface epithelial cells (OSE) were isolated from ascites specimens and from solid ovarian specimens, respectively, using methods as described by Shepherd et al. (26). Culture these cells in the appropriate medium (*see* **Section 2.1**). Perform all experiments using normal primary OSE at culture passages 2–4 and passages 1–4 for primary OvCa cells.
2. Grow and maintain established human OvCa cell lines (Hey, HeyC2, and OCC-1) in monolayer and maintained in RPMI medium (*see* **Section 2.1**).

3.2. Genomic DNA Isolation (See Note 2)

1. Culture ovarian cancer cells in 10 cm plastic plates (primary OvCa cells or established OvCa cell lines) or 6-well plates (normal OSE).
2. Isolate DNA when cells become about 90% of confluent. We use 10 cm plate culture as an example for DNA isolation in the following steps.
3. Wash the cell plates with 10 ml of 1x PBS twice and aspirate the PBS completely.
4. Add 1 ml of 1x PBS containing 1 mM EDTA to the plates and collect cells using a rubber policeman or a cell scraper and transfer the cells to 1.5 ml microcentrifuge tubes.
5. Centrifuge the cells at 3,000 rpm for 10 min to pellet the cells and aspirate the PBS.
6. Add 400 μl of DNA lysis buffer and resuspend the pellet by pipetting.
7. Place the tubes in a 55°C heat block and incubate the samples for 4 h to overnight. Make sure the cells are completely lysed and dissolved.
8. Add 400 μl of phenol:cholorfom:isoamyl alcohol (25:24:1) to the tube and mix the samples thoroughly by vortexing the tubes.
9. Centrifuge the samples at 12,000×g for 5 min. Carefully transfer the aqueous layer (top layer) to new tubes.

10. Repeat steps 8 and 9 one more time.
11. Add 400 μl of cholorfom:isoamyl alcohol (24:1) to the tube and mix the samples thoroughly by vortexing the tubes.
12. Centrifuge the samples at 12,000×*g* for 5 min. Carefully transfer the aqueous layer (top layer) to new tubes.
13. Add 800 μl of 100% ethanol to the samples and mix thoroughly by inverting the tubes (*see* **Note 3**).
14. Centrifuge the samples at 12,000×*g* for 15 min at 4°C. You should see white DNA pellets on the bottom of tubes. Decant the supernatant carefully.
15. Add 500 μl of 70% ethanol to the tubes (to remove salts) and invert the tubes to wash the DNA pellets. Centrifuge the samples at 12,000×*g* for 5 min 4°C.
16. Decant the ethanol carefully and spin the tubes briefly to bring the residual ethanol on the wall to the bottom of the tubes. Pipette the ethanol out carefully without touching the DNA pellets.
17. Air-dry the pellets for 10 min at room temperature.
18. Add 100 μl of TE pH 7.6 to dissolve the DNA (*see* **Note 4**).
19. Measure the concentration after the DNA is completely dissolved. DNA can be used immediately or stored at 4°C for future use.

3.3. Bisulfite Treatment of DNA

Bisulfite treatment of genomic DNA is conducted according to the protocol as described by Grunau et al. with minor modifications (27) (*see* **Note 5**).

1. Add 5 μg of genomic DNA to a 200 μl PCR tube. Add appropriate amount of sterile H_2O to make the total volume of 9 μl. Add 1 μl 3 M NaOH to the tube to make the final concentration of 0.3 M NaOH.
2. Incubate the sample at 42°C for 20 min using a PCR cycler to denature the DNA (*see* **Note 6**).
3. Add 120 μl of freshly prepared bisulfite solution (*see* **Section 2**) and incubate the sample at 55°C for 4 h using a PCR cycler.
4. Desalt the DNA samples using either QIAEX II (Qiagen) or Wizard DNA Clean-Up systems (Promega) according to the protocols provided by the manufacturers and resolve the DNA in 110 μl of 1 mM Tris–Cl, pH 8.0, in a 1.5 ml microcentrifuge tube.
5. Add 11 μl of a 3 M NaOH solution to the tube, mix well, and incubate the sample at 37°C for 20 min.

6. Add 22 μl of 3 M NaOAc (pH 5.2) and 400 μl 100% ethanol to the tube and mix thoroughly by hand inversion. Process immediately or store the sample at –20°C overnight.
7. Centrifuge the sample at 14,000×*g* for 15 min at 4°C. Decant the supernatant carefully.
8. Add 500 μl of 70% of ethanol to wash the pellet (to remove salts) and centrifuge the sample at 14,000×*g* for 15 min at 4°C. Decant the supernatant carefully.
9. Spin the tube briefly to bring the residual ethanol to the bottom of the tube. Pipette the ethanol out carefully without touching the DNA pellet.
10. Air-dry the DNA pellet for 10–30 min and dissolve the DNA in 30 μl of 1 mM Tris–Cl pH 8.0 (*see* **Note 7**). The bisulfite-treated DNA can be immediately used or stored at –20°C.

3.4. PCR Amplification of Promoter Sequences

3.4.1. Primer Design

Design primers that flank the promoter sequence containing the CpG sites to be analyzed. Several factors need to be taken into account in designing primers. First, the primers should not contain any CpG dinucleotide sequences to avoid biased PCR amplification of methylated or unmethylated templates. Second, since bisulfite treatment will convert cytosines to uracil, primer sequences should be changed accordingly to match DNA sequences. Third, the length of the PCR amplicon should be less than 500 bp. If the promoter is GC rich, PCR amplification efficiency for longer products is expected to be less efficient. Moreover, the efficiency of the subsequent cloning and sequencing of these PCR products will be less efficient. If necessary, more than one pair of primers should be designed to cover the region of the promoter to be analyzed.

3.4.2. PCR Reaction Mix

We used FastStart Taq DNA Polymerase from Roche Molecular Biochemicals because this system allowed us to amplify the GC-rich sequence of the human *PCSK6* promoter that proved to be difficult to be PCR amplified using regular Taq polymerase. Set up a 25 μl PCR reaction as follows: add 10.2 μl of sterile H_2O, 5 μl of DNA, 1 μl of 10 μM primers (mixed forward and reverse primers), 2.5 μl of 10x PCR reaction buffer with 20 mM $MgCl_2$, 5 μl of GC-rich solution, 1 μl of 10 mM dNTP, and 0.3 μl of FastStart Taq DNA Polymerase. Mix well and spin to bring samples down to the bottom of the tube.

3.4.3. PCR Program

Set the PCR program as follows: 1 cycle, 5 min at 95°C; 35 cycles, 30 s at 95°C, 30 s at 60°C, 45 s at 72°C; 1 cycle, 10 min at 72°C (*see* **Note 8**).

3.5. Subcloning the PCR Amplified Promoter Sequence

1. After the PCR reaction is complete, take 2 μl of reaction mixture and check the PCR reaction by running a 1.5% agarose gel using 1X TAE buffer. If the PCR is successful, you should see a single band of the expected size.
2. If the PCR is successful, then run the remaining PCR reaction in a 1% agarose gel and purify the PCR product using gel purification kit according to the manufacturer's instruction (*see* **Note 9**).
3. Elute the purified DNA, usually in 50 μl of H_2O or elution solution. However, to increase the concentration of the sample, 30 μl of H_2O or elution buffer can be used.
4. Subclone the purified PCR product into a vector capable of accepting PCR product with single 3′-deoxyadenosine overhang. For example, one can use the TOPO TA Cloning® kit or pDrive Cloning Vector kit by following the manufacturer's protocols. These vectors also allow for the blue/white screening of transformed bacterial colonies that will enhance selection of positive clones with inserts (white colonies).

3.6. Transformation of Bacteria and Miniprep of Plasmid DNA

1. Conduct transformation and blue/white screening by following the protocol provided by the commercial kits (*see* **Note 10**).
2. Pick up a dozen of white colonies and culture them in 2–5 ml of LB medium overnight at 37°C.
3. Prepare plasmid DNA using kits according to the protocol provided by manufacturers.
4. Measure the concentration of plasmid DNA.
5. Conduct an enzyme digestion to ensure that there is an insert of the expected size in each of the plasmid (*see* **Note 11**). The insert of the plasmid can be sequenced either in-house or sent out for sequencing (*see* **Note 12**).

3.7. Calculation of Methylation

The cytosine residues in CpG dinucleotides will be converted to uracil by sulfite treatment, and the uracil will be converted to thymidine in the following PCR procedure. However, if the cytosine is methylated, it will remain unchanged. Therefore, the presence of a cytosine in a CpG dinucleotide will indicate the methylation of that cytosine (*see* **Note 13**). Methylation status is expressed as percentage of methylated cytosines per total CpG dinucleotides (methylation (%) = methylated cytosines/total cytosines in CpG dinucleotides × 100) (**Fig. 12.1**).

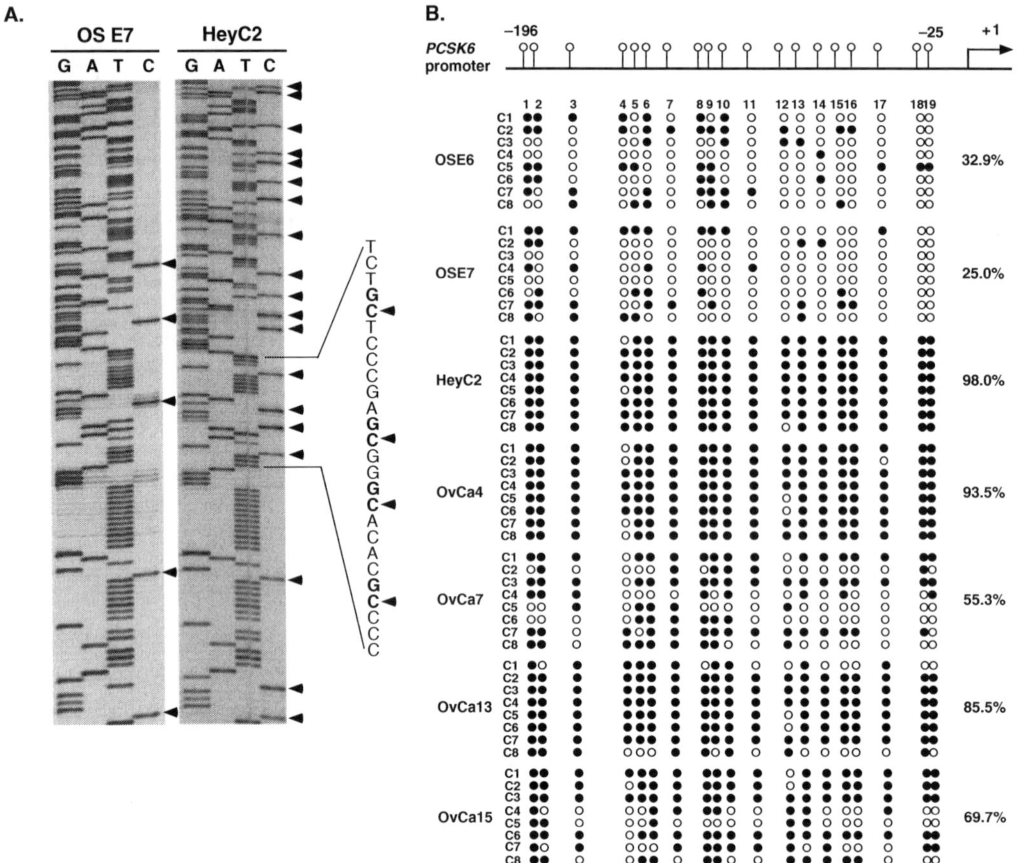

Fig. 12.1. Methylation status of the endogenous *PCSK6* promoter. Methylation status of 19 CpG dinucleotides in the *PCSK6* promoter (–196/–25, relative to translation start site) was determined using bisulfite genomic sequencing. (**a**) Representative bisulfite sequencing gel indicates complete conversion of non-CpG cytosines in both normal OSE and HeyC2 cells. Methylated cytosines in CpG dinucleotides remained unchanged (*arrowhead*). An example of the wild-type *PCSK6* DNA sequence is shown with *arrowheads* indicating methylated CpG nucleotides, while non-CpG cytosine nucleotides have been converted to thymidine. (**b**) Summary of bisulfite genomic sequencing in normal OSE (OSE6 and OSE7), HeyC2, and primary OvCa (OvCa4, 7, 13, and 15) cells. Genomic DNA of eight individual clones (C1–8) from each cell sample was sequenced; the methylation status is summarized as the percentage of methylated cytosine in CpG dinucleotides. Each *circle* is a CpG dinucleotide; *closed circle*, methylated; *open circle*, un-methylated. The relative position of 19 CpG dinucleotides in the *PCSK6* gene (–196/–25) is schematically represented at the top of the figure [Modified from Fu et al. (19)].

3.8. Treatment of Cultured OvCa Cells with DAC and TSA

3.8.1. OvCa Cell Lines

1. Seed OvCa cell lines (HeyC2 and OCC-1) at 3 × 10^5 cells in 10 cm dishes and culture overnight.
2. Treat cells with 1 or 10 μM DAC for 3 days, feeding them fresh medium containing DAC every day.
3. Apply TSA at 0.1 or 1 μM alone or in combination with DAC for the last 18 h of culture (*see* **Note 14**).

3.8.2. Primary OvCa Cells

1. Treat primary OvCa cells with 10 μM DAC for 6 days (due to their slow rate of doubling compared to OvCa cell lines).
2. Include 1 μM TSA for the last 18 h of culture.

3.9. Examination of Induced Gene Expression by RT-PCR

1. After the treatment, isolate total RNA from these cells for RT-PCR analysis of *PCSK6* and *PCSK7* mRNA expression.
2. Terminate PCR reaction after 20 cycles.
3. Detect PCR products by Southern blot using radiolabeled *PCSK6* or *PCSK7* cDNA probes (*see* **Note 15**).
4. Visualize signals by autoradiography; scan and quantify them using NIH Image 1.62.

GAPDH mRNA was used as a positive control for PCR and used to normalize *PCSK6* and *PCSK7* expression. It should be noted that there was no change in *PCSK7* expression with these treatments, indicating that expression of this gene is not modified by epigenetic alterations in human OvCa samples. A synergistic induction of *PCSK6* expression by DAC and TSA in OvCa cells was observed and shown in **Fig. 12.2**.

4. Notes

1. FastStart Taq DNA Polymerase (Roche, 12 158 264 001) optimizes amplification of GC-rich sequences and generates single 3′-deoxyadenosine overhang of PCR amplicons that is required for the following subcloning of the PCR products.
2. Filtered tips and sterile microcentrifuge tubes should be used throughout the protocol to avoid cross-contamination between samples because any contamination can be amplified by the PCR procedure.
3. There is no need to add sodium acetate because there is enough salt from the lysis buffer.
4. Use appropriate amount of TE to dissolve DNA, as excess TE will result in too diluted a sample.
5. Commercial kits for bisulfite treatment are available from several suppliers. If you prefer to use these kits, you can conduct bisulfite treatment by following the instruction manuals.
6. Because the reaction volume is only 10 μl, using a PCR cycler with a heated lid will avoid the evaporation and change of the reaction volume.

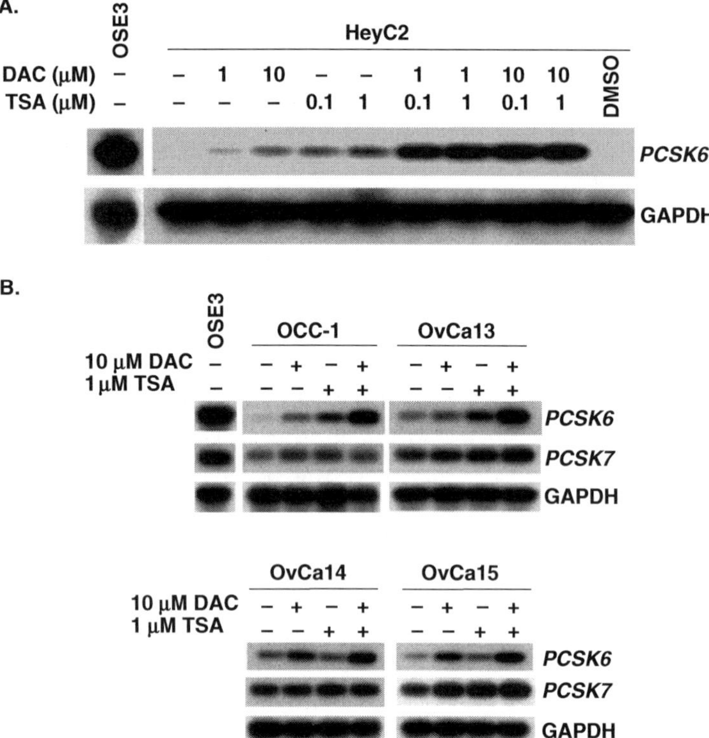

Fig. 12.2. Induction of *PCSK6* gene expression by treatment with a demethylating agent and/or a histone deacetylase inhibitor. (**a**) HeyC2 cells were treated with DAC (1 or 10 μM) or TSA (0.1 or 1 μM) or combination of various doses of DAC and TSA. Cells were cultured in the medium containing DAC for 3 days, with medium changes every day. TSA was applied for the last 18 h. *PCSK6* expression was examined using Southern analysis of RT-PCR products (isolated after 20 PCR cycles) using the full-length *PCSK6* cDNA as a probe. DMSO was used as a vehicle control. Expression in untreated normal OSE (OSE3) was used as a positive control. (**b**) OCC-1 and three primary OvCa cell samples (OvCa 13–15) were untreated (–) or treated (+) with 10 μM DAC, 1 μM TSA or both, and *PCSK6*, *PCSK7*, or *GAPDH* expression was examined. Expression in untreated normal OSE (OSE3) was used as a positive control [Modified from Fu et al. (19)].

7. Bisulfite-treated DNA is not stable in dH_2O, so it should be dissolved in 1 mM Tris–HCl, pH 8.0 if it is not used immediately.
8. PCR cycle numbers can be from 30 to 40 cycles depending on the amount of starting DNA; annealing temperature depends on the melting temperature (T_m) of primers, usually 3–5° lower than the T_m; and elongation time depends on the length of PCR product, usually 1 min for 1 kb.
9. Gel purification of PCR product will increase the efficiency of the subsequent step of cloning PCR product to cloning vectors.

10. Usage of high-efficiency competent bacterial cells will enhance colony number.
11. In the commercial cloning vectors, there are usually two EcoRI sites flanking the insert; therefore, you can use EcoRI to cut the insert out. However, if there are internal EcoRI site(s) in the insert, you have to choose other enzymes for digestion. You may need to conduct digestion using two different enzyme sites flanking the insert.
12. It should be noted that for greatest accuracy several clones from independent experiments must be analyzed to provide the truest representation of the methylation status of the DNA region under examination. In many commercial cloning vectors, T7 promoter and M13 priming sites are present that allow bidirectional sequencing.
13. The success of bisulfite sequencing approach to examine DNA methylation depends on the complete conversion of C to U by the bisulfite treatment. If the bisulfite treatment is successful, there should not be Cs in non-CpG sites. Therefore, the presence of Cs in non-CpG sites will indicate the incomplete bisulfite treatment.
14. A dose response and a time course experiment should be conducted to determine the appropriate concentrations of DAC and TSA to be used for your cells and the durations of treatment to reactivate genes of interest. An equal volume of DMSO was used as a vehicle control for DAC and TSA treatment.
15. Although we used a RT-PCR followed by southern blot approach to quantify the expression of *PCSK6* and *PCSK7* genes in our experiments, we would recommend the real-time RT-PCR technique for the quantitative analysis of the re-expression of silenced genes.
16. An alternative technique to examine DNA methylation is methylation-specific PCR (MSP) (28). MSP requires sodium bisulfite conversion of DNA, followed by two sequential PCR reactions amplifying the same DNA target regions containing the CpG sites of interest. The initial PCR reaction uses primers specific for methylated DNA and the second PCR reaction uses primers specific for unmethylated DNA. The presence and the amount of PCR amplicons from each reaction, visualized by agarose gel electrophoresis, will reflect the methylation status of target sites. MSP is easy to perform, but it provides only qualitative data (20).
17. A common tool to examine chromatin modifications, such as histone acetylation and methylation, is the chromatin immunoprecipitation (ChIP) assay using antibodies against

acetylated or methylated histone proteins. ChIP is either followed by quantitative PCR (ChIP-qPCR) analysis of the enriched DNA to analyze the histone modifications on individual promoters or followed by microarray (ChIP-chip) or by sequencing (ChIP-seq) for genome-wide analysis of chromatin modifications (29, 30).

Acknowledgments

This work was supported by funds to MWN from the Canadian Institute of Health Research regional partnership grant program (ROP-91758) partnered with the Nova Scotia Health Research Foundation (MED-Matching-2008-4881) and the Dalhousie Cancer Research Program.

References

1. Jones, P. A., and Baylin, S. B. (2007) The epigenomics of cancer *Cell* **128**, 683–92.
2. Lopez, J., Percharde, M., Coley, H. M., Webb, A., and Crook, T. (2009) The context and potential of epigenetics in oncology *Br J Cancer* **100**, 571–7.
3. Balch, C., Fang, F., Matei, D. E., Huang, T. H., and Nephew, K. P. (2009) Minireview: Epigenetic changes in ovarian cancer *Endocrinology* **150**, 4003–11.
4. Feinberg, A. P. (2007) Phenotypic plasticity and the epigenetics of human disease *Nature* **447**, 433–40.
5. Nan, X., Ng, H. H., Johnson, C. A., Laherty, C. D., Turner, B. M., Eisenman, R. N., and Bird, A. (1998) Transcriptional repression by the methyl-CpG-binding protein MeCP2 involves a histone deacetylase complex *Nature* **393**, 386–9.
6. Jones, P. L., Veenstra, G. J., Wade, P. A., Vermaak, D., Kass, S. U., Landsberger, N., Strouboulis, J., and Wolffe, A. P. (1998) Methylated DNA and MeCP2 recruit histone deacetylase to repress transcription *Nat Genet* **19**, 187–91.
7. Zhang, Y., Ng, H. H., Erdjument-Bromage, H., Tempst, P., Bird, A., and Reinberg, D. (1999) Analysis of the NuRD subunits reveals a histone deacetylase core complex and a connection with DNA methylation *Genes Dev* **13**, 1924–35.
8. Harikrishnan, K. N., Chow, M. Z., Baker, E. K., Pal, S., Bassal, S., Brasacchio, D., Wang, L., Craig, J. M., Jones, P. L., Sif, S., and El-Osta, A. (2005) Brahma links the SWI/SNF chromatin-remodeling complex with MeCP2-dependent transcriptional silencing *Nat Genet* **37**, 254–64.
9. Szyf, M. (2009) Epigenetics, DNA methylation, and chromatin modifying drugs *Annu Rev Pharmacol Toxicol* **49**, 243–63.
10. Bassi, D. E., Mahloogi, H., and Klein-Szanto, A. J. (2000) The proprotein convertases furin and PACE4 play a significant role in tumor progression *Mol Carcinog* **28**, 63–9.
11. Bassi, D. E., Fu, J., Lopez de Cicco, R., and Klein-Szanto, A. J. (2005) Proprotein convertases: "master switches" in the regulation of tumor growth and progression *Mol Carcinog* **44**, 151–61.
12. Jemal, A., Siegel, R., Ward, E., Hao, Y., Xu, J., Murray, T., and Thun, M. J. (2008) Cancer statistics *CA Cancer J Clin* **58**, 71–96.
13. Scully, R. E., Young, R. H., and Clement, P. B. (1996) Tumors of the ovary, maldeveloped gonads, fallopian tube, and broad ligament. In *Atlas of Tumor Biology*, ed. Rosai, J. Washington, DC: Armed Forces Institute of Pathology.
14. Scully, R. E. (1995) Pathology of ovarian cancer precursors *J Cell Biochem Suppl* **23**, 208–18.
15. Horiuchi, A., Itoh, K., Shimizu, M., Nakai, I., Yamazaki, T., Kimura, K., Suzuki, A., Shiozawa, I., Ueda, N., and Konishi, I.

(2003) Toward understanding the natural history of ovarian carcinoma development: A clinicopathological approach *Gynecol Oncol* **88**, 309–17.

16. Dubeau, L. (1999) The cell of origin of ovarian epithelial tumors and the ovarian surface epithelium dogma: Does the emperor have no clothes? *Gynecol Oncol* 72, 437–42.
17. Auersperg, N., Wong, A. S., Choi, K. C., Kang, S. K., and Leung, P. C. (2001) Ovarian surface epithelium: Biology, endocrinology, and pathology *Endocr Rev* **22**, 255–88.
18. Fathalla, M. F. (1971) Incessant ovulation–a factor in ovarian neoplasia? *Lancet* **2**, 163.
19. Fu, Y., Campbell, E. J., Shepherd, T. G., and Nachtigal, M. W. (2003) Epigenetic regulation of proprotein convertase PACE4 gene expression in human ovarian cancer cells *Mol Cancer Res* **1**, 569–76.
20. Ammerpohl, O., Martin-Subero, J. I., Richter, J., Vater, I., and Siebert, R. (2009) Hunting for the 5th base: Techniques for analyzing DNA methylation *Biochim Biophys Acta* **1790**, 847–62.
21. Clark, S. J., Statham, A., Stirzaker, C., Molloy, P. L., and Frommer, M. (2006) DNA methylation: Bisulphite modification and analysis *Nat Protoc* **1**, 2353–64.
22. Clark, S. J., Harrison, J., Paul, C. L., and Frommer, M. (1994) High sensitivity mapping of methylated cytosines *Nucleic Acids Res* **22**, 2990–7.
23. Wu, J. C., and Santi, D. V. (1985) On the mechanism and inhibition of DNA cytosine methyltransferases *Prog Clin Biol Res* **198**, 119–29.
24. Sheikhnejad, G., Brank, A., Christman, J. K., Goddard, A., Alvarez, E., Ford, H., Jr., Marquez, V. E., Marasco, C. J., Sufrin, J. R., O'Gara, M., and Cheng, X. (1999) Mechanism of inhibition of DNA (cytosine C5)-methyltransferases by oligodeoxyribonucleotides containing 5, 6-dihydro-5-azacytosine *J Mol Biol* **285**, 2021–34.
25. Bolden, J. E., Peart, M. J., and Johnstone, R. W. (2006) Anticancer activities of histone deacetylase inhibitors *Nat Rev Drug Discov* **5**, 769–84.
26. Shepherd, T. G., Theriault, B. L., Campbell, E. J., and Nachtigal, M. W. (2006) Primary culture of ovarian surface epithelial cells and ascites-derived ovarian cancer cells from patients *Nat Protoc* **1**, 2643–9.
27. Grunau, C., Clark, S. J., and Rosenthal, A. (2001) Bisulfite genomic sequencing: Systematic investigation of critical experimental parameters *Nucleic Acids Res* **29**, E65–5.
28. Herman, J. G., Graff, J. R., Myohanen, S., Nelkin, B. D., and Baylin, S. B. (1996) Methylation-specific PCR: A novel PCR assay for methylation status of CpG islands *Proc Natl Acad Sci USA* **93**, 9821–6.
29. Hon, G. C., Hawkins, R. D., and Ren, B. (2009) Predictive chromatin signatures in the mammalian genome *Hum Mol Genet* **18**, R195–201.
30. Park, P. J. (2009) ChIP-seq: Advantages and challenges of a maturing technology *Nat Rev Genet* **10**, 669–80.

Chapter 13

Genetic and Functional Characterization of PCSK1

Hélène Choquet, Pieter Stijnen, and John W.M. Creemers

Abstract

PC1/3 is a neuroendocrine-specific member of the mammalian subtilisin-like proprotein convertase family. This seven-member family is involved in the endoproteolytic cleavage of a large number of precursor proteins including prohormones, proneuropeptides, zymogens, and proreceptors. PC1/3 is synthesized as a zymogen, proPC1/3, and its propeptide is rapidly and autocatalytically cleaved in the endoplasmic reticulum. The mature protein is sorted and stored in dense-core secretory vesicles, together with its substrates. Compound-inactivating mutations in the *PCSK1* gene, which encodes PC1/3, cause monogenic obesity. Furthermore, the contribution of two common nonsynonymous variants in *PCSK1* to polygenic obesity risk has recently been established. Additional rare variants have been identified in non-consanguineous extremely obese Europeans but functional characterization has not yet been described. Sequencing efforts of larger cohorts of obese patients might reveal more variants conferring risk of obesity.

Key words: *PCSK1*, PC1/3, obesity, convertase, endoproteolytic processing, neuropeptides, hormones.

1. Introduction

PCSK1, together with *leptin*, was the first human gene associated with monogenic obesity to be discovered (1). Although homozygous or compound heterozygous loss-of-function mutations in *PCSK1* are rare and have so far been described only in three patients, common nonsynonymous variants have recently been found to make a significant contribution to polygenic obesity (1–4). Additionally, rare heterozygous mutations have been identified in European subjects contributing to severe forms of obesity (Creemers et al., manuscript in preparation). Current developments in high-throughput sequencing make the analysis of the

M. Mbikay, N.G. Seidah (eds.), *Proprotein Convertases*, Methods in Molecular Biology 768,
DOI 10.1007/978-1-61779-204-5_13, © Springer Science+Business Media, LLC 2011

entire *PCSK1* gene in cohorts of obese patients feasible and will firmly establish the contribution of *PCSK1* variants to obesity. The maturation and activity of a novel PC1/3 allele must subsequently be assessed to determine its potential impact. Here we describe a convenient method to sequence the *PCSK1* gene and functionally characterize the genetic variants.

2. Materials

2.1. Sequencing of PCSK1 Gene in Obese Patients

2.1.1. Instrumentation and Software

1. Vacuum pump, mechanical purification by filter (Millipore, Billerica, USA).
2. 3730XL DNA sequencer analyzer (Applied Biosystems, ABI, Foster City, CA., USA).
3. Software for sequences analysis. We use the VariantR software (Applied Biosystems, ABI, Foster City, CA, USA).

2.1.2. Reagents and Consumables

All reagents and consumables should be of standard molecular biology grade. Use sterile distilled or deionized water:

1. PCR primers.
2. PCR reagent kit containing FastStart Taq, 10× PCR buffer + $MgCl_2$, 10× PCR buffer without $MgCl_2$, 25 mM $MgCl_2$, 5× GC-RICH sol. (Roche).
3. Deoxyribonucleotide (dNTP) set, 100 mM for each dNTP (dATP, dCTP, dGTP, dTTP) (Promega, Madison, USA).
4. Big Dye Terminator Cycle Sequencing Ready Reaction Kit (Applied Biosystems, ABI, Foster City, CA, USA).
5. Sephadex G-50 Fine (GE Healthcare, Uppsala, Sweden).
6. 96-Well plates (ABgene).
7. 96-Well MultiScreen plates (Millipore, Billerica, MA, USA).
8. 96-Well MultiScreen filter plates for high-throughput separations (Millipore).

2.2. PC1 Activity Assay

1. Reaction buffer stock solution: To 5 ml of 1 M sodium acetate, pH 5.5, add 5 ml of 1% Brij-30, 5 ml of 50 mM $CaCl_2$, 5 ml of 10× inhibitor cocktail (*see* **Note 1**).
2. 96-Well microtiter plate fluorimeter, reading at 380 nm excitation, 460 nm emission (or similar wavelengths).
3. Amino methylcoumarin (amc) standard (Peptides International).
4. 2 mM pERTKR-amc substrate stock (Peptides International) in water.

5. 1 M Sodium acetate, pH 5.5 (keep at 4°C).
6. 1% Brij-30
7. 10× Inhibitor cocktail (required when the PC1 enzyme is not being assayed in purified form): 10 μM E-64, 1 μM leupeptin, 10 μM pepstatin, and 1 mM TPCK (all from Sigma). Store the cocktail aliquoted at 20°C.
8. Polypropylene 96-well microtiter plates (black, flat-bottomed; Costar).
9. SealPlate (Excel Scientific #100-SEAL-PLT).

3. Methods

3.1. Sequencing of PCSK1 Gene in Obese Patients

1. Design PCR primers flanking the coding regions of interest using an online software (*see* **Note 2**).
2. Test each PCR primer pair in an in silico PCR to verify that a specific DNA fragment is being amplified (*see* **Note 3**).
3. Resuspend the PCR primers.
4. Amplify DNA samples by PCR according to an optimized protocol (**Table 13.1**).
5. Perform the PCR reactions in a thermocycler using an initial 4-min denaturation step at 95°C followed by 40 cycles of 30 s at 95°C and 30 s at melting temperature and 45 s at 72°C.
6. Verify the success of the amplification on a 2% agarose gel for a subset of samples.

Table 13.1
DNA amplification reaction protocol

Reagent	Volume per reaction (μL)
DNA, 10 ng/μL	3
$MgCl_2$, 25 mM	2
Buffer, 10×	2.5
dNTP, 4 mM	2
Fast Taq	0.2
Primer forward, 10 μM	0.8
Primer reverse, 10 μM	0.8
Total volume	25

7. Add 75 μl of H_2O in each well of a 96-well MultiScreen plate; then add the PCR products (approximately 25 μl).
8. Put the plate on the vacuum pump to make mechanical purification by filter for 15 min.
9. Add 20–25 μl of H_2O to resuspend the PCR products.
10. Put the plate on agitation in a vortex for 10 min.
11. Recup the PCR products purified (PPP) and eventually for a quality control, check on a 2% agarose gel for a subset of the samples.
12. Process to the sequence reaction according to an optimized protocol (**Table 13.2**).
13. Perform the sequence reaction in a thermocycler using an initial 1-min denaturation step at 96°C followed by 35 cycles of 10 s at 96°C, 5 s at 50°C, and 2 min 30 s at 60°C.
14. Add Sephadex G-50 Fine in each well of a 96-well MultiScreen filter plate for high-throughput separations, then add 200 μl of H_2O, next put this plate minimum 2 h at 4°C.
15. Centrifuge the 96-well MultiScreen filter plate for 10 min at speed 910 rcf (to remove water).
16. Pipette the sequence reaction products and put them on the 96-well MultiScreen filter plate.
17. Centrifuge the 96-well MultiScreen filter plate for 10 min at speed 910 rcf.
18. Scan the 96-well MultiScreen filter plate on the 3730XL DNA sequencer analyzer.
19. Use a software for sequence analysis and to detect new mutations.

Table 13.2
Sequence reaction protocol

Reagent	Volume per reaction (μL)
PPP	3
Big Dye buffer	3.75
Big Dye v.3.1	0.5
Primer forward or reverse, 10 μM	0.16
Total volume	20

3.2. PC1 Activity Assay (See Note 4)

1. Pipette 20 μl of reaction buffer stock solution into a 96-well microtiter plate. Perform experiments at least in duplicate or triplicate.
2. Add sample volumes of up to 25 μl (*see* **Notes 5–6**).
3. Pipette known concentrations of free amc in duplicate in separate wells for construction of a standard curve.
4. Add distilled water to reach a total reaction volume of 45 μl.
5. Add 5 μl of pERTKR-amc substrate to each well. Mix gently tapping the side of the plate.
6. Cover the plate tightly with SealPlate to prevent evaporation and incubate at 37°C.
7. Read the fluorescence of the samples with the microtiter fluorimeter at 5–15 min intervals. Return the plate to the incubator after each reading if the plate reader cannot be set at 37°C. Take readings for up to 2 h (*see* **Note 7**).
8. Plot fluorescence against time, first subtracting the fluorescence in the blank, and estimate PC1 enzyme activity by reference to the standard curve. Use the linear portion of the activity curve for activity estimations (i.e., after the lag phase) (*see* **Note 8**).

4. Notes

1. The final concentration is 0.1 M sodium acetate, pH 5.5, 0.1% Brij-30, 1× inhibitor mix, and 5 mM $CaCl_2$. The inhibitor cocktail is not necessary if purified PC1 enzyme is used. Without inhibitor cocktail, the reaction buffer stock solution can be stored at 4°C for a prolonged period of time (up to a year and longer) if 0.02% azide is added. With inhibitor cocktail the solution should be made fresh and kept on ice. Do not use commercially available inhibitor cocktails as they usually contain serine protease inhibitors and chelators like EDTA.
2. The length of the PCR fragments should be not too long, optimally around 200–700 bp. NCBI/Primer-BLAST is freely available on the Internet (http://www.ncbi.nlm.nih.gov/tools/primer-blast/).
3. The online software for in silico PCR can be found at http://genome.ucsc.edu/cgi-bin/hgPcr.
4. Recombinant PC1/3 can be conveniently collected from conditioned medium of transiently transfected cells (e.g.,

HEK293T, Neuro2A, βTC3, CHO) without further purification. For this purpose, incubate the cells 8–16 h in serum-free medium. Activity can be normalized for PC1/3 expression by means of Western blotting. To this end, precipitate 1 volume (e.g., 250 μl) of conditioned medium with 4 volumes methanol at –20°C for >2 h, centrifuge for 15 min at maximum speed in a benchtop centrifuge, and resuspend the dried pellet in sample buffer.

5. When intracellular PC1/3 is analyzed, immunopurification is recommended. Lysis of the cells should be performed with a mild non-denaturing detergent like 1% Brij-30 or 1% Triton X-100. A PC1/3-specific antibody can be used, but in our hands, an epitope-tagged construct (e.g., FLAG: Asp-Tyr-Lys-Asp-Asp-Asp-Asp-Lys) gives best results (1–4). In this construct, the epitope tag is inserted between the propeptide and the catalytic domain, so all protein forms of PC1/3 will be immunoprecipitated. Normalization can be performed by using 10–50% of the immunoprecipitate for Western blotting and the remaining 50–90% for the activity study. Before use, make sure to remove all residual fluid from the beads using a syringe and premix 25 μl of water with 20 μl of reaction buffer before addition.
6. Appropriate controls in the same buffer should be included. This allows estimation of the intrinsic fluorescence/water hydrolysis of the substrate alone. Typically, 25 μl of conditioned culture medium is assayed and plain medium is used in control wells.
7. The amount of substrate hydrolyzed should be kept below 10% of the total substrate present for accurate quantitation of the enzyme activity, but an activity of at least 20 times the blank should be achieved.
8. Linear rates of hydrolysis are often not achieved until a lag period of ~30 min after the start of the reaction. Full-length PC1 exhibits much longer rates of linear hydrolysis than do carboxyterminally truncated forms of PC1.

Acknowledgments

This work was supported by GOA2008/16, "FWO Vlaanderen," Le Conseil Régional Nord Pas de Calais/FEDER, and the Agence Nationale de la Recherche.

References

1. Jackson, R. S., Creemers, J. W., Ohagi, S., Raffin-Sanson, M. L., Sanders, L., Montague, C. T., Hutton, J. C., and O'Rahilly, S. (1997) Obesity and impaired prohormone processing associated with mutations in the human prohormone convertase 1 gene *Nat Genet* **16**, 303–6.
2. Benzinou, M., Creemers, J. W., Choquet, H., Lobbens, S., Dina, C., Durand, E., Guerardel, A., Boutin, P., Jouret, B., Heude, B., Balkau, B., Tichet, J., Marre, M., Potoczna, N., Horber, F., Le Stunff, C., Czernichow, S., Sandbaek, A., Lauritzen, T., Borch-Johnsen, K., Andersen, G., Kiess, W., Korner, A., Kovacs, P., Jacobson, P., Carlsson, L. M., Walley, A. J., Jorgensen, T., Hansen, T., Pedersen, O., Meyre, D., and Froguel, P. (2008) Common nonsynonymous variants in PCSK1 confer risk of obesity *Nat Genet* **40**, 943–5.
3. Farooqi, I. S., Volders, K., Stanhope, R., Heuschkel, R., White, A., Lank, E., Keogh, J., O'Rahilly, S., and Creemers, J. W. (2007) Hyperphagia and early-onset obesity due to a novel homozygous missense mutation in prohormone convertase 1/3 *J Clin Endocrinol Metab* **92**, 3369–73.
4. Jackson, R. S., Creemers, J. W., Farooqi, I. S., Raffin_Sanson, M. L., Varro, A., Dockray, G. J., Holst, J. J., Brubaker, P. L., Corvol, P., Polonsky, K. S., Ostrega, D., Becker, K. L., Bertagna, X., Hutton, J. C., White, A., Dattani, M. T., Hussain, K., Middleton, S. J., Nicole, T. M., Milla, P. J., Lindley, K. J., and O'Rahilly, S. (2003) Small-intestinal dysfunction accompanies the complex endocrinopathy of human proprotein convertase 1 deficiency *J Clin Invest* **112**, 1550–60.

Section IV

Experimental Models

Chapter 14

Prohormone Processing in Zebrafish

Michael G. Morash, Kelly Soanes, and Younes Anini

Abstract

Proprotein convertases (PCs) are secretory proteolytic enzymes that activate precursor proteins into biologically active forms by limited proteolysis at one or multiple internal sites. PCs are implicated in the processing of multiple protein precursors, including hormones, proteases, growth factors, angiogenic factors, and receptors. PCs have been linked recently to various pathologies such as Alzheimer's disease, tumorigenesis, and infections. The zebrafish has emerged as an attractive model for studying the role of PCs not only in substrate production but also in development. Herein we describe methods that are used to characterize DNA sequences of PCs in zebrafish, as well as to evaluate the ontogeny and tissue distribution of their transcripts. We also provide information on the morpholino-mediated knockdown of proprotein convertases.

Key words: Zebrafish, morpholino, processing, expression.

1. Introduction

Proprotein convertases (PCs), or proprotein convertases subtilisin/kexins (PCSKs), are calcium-dependent subtilases that proteolytically convert secretory precursor proteins to bioactive polypeptides. So far nine PCs have been discovered, of which seven are yeast kexin homologues and cleave peptide bonds C-terminal to basic residues [for a comprehensive review, see (1)]. Two other subtypes, namely subtilisin kexin isozyme-1 (SKI-1)/ site 1 protease (S1P)/PCSK8 (2) and neural apoptosis regenerative convertase-1 (NARC-1)/PCSK9 (3), which cleave peptide bonds C-terminal to a non-basic amino acid, were discovered later. Typically, kexin-like PCs cleave peptide bonds at the general motif (K/R)-$(X)_n$-(K/R)↓, where $n = 0, 2, 4$, or 6 and X is not Cys. PCs are found in all cells in varying combinations and

M. Mbikay, N.G. Seidah (eds.), *Proprotein Convertases*, Methods in Molecular Biology 768, DOI 10.1007/978-1-61779-204-5_14, © Springer Science+Business Media, LLC 2011

amounts, and many of them are crucial for normal development and physiology. At the cellular level, PCs are upstream participants in pathways controlling cell growth and differentiation [for a review, see (4)].

The zebrafish (*Danio rerio*) has become a popular model for developmental biologists due to its amenability to a variety of techniques. Primarily, external egg fertilization and a short embryonic development (~5 days until free feeding) make investigation of systems throughout development possible. The zebrafish is emerging as a unique animal model to investigate the physiological and pathophysiological roles of PCs. Four of the proprotein convertases have been described in zebrafish to date. The neuroendocrine proprotein convertases PC1 and PC2 were cloned and sequenced from zebrafish (5). Both PC1 and PC2 display high degrees of similarity to their counterparts in other species, including the conservation of the catalytic triad and other essential residues. The brain contained the highest expression levels of both PC1 and PC2, and both transcripts were also detectable in the fore, mid, and distal gut. PC1 and PC2 were detectable at 4.5 h post-fertilization, and while PC1 expression increased throughout development, PC2 expression remained unchanged until increasing substantially at 5 days post-fertilization (5). The PC1 co-factor proSAAS has also been identified in zebrafish (6). Although zebrafish proSAAS protein shows an overall amino acid sequence identity of only 30% with mouse proSAAS, two 14–16 residue hydrophobic segments (predicted to form alpha-helices) and two 9–11 residue sequences containing basic convertase cleavage sites are highly conserved, indicating that this sequence may be of functional importance. Zebrafish proSAAS exhibited neural and endocrine distributions. Also, the zebrafish homologue of PCSK9 has been characterized (7). PCSK9 displays an expression profile in the CNS similar to that of mice, where it is involved in organization of cerebellar neurons. Knockdown of PCSK9 in zebrafish results in embryonic death. Finally, two homologues of furin have been described in zebrafish; they are involved in craniofacial patterning (8).

In order to assess the function of proprotein convertases in zebrafish, it is first necessary to obtain their entire sequence, determine their tissue expression levels, and then disrupt their expression.

2. Materials

2.1. Animal Handling and Tissue Extraction

1. Fish food: Nutrafin freeze-dried brine shrimp and Nutrafin staple flake food (Rolf C. Hagen, Inc., Montreal, PQ, Canada).

2. 5 gallon fish tanks at 28°C on a 14:10 light:dark cycle and 2 L breeding tanks with dividers and egg traps.
3. 1X E3 medium: 5 mM NaCl, 0.17 mM KCl, 0.33 mM $CaCl_2$, 0.33 mM $MgSO_4$, made up as 60X stock.
4. 0.04% (w/v) MS-222 (ethyl 3-aminobenzoate methane-sulfonate salt (Tricaine); Sigma Chem. Co., Ontario, Canada).
5. RNAlater (Qiagen).
6. RNase away (Invitrogen).
7. 100 × 15 mm polystyrene Petri dishes.
8. Plastic tea strainer.
9. Polyethylene spray bottle.

2.2. RNA Isolation

1. Tissue homogenizer or 23 gauge needle and 1 ml syringes, RNase free.
2. RNase- and DNase-free tubes, pipette tips.
3. UV spectrophotometer.
4. Thermal cycler.
5. Refrigerated microcentrifuge.

We have successfully used two RNA extraction protocols using adult zebrafish tissues and embryos. Thus they will both be discussed herein.

2.2.1. Protocol #1: Trizol

1. Trizol reagent (Invitrogen, ON). Store at 4°C (*see* **Note 1**).
2. Chloroform (Sigma), Store at room temperature (*see* **Note 1**).
3. Diethyl pyrocarbonate (DEPC, Sigma).
4. DPEC-treated distilled deionized water (ddH_2O): 0.5 ml of DEPC into 500 ml of ddH_2O, shake vigorously for 5 min. Allow to sit in fume hood for at least 1 h, then autoclave (*see* **Note 1**).
5. 100% isopropanol and 70% ethanol (made in DEPC-treated ddH_2O) from Sigma stored at –20°C.
6. DNaseI (50–375 U/μl) (Invitrogen).
7. 10X DNaseI buffer: 200 mM Tris-HCl (pH 8.4), 20 mM $MgCl_2$, and 500 mM KCl.

2.2.2. Protocol #2: Bio-Rad

1. Aurum Total RNA Fatty and Fibrous Tissue Kit (Bio-Rad) stored at room temperature, except PureZOL reagent at 4°C.
2. Reconstituted DNaseI solution (included in kit).
3. Elution buffer prewarmed to 70°C (included in kit).

2.3. Determining RNA Integrity

1. 2X RNA loading dye (MBI Fermentas): 95% formamide, 0.025% SDS, 0.025% bromophenol blue, 0.025% xylene cyanol FF, 0.025% ethidium bromide, 0.5 mM EDTA.
2. MOPS running buffer: 0.4 M 3-*N*-morpholinopropanesulfonic acid (MOPS), 0.1 M sodium acetate, and 0.01 M EDTA.
3. Formaldehyde agarose gel: per 100 ml, 1 g low-melting point agarose, 18 ml of 37% formaldehyde, 10 ml of MOPS running buffer, and 72 ml of ddH_2O.
4. Electrophoresis equipment in a fume hood.
5. Gel imaging system and densitometry software.

2.4. cDNA Synthesis

1. Superscript II reverse transcriptase (200 U/μl, Invitrogen).
2. 5X first-strand buffer: 250 mM Tris-HCl, pH 8.3 at room temperature; 375 mM KCl, 15 mM $MgCl_2$, and 0.1 M DTT.
3. 250 ng of random decamers/μl DEPC ddH_2O.
4. 10 mM dNTP.
5. Optional RNaseOUT 40 U/μl (Invitrogen) (*see* **Note 2**).

2.5. PCR and Rapid Amplification of cDNA Ends

1. Platinum Taq DNA polymerase kit (Invitrogen). In each reaction: 1X buffer [20 mM Tris-HCl (pH 8.4), 50 mM KCL], 0.2 mM dNTP mix, 1.5 mM $MgCl_2$, 0.2 μM forward/reverse primers, 1.0 units of Taq polymerase, 2 μl of cDNA template, and ddH_2O to 50 μl.
2. FirstChoice RLM-RACE Kit for 3′- and 5′-rapid amplification of cDNA ends (Ambion).
3. Oligonucleotides (oligos) for the following applications: detection, full-length transcripts, two nested reverse primers near the 5′-end of the transcripts for RACE.
4. 10X Tris borate EDTA (TBE): 108 g Tris base, 55 g boric acid, 20 ml of 0.5 M EDTA, ddH_2O to 1 l.
5. Agarose gel: 1 g low-melting point agarose, 10 ml of 10X TBE, ddH_2O to 100 ml, microwave for 2 min, allow to cool to 60°C, cast.
6. 1X TBE running buffer: 100 ml 10X TBE and 900 ml ddH_2O.
7. 10X DNA loading buffer: dissolve 250 mg of bromophenol blue in 33 ml of 150 mM Tris pH 7.6, add 60 ml glycerol and 7 ml H_2O. Store at room temperature.

2.6. DNA Sequencing, Sequence Analysis, and Alignment

1. DNA sequencing facility.
2. Multiple sequence alignment program, such as Clustal W2 (http://www.ebi.ac.uk/Tools/clustalw2/index.html).
3. DNA sequence manipulation software (example: http://www.generunner.net/).
4. BLAST algorithm (http://blast.ncbi.nlm.nih.gov/Blast.cgi).

2.7. Quantitative Real-Time PCR

1. 2X iQ SYBR Green Supermix (Bio-Rad).
2. Oligos for qPCR.
3. Real-time PCR thermal cycler.

2.8. Morpholino Knockdown

Adapted from (9).

1. Kwik-Fil borosilicate glass capillaries (World Precision Instruments) (*see* **Note 3**).
2. Standard Fisherbrand microscope slides (or equivalent).
3. E3 media (*see* above and **Note 4**).
4. Phenol red (Sigma).
5. 5.0 ml graduated LDPE Ultident brand transfer pipettes.
6. Gene-specific morpholino (MO) antisense oligos and mispair oligo controls (*see* **Note 5**).
7. Mineral oil for bolus calibration.
8. mMessage mMachine T7 kit (Ambion).
9. Complete cDNA clone of gene of interest with T7 or T3 upstream promoter (Open Biosystems).
10. RNeasy MinElute Cleanup Kit (Qiagen).
11. RNase-Free DNase Set (Qiagen).
12. SuperScript® III Reverse Transcriptase kit (Invitrogen).
13. Taq DNA Polymerase with ThermoPol Buffer (New England BioLabs).
14. dNTP kit (Fermentas).
15. Gene-specific primers to detect MO-mediated alternative splicing (IDT).
16. Distilled water (dH_2O).
17. Optional: Commercially available injection molds for creating agarose injection dishes.

2.0.1. Equipment

1. Programmable microinjector (e.g., Epindorph FemtoJet Express).
2. Micromanipulator (e.g., Narishege MM-151).
3. Micropipette needle puller (e.g., Sutter Instrument Co.).

4. Stereomicroscope (Nikon or Leica) with transmitted light base, oblique illumination control is desirable (*see* **Note 6**).
5. Standard bench top incubator capable of maintaining a 28°C temperature (e.g., Sheldon Economy Incubators).
6. Standard dry bath incubator (Fisher Scientific).
7. 1 × 0.01 mm stage micrometer (Fisher Scientific).
8. Standard thermocycler (PerkinElmer).

3. Methods

3.1. Fish Maintenance, Tissue Harvesting, and Embryo Collection

1. Maintain the broodstock according to standard culture conditions: 28°C, 14:10 light:dark cycle, following appropriate animal care guidelines. Zebrafish were housed at 30 adult (20–30 mm in length) per 5 gal tank (*see* **Note** 7).
2. For tissue extraction, individual fish were placed in a 0.04% tricaine bath (in 100 ml of system water) until the operculi stop moving.
3. Dissect fish and place removed tissues of interest into 1 ml RNA later, and stored at –20°C for no more than 24 h.
4. Egg collection: late Egg collection: late afternoon, the day before the injections, remove an equal number of similarly sized, well-fed males and females of the appropriate genotype from the stock tanks using a clean net and place them in the breeding traps filled with fresh fish water. Separate the males and females with the plastic dividers provided.
5. Maintain the breeding traps at 28°C following the same 14:10 light:dark cycle. To maintain water quality in the breeding traps and therefore optimal spawning use a maximum of four fish per 2 L breeding trap.
6. The next day once the lights are on pull the dividers and allow fish to spawn naturally (*see* **Note 8**). Make a note of the time the dividers were pulled, consider this time 0. Once eggs are deposited (15–30 min), transfer fish back to housing tanks and transfer embryos by pouring out contents of breeding tank into a tea strainer. Rinse well with 1X E3 media and transfer embryos to a 100 mm Petri dish with 1X E3, so that eggs are stored at 100 embryos per plate.
7. For RNA isolation, unfertilized and fungal eggs are sorted out in E3 media, and 30–40 4 & 24 h, 30 48 & 72 h and 25 96 & 120 h embryos were used per group, embryos were used immediately for RNA extraction.

3.2. RNA Isolation (see Note 9)

Two techniques were used for RNA isolation. Each will be described separately.

3.2.1. Trizol Method

1. 100 mg of tissue or the appropriate number of embryos receive 1 ml of Trizol on ice. Homogenize tissues mechanically, entirely, by the use of a tissue homogenizer in 13 ml tubes or by repeated passage through a 23 gauge needle in a 1.5-ml tube. Transfer samples to fresh 1.5 ml tubes. Change needles and syringes after every tissue. Clean the homogenizer with ddH_2O containing RNase away three times between every sample, wipe with a tissue, and inspect for cleanliness after each wash. During processing, keep already homogenized samples at –20°C for no more than 1 h.
2. Once all samples are prepared, thaw frozen samples at room temperature, and then incubate all samples at room temperature for 5 min.
3. Add 200 μl of chloroform to each sample, shake vigorously by hand for 1 min, and incubate at room temperature for 2 min 30 s.
4. Centrifuge samples for 15 min, 3,000 rpm, at 4°C.
5. The material will separate into two layers. Gently remove the tubes, and using a 200 μl pipette tip, transfer the top (aqueous) layer (approximately 550 μl) to a fresh 1.5 ml tube, avoiding the interface and the organic (red) phase (*see* **Note 10**).
6. Add 550 μl of isopropanol to the aqueous phase, mix by inverting, and incubate at room temperature for 15 min (*see* **Note 11**). Spin 14,000 rpm, 15 min, at 4°C.
7. The RNA pellet will have formed on the bottom of the tube (*see* **Note 12**). Discard supernatant, add 1 ml of –20°C, 70% ethanol to the tube, and loosen the pellet from the bottom with gentle pipetting. It is not necessary to break up the pellet completely.
8. Spin for 10 min, 14,000 rpm, at 4°C. Remove ethanol and air-dry pellet until tube is dry (*see* **Note 13**). Resuspend the pellet in 10–20 μl of DEPC-treated water.
9. Dilute RNA 1/100 in DEPC ddH_2O. Determine $A_{260/280}$. A ratio of 1.8 or greater should be obtained.
10. A DNaseI digestion is required. Prepare DNaseI digestions so that 1 μg of RNA will be treated per each 10 μl of digestion reaction. Dilute 10X DNaseI buffer in DEPC ddH_2O, add RNA, and finally, add 1 μl of DNaseI (20,000 U). Treat at room temperature for

15 min; add 1 μl of EDTA solution per microgram of RNA prep. Heat at 65°C for 10 min, and then store on ice.

3.2.2. Bio-Rad Method

1. The homogenizing steps are identical to that of Trizol, except using the PureZol reagent, including the 5 min room temperature incubation step.
2. Add 0.2 ml of chloroform, shake 15 s, and incubate again for 5 min at room temperature.
3. Centrifuge >12,000×*g*, 15 min, at 4°C.
4. Carefully remove the aqueous top layer, avoiding the organic (red) layer and the interface (*see* **Note 10**). Add an equal volume (~550 μl) of 70% ethanol, mix thoroughly.
5. Pipette 700 μl of the RNA prep into an assembled RNA-binding column and collection tube. Spin 60 s, room temperature, 12,000×*g*. Discard flow through and repeat with remaining prep.
6. Discard flow through, add 700 μl of low-stringency wash buffer, spin 30 s, >12,000×*g*.
7. Add 80 μl of diluted DNaseI to the column and incubate 15 min at room temperature, centrifuge column for 30 s, room temperature, 12,000×*g*.
8. Wash with 700 μl of high-stringency wash, spin 30 s, room temperature, 12,000×*g*.
9. Wash with 700 μl of low-stringency wash, spin 30 s, room temperature, 12,000×*g*. Centrifuge an additional 2 min, room temperature, 12,000×*g* to remove any excess wash solution.
10. Transfer column to a collection tube, add 30–40 μl of 70°C elution buffer to the column, incubate 1 min at room temperature, spin 2 min, at room temperature, 12,000×*g*.
11. Dilute RNA 1/100 in DEPC ddH_2O. Determine $A_{260/280}$. A ratio of 1.8 or greater should be obtained.

3.3. Determining RNA Integrity

1. Mix 1 μg of RNA sample (diluted in DEPC ddH_2O if necessary) with an equal volume of loading buffer, heat for 10 min, 70°C, then chill on ice (*see* **Note 14**).
2. Prepare formaldehyde agarose gel. Dissolve 1 g agarose into 72 ml of ddH_2O by boiling for 1–2 min. Allow agarose to cool slightly, but remember both the MOPS buffer and formaldehyde are at room temperature and will decrease the temperature of the agarose further. Add 10 ml MOPS buffer, swirl to dissolve, then add 18 ml of formaldehyde in a fume hood, and cast gel. Cover gel with 1X MOPS running buffer, load RNA samples in RNA loading buffer into each well, and

then run samples on gel at 100 V until decent separation of the dye fronts occurs.

3. Visualize and assess ratio of 28S to 18S using densitometry. Ratio should be ~2 (*see* **Note 14**).

3.4. cDNA Synthesis

1. Dilute RNA to a final concentration of 0.1 μg/μl in DEPC ddH_2O. 10 μl of diluted RNA prep receives 1 μl of 10 mM dNTP mix and 250 ng of random decamers (*see* **Note 15**). Heat solution to 65°C for 5 min and snap chill on ice.
2. For each 12 μl of RNA/dNTP/primer mix, add 4 μl of 5X buffer, 2 μl of DTT, and 1 μl of DEPC ddH_2O or RNase out (*see* **Note 2**).
3. Heat to 42°C for 2 min, and then add 1 unit of reverse transcriptase. Incubate at 42°C for 50 min, followed by 70°C for 15 min.

3.5. PCR and RACE

1. Using expression primers (*see* **Note 16**), perform PCR on various tissues to determine which tissues express the desired PC. Reaction consists of 2 μl cDNA prep, 1 μl of 10 mM primers, $MgCl_2$, buffer, dNTPs, Taq, and water to 25 μl. Run 35 cycles. Mix 1 μl of 10X DNA loading dye with 9 μl of PCR product, and load sample into the prepared agarose gel (*see* **Section 2.5**). Include a DNA ladder for size determination. Apply 100 V of current to the gel until decent separation occurs; visualize to ensure size and presence (*see* **Note 16**).
2. After identifying a tissue with expression and using the available genome sequence to design primers that will amplify the entire open reading frame, PCR the entire coding sequence using the PCR reaction described above (*see* **Note 8**).
3. Perform RACE to identify 5′-start site and upstream coding sequence.

3.6. DNA Sequencing, Sequence Analysis, and Alignment

1. Submit the full-length cDNA and the RT-PCR amplicons for DNA sequencing following the sequencing facility's protocol.
2. Once the DNA sequences are obtained, perform a multiple sequence alignment to develop a consensus sequence.
3. BLAST the consensus against the genome in order to identify the exons. Then design and test primers for qPCR, and sequence resulting amplicons (*see* **Note 16**).

3.7. qRT-PCR

1. Prepare a premix containing enough reaction mix for four to five more reactions than you are running (per reaction; 12.5 μl Bio-Rad SYBR Green 2X premix, 8.5 μl ddH_2O, 1 μl of each 10 mM primer) (*see* **Note 16**).

2. Add 23 μl of premix to each well of the PCR plate. Add 2 μl of template to each reaction mix. Briefly spin PCR plate 200–300 rpm, 15 s, 4°C to collect all reagents at the bottom of wells. Include serial dilutions of cDNA from known expressing tissue (5- to 10-fold) in order to determine primer efficiency.
3. Run qPCR reaction on a real-time thermal cycler. Use provided software for analysis (*see* **Note 17**).

3.8. Morpholino Knockdown Studies

1. The day prior to experimentation, set up breeding pairs as described above. Proceed with egg collection the day of the experiment as described.
2. From the prepared stock of the MO antisense oligos create the appropriate dilutions in dH_2O (*see* **Note 18**). If desired, the injection tracking dye, phenol red, can be added to the MO solution (0.05%).
3. While fish are spawing, pull needles and load 2 μl of the appropriate MO dilution (*see* **Notes 19 and 20**) onto the Kwik-Fil needles and allow time to fill (*see* **Note 3**).
4. Prepare the injection station. Break the end of the filled needle with a pair of forceps to create an appropriate sized bore (*see* **Note 20**). Position the needle in the holder of the microinjector and connect to the micromanipulator and calibrate the needle (*see* **Note 21**).
5. Prepare the injection dish and load embryos using plastic transfer pipettes (**Note 22**).
6. Between the one- and four-cell stages deliver the appropriately sized bolus of MO antisense oligo near the yolk/cell interface (**Note 23**).
7. Wash embryos off the injection plate into a new Petri dish with E3. Label with the appropriate genotype and experimental conditions, incubate at 28°C and monitor development (*see* **Note 24**).
8. Continue to remove unfertilized and dead embryos with transfer pipettes to prevent toxins produced from dying embryos from adversely affecting development of the remaining embryos in the dish.
9. Repeat the injections using the control MOs at the same concentration(s) used for the gene-specific antisense oligos.
10. Observe and record the percent survival and morphant phenotypes from each experiment, and use the data to optimize the effective MO concentration for knockdowns.

11. Confirm MO-mediated gene-specific knockdowns through Western analysis, whole mount immunohistochemistry, RT-PCR, or mRNA rescue (*see* **Note 25**).

4. Notes

1. DEPC is a carcinogen. Observe caution when using. Trizol and chloroform are harmful chemicals. Avoid skin contact and carry out procedures in a fume hood.
2. RNase OUT is an RNA inhibitor which is recommended when using the Superscript II. We have found it not necessary.
3. Kwik-Fil borosilicate glass capillaries are recommended for ease and speed of loading.
4. 1×10^{-5}% methylene blue can be added to the E3 as a fungicide; however, it is not necessary if the plates are frequently cleaned and the density of embryos is maintained at less than 100 embryos/plate. Transferring embryos to fresh Petri dishes daily also reduces fungal growth. Methylene blue will interfere with fluorescence microscopy. To ensure there is no adverse effect on development due to overcrowding no more than 100 embryos should be held in a Petri dish in approximately 20 ml of E3 overnight. Dishes should be cleaned frequently by removing dead embryos. To control fungal growth surviving embryos should be transferred to new plates with fresh E3, daily.
5. Use Gene Tools MO antisense design service to create the appropriate gene-specific translational, splice-blocking and mispair control MOs; see http://www.gene-tools.com/.
6. Although a simple transmitted light base on a basic stereomicroscope is sufficient for microinjection setup, a stereomicroscope with some level of controlled oblique illumination makes visualization of the embryos is much easier.
7. In our hands an AB/Tubingen hybrid line has significantly higher fecundity, fertility, and breeding longevity on average than either wild-type AB or Tubingen lines. Due to the elevated nutritional requirements for spawning each adult should not be bred more than once per week.
8. To control the time of spawning breeding traps can be kept in the dark until embryos are required. To establish multiple rounds of injections on the same day separate breeding traps can be kept in the dark prior to pulling the dividers and initiating spawning.

9. RNA is easily degraded and contaminated. All materials should be cleaned with RNase away. Ideally, pipettors and other equipment should be dedicated to RNA work. All tips and tubes must be RNase free. RNase is *not* inhibited by autoclaving. RNA-later states that it can be archived at –80°C indefinitely, but ideally, RNA should be used as soon as possible to limit the possibility of degradation.
10. It is better to leave some aqueous phase behind then to accidentally remove any contaminating layers. The organic layer can be saved for downstream protein isolation (see the included Trizol protocol).
11. Examine the sample; it is common in liver and pancreas extracts to have a precipitate form. We assume this is glycogen. After the 15 min at room temperature, spin the tube at 500 rpm for 30 s to collect this precipitate, transfer supernatant to a fresh tube and proceed.
12. Be consistent in the way in which you orient the tubes in the rotor, so the pellet will always form on the same side of the tube. Small pellets can be difficult to find.
13. Do not overdry pellet. As the pellet dries, it will become clear (from white). Once this occurs, resuspend the pellet. Overdrying it will decrease the solubility significantly.
14. In order to minimize pipetting error when loading, use at least 10 μl of total volume, making the remaining volume with DEPC-treated ddH_2O. No more than 1 μg of RNA should be loaded per well, in order to minimize saturation of the 28S and 18S bands. The ratio of these bands (28S:18S) should be approximately 2. For a comprehensive discussion of the subject, see Ambion (http://www.ambion.com/techlib/tn/111/8.html).
15. In this step, both random primers (hexamers and decamers) and/or oligo dT can be used. While better cDNA yields are obtained with random primers, oligo dT will give a more complete library of transcripts. We suggest starting with random primers, and then using oligo dT if there are amplification difficulties. Primer design is an essential step in PCR, and there are considerations for each desired type of PCR reaction. Given the partially complete genome sequence available for zebrafish, the development of "detection" primer sets is the first step. Ideally, based on the predicted cDNA sequence, the product should be between 300 and 500 bp, and the annealing temperatures should ideally be 60–64°C. Also, ideally, PCR products should be present at 30–35 cycles to limit the possibility of spurious products. Multiple tissues should be tested to determine expressing tissues and possibly identify

alternative transcripts. For extension times, typically 1 min/kb of desired product is sufficient. For full-length cDNA PCR, use the available sequence as a starting point, although if there are assembly errors, internal primers will need to be created. The "ideal" primer (60–62°C) is not always possible; primers must be designed in order to minimize primer dimers, inter- and intra-primer binding, and hairpin loops. Nested RACE primers can also be created at this point according to the RACE kit manufacturer's instructions. Sequencing primers for a particular strand should be spaced no more than 500 bp apart, in order to ensure enough sequence read for overlap. Submit multiple samples for sequencing, to discount sequencing reaction errors. Generate a consensus sequence from the sequencing results.

16. qPCR primer selection: Primers need to be designed that span introns or intron–exon junctions. This is in order to eliminate the possibility of amplifying chromosomal DNA. Primers have annealing temperatures of at least 60°C and must meet the criteria outlined above. Additionally, amplicons should not exceed 200 bp, in order to minimize SYBR green saturation. Primers need to be tested initially against serial dilutions of cDNA from a known expressing tissue in order to calculate amplification efficiency. Efficiency will dictate the analysis method used (*see* **Note 17**).
17. Analysis of qPCR data: An exhaustive discussion about qPCR data analysis is not the aim of this chapter. For more information, *see* (10).
18. Gene Tools recommend MOs be reconstituted in dH_2O to create a stock of 1 mM or less and stored at room temperature to avoid precipitation. We maintain small aliquots of the MO stock at –20°C for long-term storage with no observable changes in activity.
19. The utility of any given MO will only be determined empirically in conjunction with appropriate controls through a range of concentrations. The identification of an effective MO, for a single gene, may require testing multiple antisense and mispair control MOs.
20. The ideal taper of the needle must be determined empirically by mock injections on embryos. Manipulating the needle puller program can create a range of needles for many applications. Needles must be rigid enough to puncture the chorion and tissues but thin enough to limit the damage to the cellular membranes caused by the injection. Needles that are too thin, however, are also more likely to plug. Once an appropriate program has been established a

number of needles can be prepared and stored by mounting on Plasticine in a closed container to avoid damage.

21. Each needle must be calibrated to determine the volume of the bolus delivered at a particular microinjection setting. The calibration of the needle occurs prior to injection with the MO being injected. The needle can be loaded first and then the tip broken using forceps to produce an appropriate bore size. The needle is then used to inject the MO into mineral oil on a slide micrometer. The volume of the bolus is calculated based on the diameter of the bolus determined from the slide micrometer. The injection settings are adjusted until the bolus delivered is of the desired volume. The value is then applied to all injections using the same needle and microinjection settings. If the needle or settings are changed the needle must be recalibrated.
22. Although commercial injection molds are available for creating agarose injection dishes a simple injection plate can be created by placing a standard Fisherbrand microscope (or equivalent) slide in a 10 × 150 mm Petri dish. The edge of the slide is used as backing for injecting the embryos. Once distributed along the edge of the slide the embryos are held in place by wicking away excess E3.
23. Embryos can be injected with MO antisense oligos until the four-cell stage; however, the later the injection the greater the chance of mosaicism. Oligos delivered into the yolk under the cells will be drawn into the cells by ooplasmic streaming through the pores between the yolk and the cells.
24. After injection, monitor the development of the morphants regularly. Look for alterations in tail formation (bifurcation, truncation, failure to separate from yolk ball) and head development [eyes, brain (formation and cloudiness), and symmetry]. Later, look for blood flow irregularities, chromatophore changes, heart beat, coiling, etc.
25. In the absence of gene-specific antibodies, the effectiveness of splice-blocking MO can be evaluated through the identification of alternatively spliced mRNA by RT-PCR. Although a standard MO antisense oligo control is available through Gene Tools, the gene-specific mismatch control oligo is the most appropriate antisense oligo injection control. Morphant phenotype rescue by injection of gene-specific mRNA is another acceptable confirmation of MO-mediated knockdowns.

Acknowledgments

Research in Dr. Anini's laboratory is supported by an operating grant from the Canadian Institutes of Health Research (CIHR) and an infrastructure grant from the Canada foundation of Innovation (CFI).

References

1. Seidah, N. G., and Chretien, M. (1999) Proprotein and prohormone convertases: A family of subtilases generating diverse bioactive polypeptides *Brain Res* **848**, 45–62.
2. Seidah, N. G., Mowla, S. J., Hamelin, J., Mamarbachi, A. M., Benjannet, S., Toure, B. B., Basak, A., Munzer, J. S., Marcinkiewicz, J., Zhong, M., Barale, J. C., Lazure, C., Murphy, R. A., Chretien, M., and Marcinkiewicz, M. (1999) Mammalian subtilisin/kexin isozyme SKI-1: A widely expressed proprotein convertase with a unique cleavage specificity and cellular localization *Proc Natl Acad Sci USA* **96**, 1321–6.
3. Seidah, N. G., Benjannet, S., Wickham, L., Marcinkiewicz, J., Jasmin, S. B., Stifani, S., Basak, A., Prat, A., and Chretien, M. (2003) The secretory proprotein convertase neural apoptosis-regulated convertase 1 (NARC-1): Liver regeneration and neuronal differentiation *Proc Natl Acad Sci USA* **100**, 928–33.
4. Seidah, N. G., Mayer, G., Zaid, A., Rousselet, E., Nassoury, N., Poirier, S., Essalmani, R., and Prat, A. (2008) The activation and physiological functions of the proprotein convertases *Int J Biochem Cell Biol* **40**, 1111–25.
5. Morash, M. G., MacDonald, A. B., Croll, R. P., and Anini, Y. (2009) Molecular cloning, ontogeny and tissue distribution of zebrafish (Danio rerio) prohormone convertases: PCSK1 and PCSK2 *Gen Comp Endocrinol* **162**, 179–87.
6. Kudo, H., Liu, J., Jansen, E. J., Ozawa, A., Panula, P., Martens, G. J., and Lindberg, I. (2009) Identification of proSAAS homologs in lower vertebrates: Conservation of hydrophobic helices and convertase-inhibiting sequences *Endocrinology* **150**, 1393–9.
7. Poirier, S., Prat, A., Marcinkiewicz, E., Paquin, J., Chitramuthu, B. P., Baranowski, D., Cadieux, B., Bennett, H. P., and Seidah, N. G. (2006) Implication of the proprotein convertase NARC-1/PCSK9 in the development of the nervous system *J Neurochem* **98**, 838–50.
8. Walker, M. B., Miller, C. T., Coffin Talbot, J., Stock, D. W., and Kimmel, C. B. (2006) Zebrafish furin mutants reveal intricacies in regulating Endothelin1 signaling in craniofacial patterning *Dev Biol* **295**, 194–205.
9. Nusslein-Volhard, C., and Dahm, R. (2002) *Zebrafish: A Practical Approach.* Oxford University Press, New York, NY and Cary, NC.
10. Bustin, S. A., Benes, V., Garson, J. A., Hellemans, J., Huggett, J., Kubista, M., Mueller, R., Nolan, T., Pfaffl, M. W., Shipley, G. L., Vandesompele, J., and Wittwer, C. T. (2009) The MIQE guidelines: Minimum information for publication of quantitative real-time PCR experiments *Clin Chem* **55**, 611–22.

Chapter 15

Use of Zebrafish and Knockdown Technology to Define Proprotein Convertase Activity

Babykumari P. Chitramuthu and Hugh P.J. Bennett

Abstract

The Zebrafish (*Danio rerio*) is a powerful and well-established tool used extensively for the study of early vertebrate development and as a model of human diseases. Zebrafish genes orthologous to their mammalian counterparts generally share conserved biological function. Protein knockdown or overexpression can be effectively achieved by microinjection of morpholino antisense oligonucleotides (MOs) or mRNA, respectively, into developing embryos at the one- to two-cell stage. Correlating gene expression patterns with the characterizing of phenotypes resulting from over- or underexpression can reveal the function of a particular protein. The microinjection technique is simple and results are reproducible. We defined the expression pattern of the proprotein convertase PCSK5 within the lateral line neuromasts and various organs including the liver, gut and otic vesicle by whole-mount in situ hybridization (ISH) and immunofluorescence (IF). MO-mediated knockdown of zebrafish PCSK5 expression generated embryos that display abnormal neuromast deposition within the lateral line system resulting in uncoordinated patterns of swimming.

Key words: Zebrafish, microinjection, morpholino, PCSK5, lateral line neuromast, in situ hybridization, immunofluorescence.

1. Introduction

The zebrafish has emerged as a powerful model to study early developmental events and is being increasingly used to determine mechanisms underlying human disease (1–3) and as a tool in the discovery of potential therapeutics (4, 5). Like all teleosts zebrafish have the ability to regenerate damaged or amputated tissue including the heart and nervous system (6, 7). All organ primordia are present together with partial cytodifferentiation within

M. Mbikay, N.G. Seidah (eds.), *Proprotein Convertases*, Methods in Molecular Biology 768,
DOI 10.1007/978-1-61779-204-5_15,

48 h post-fertilization (hpf) with full organ development occurring within 7 days post-fertilization (dpf) with exception of the reproductive organs (zebrafish only become sexually mature after 3 months). The embryos are optically clear, facilitating unhindered real-time observation during organogenesis and assessment of gene expression by whole-mount in situ and immunohistochemical analysis. They develop ex utero permitting analysis of gene function through all stages of embryonic development. We have established the zebrafish animal model in our laboratory as a tool for gene function discovery (8, 9). Wild-type zebrafish are kept on a 14 h/10 h light/dark cycle at 28.5°C in a laboratory aquarium (10).

To define the function of PCSK5 during development we first characterized its expression pattern by whole-mount in situ hybridization (ISH) and immunofluorescence (IF). The ISH technique allows for the determination of the temporal and spatial expression of particular genes. The current protocol describes ISH of whole-mount zebrafish embryos using digoxigenin-labelled antisense RNA probes synthesized by in vitro transcription of cloned and linearized template of PCSK5 with digoxigenin-linked nucleotides. Staged embryos are dechorionated, fixed, permeabilized and post-fixed before being incubated with specific probes. After washing away excess probe, embryos are incubated with alkaline phosphatase-conjugated anti-digoxigenin antibody. PCSK5 expression is detected using a chromogenic substrate for alkaline phosphatase. The whole procedure takes only 3–4 days depending on the level of expression. The IF technique allows for the detection of the protein of interest. In this technique the dechorionated, fixed, permeabilized and post-fixed embryos were incubated with PCSK5 antibody. After extensive washing in PBST, embryos were incubated with appropriate secondary antibody conjugated with Alexa488. PCSk5 immunoreactivity was observed using a GFP filter-equipped microscope.

We employed a microinjection technique (11) to knockdown PCSK5 expression in order to define its biological function. In our method, a microcapillary filament is filled with MO solution and attached to the holder of the microinjector that forces the solution out of the filament under air pressure. A small amount of solution is then expelled into the cytoplasm and the injected embryos are incubated at 28.5°C to develop further. The injected MO is evenly distributed through all cells of the developing embryo. Within each cell, the MO binds to target mRNA with high affinity and specificity thereby blocking the initiation of translation. The functional roles of the protein products can be assessed by following morphological, physiological, or molecular changes using marker gene approaches following over- or under-expression.

2. Materials

2.1. Whole-Mount In Situ Hybridization

2.1.1. Digoxigenin-Labelled RNA Probe Synthesis

1. cDNA or genomic DNA.
2. Taq polymerase (Invitrogen).
3. TopoTA cloning kit-dual promoter pCRII-TOPO (Invitrogen).
4. HQ Mini Plasmid Purification kit (Invitrogen).
5. Agarose (Roche, Laval, QC).
6. *Not*I and *Hin*dIII restriction enzymes (Invitrogen).
7. Buffer-saturated phenol ultrapure (Invitrogen): Store at 4°C. Eyes, skin and respiratory tract irritant and suspected carcinogen. Care should be taken while handling.
8. Chloroform (Fisher): Toxic and suspected carcinogen. Work under fume hood.
9. RNase-free DNAse I (Roche, Laval, QC): Prepare small aliquots and store at –20°C.
10. DIG-RNA Labelling Mix with Sp6, T7 and T3 RNA polymerase (Roche, Laval, QC).
11. RNase inhibitor (Roche, Laval, QC).
12. 3.0 M sodium acetate (pH 5.5, Fisher).
13. 100% ethanol (Commercial Alcohols).
14. 75% ethanol.
15. Diethyl pyrocarbonate (DEPC, Sigma): Add 1 ml of DEPC in 1 l of water, incubate at 50°C overnight, autoclave for 20 min and bring to room temperature before use. DEPC is an eye, skin and respiratory irritant. Avoid contact with skin and eyes. Use safety glasses, gloves and mask.
16. NaOH (Fisher).
17. EDTA 0.5 M (pH 8.0): Dissolve 186.1 g of EDTA in 800 ml of water. Add 15 g of NaOH pellets and make up to 1 l by adding water when all pellets are dissolved.

2.1.2. Embryo Preparation

1. Instant Ocean Salt (Aquarium Systems Inc, Mentor, OH): Prepare 100X egg water by dissolving 6 g of Instant Ocean Salt in 1 l of water. Combine 1 part of salt with 99 parts of water to make a working solution.
2. Phosphate buffered saline (PBS, Fisher): Reconstitution in 1 l of water results in 10X PBS. Autoclave before storage at room temperature. Prepare working solution by dilution of 1 part with 9 parts of water.
3. Paraformaldehyde (PFA, Sigma): Prepare a 4% (w/v) solution in PBS fresh for each experiment. The solution may

need to be carefully heated (use a stirring hot plate in a fume hood) to dissolve, then cool to room temperature and store at –20°C as 50 ml aliquots for later use. PFA is toxic. Use safety glasses, gloves and dust mask.

4. Phenylthiocarbomide (PTC, Sigma): 0.3% (w/v) in egg water. PTC is highly toxic. Use safety glasses, gloves and dust mask.
5. Methanol (Fisher).
6. Dumont (Watchmaker's) Forceps pattern no. 5 (Fine Science Tools, North Vancouver, BC).

2.1.3. Whole-Mount In Situ Hybridization

1. Six-well cell culture cluster (Costar).
2. Baskets are made of nylon mesh and Eppendorf tubes (*see* **Fig. 15.1**). To make the baskets, cut Eppendorf tubes (1.5 ml) with or without rims to remove the conical end. Cut nylon mesh into small pieces corresponding to the size of the cut ends of the Eppendorf tubes. Place tubes with the nylon mesh covering the cut end on an electrical hot plate until both the tube and nylon mesh stick together (carry out in fume hood). Cut off the excess mesh from the baskets and store them in 100% methanol until use (**Fig. 15.1**).
3. Methanol (Fisher).
4. Phosphate buffered saline (PBS, Fisher).
5. Phosphate buffered saline Tween (PBST): Prepare PBS as outlined in **Section 2.1.2**. Polyoxyethylenesorbitan monolaurate (Tween 20, 10% (v/v)) (Sigma) stock solution is prepared by diluting 10 ml of Tween 20 with 90 ml of sterile water. After complete homogenization of the solution,

Fig. 15.1. Eppendorf-nylon mesh baskets used to hold staged embryos.

it is stored at room temperature. Protect this solution from light. Combine 1 part of 10X PBS, 1 part of 10% Tween with 8 parts of sterile water to make 1X PBST.

6. Proteinase K (Fermentas, Burlington, ON).
7. Paraformaldehyde (PFA) (Sigma): prepare PBS as outlined in **Section 2.2**.
8. Acetic anhydride (Sigma).
9. Triethanolamine (Sigma).
10. Acetylation mix: combine 125 μl of triethanolamine and 27 μl of acetic anhydride in 10 ml sterile water.
11. Formamide, high purity grade (Sigma): Formamide is highly corrosive on contact with skin or eyes. It is also a teratogen. Handle with proper safety attire including gloves and goggles.
12. AG501-X8 Resin (Bio-Rad, Mississauga, ON).
13. Deionized formamide: Add formamide to 10 g/l of AG501-X8 and stir slowly for 30 min and repeat once. Filter the solution to remove the resin, aliquot into 20 ml, and store in the dark at –20°C.
14. Citric acid monohydrate (Sigma): Prepare 1 M solution. Filter and store at 4°C.
15. Heparin sodium salt (Sigma): Prepare stock solution as 50 mg/ml and store at 4°C.
16. Saline-sodium citrate buffer (SSC) (Sigma): Dissolve powdered blend in 1 l of water to make 20X SSC stock solution and store at room temperature.
17. tRNA from bakers yeast type X, lyophilized powder (Sigma): Resuspend in water at a concentration of 50 mg/ml and store at –20°C as small aliquots.
18. Hybridization mix (HM): 50% deionized formamide, 5X SSC, 0.1% Tween 20, 50 μg/ml of heparin, 500 μg/ml of RNase-free tRNA, adjust pH 6.0 by adding citric acid (460 μl of 1 M citric acid solution for 50 ml of HM). Prepare mix without tRNA and then add tRNA only to make volume needed for pre-hybridization and hybridization.
19. NaCl (Fisher): Prepare 5 M stock solution and store at room temperature.
20. Tris-HCl (Fisher): 1 M stock solution Tris-HCl pH 9.5: Dissolve 157.1 g of Tris base in 900 ml of water, adjust pH to 9.5 by adding HCl, and then adjust the volume to 1 l by adding water.
21. $MgCl_2$ (Fisher): Prepare 1 M stock and store at room temperature.

22. Levamisole hydrochloride (Sigma): 100 mM prepare fresh when needed.
23. Alkaline Tris (pre-staining) buffer: 100 mM Tris-HCl, pH 9.5, 50 mM $MgCl_2$, 100 mM NaCl, 0.1% Tween 20 (v/v) and 1 mM levamisol.
24. *N*, *N*-Dimethylformamide anhydrous (DMF, Sigma): Irritant, toxic, combustible and suspected teratogen. Handle with proper safety attire including gloves and goggles.
25. Nitroblue tetrazolium (NBT) (Sigma): Dissolve 50 mg of NBT in 0.7 ml of *N*, *N*-dimethylformamide and 0.3 ml of sterile water and store at –20°C. Protect this solution from light.
26. 5-Bromo 4-chloro 3-indolyl phosphate (BCIP) (Sigma): Dissolve 50 mg of BCIP in 1 ml of *N*, *N*-dimethylformamide anhydrous and store at –20°C. Protect this solution from light.
27. Albumin from bovine serum, purified fraction V (BSA) (Sigma): Prepare 100 mg/ml and store at –20°C.
28. Sheep anti-digoxigenin-AP Fab fragments (Roche Diagnostics, Laval, QC): Store at 4°C.
29. Blocking buffer: 1X PBST, 2% calf serum (vol/vol), 2 mg/ml BSA.
30. Staining buffer: Add 23 μl of 50 mg/ml NBT and 13 μl of 50 mg/ml BCIP in 4 ml of alkaline Tris (pre-staining buffer). Prepare fresh when needed.
31. Stop solution: 1 PBS, pH 5.5, 1 mM EDTA, 0.1% Tween 20 (v/v).
32. Glycerol (Sigma).

2.2. Whole-Mount Immunofluorescence

Same as listed for in situ hybridization (*see* **Section 2.1.2** for in situ technique).

2.2.1. Embryo Preparation

2.2.2. Whole-Mount Immunofluorescence

1. Six-well cell culture cluster (Costar).
2. Methanol (Fisher).
3. Phosphate buffered saline (PBS, Fisher).
4. Phosphate buffered saline Tween (PBST): prepare PBS as outlined under **Section 2.3**.
5. Proteinase K (Fermentas, Burlington, ON).
6. Paraformaldehyde (PFA) (Sigma): prepare PBS as outlined under **Section 2.2**.
7. Fetal calf serum (Invitrogen).

8. Dimethyl sulfoxide (DMSO) (Sigma).
9. Primary antibodies: Peptides (D-Y-K-P-S-Y-P-G-T-V-Q-S-S-M-C) and (T-E-T-M-D-Q-T-H-D-G-S-F-N-D-C) corresponding to the predicted amino-terminal sequences of mature PCSK5.1 and PCSK5.2, respectively (following cleavage of the propeptide; carboxyl-terminal cysteine residue added to facilitate conjugation), were synthesized by solid phase chemistry and polyclonal antibodies were raised in rabbits by immunization with peptides conjugated to keyhole limpet hemocyanin via the carboxyl-terminal cysteine. Antisera were purified on peptide affinity columns prior to use. Peptide synthesis, antigen conjugation, antibody production and purification were carried out at the Sheldon Biotechnology Centre, McGill University, Montreal, Quebec, Canada.
10. Secondary antibody: Alexa Fluor 488 donkey anti-rabbit IgG antibody.
11. Blocking buffer: 1X PBST, 5% calf serum (v/v) and 1% DMSO.

2.3. Microinjection

2.3.1. Embryo Preparation

1. Wild-type zebrafish are purchased from Aquatica Tropicals (Florida) and maintained on a 14 h/10 h light/dark cycle at 28.5°C in a laboratory aquarium (Allentown Caging Equipment Co.). A zebrafish facility of approximately 200 fish produces hundreds of embryos daily by natural spontaneous spawning.
2. Zebrafish Housing system (Aquaneering Inc., San Diego, CA).
3. Instant Ocean Salt (Aquarium Systems Inc, Mentor, OH): Prepare 100X egg water by dissolving 6 g of Instant Ocean Salt in 1 l of water. Combine 1 part of salt with 99 parts of water to make a working solution.
4. Nylon netting (Safari Animal Centre, Pointe-Claire, QC) to catch and transfer fish between tanks.
5. Petri dishes.

2.3.2. Microinjection

1. 0.5% phenol red in DPBS (Sigma).
2. Fluorescein isothiocyanate (FITC) (Sigma).
3. 1X Danieau buffer: 58 mM NaCl, 0.7 mM KCl, 0.4 mM $MgSO_4$, 0.6 mM $Ca(NO_3)2$ and 5.0 mM HEPES pH 7.6.
4. Morpholino (MO) oligonucleotides are directed against the initiator ATG sequence of the PCSK5.1 gene (Gene tools, Philomath, OR). MOs are diluted in 1X Danieau buffer with 0.1% phenol red and 0.05% FITC-dextran. Phenol red facilitates the estimation of the volume of DNA

injected FITC and also indicates the efficiency of incorporation of MO inside the developing embryo.

5. Microinjection Station to hold embryos during microinjection. Pour about 40 ml of 2% agarose in a Petri dish (150 mm × 15 mm) and let it solidify. Prepare furrow by making wedge-shaped cut using a sterile scalpel. Two furrows can be made per dish.
6. Dissecting microscope (Leica, MS 5).
7. Microinjection pipettes: Microinjection pipettes are prepared from standard-walled glass capillaries with an inner filament (1.0 mm × 0.58 mm, HRS Scientific, Montreal, QC). Microinjection pipettes (needle) are drawn using a needle puller (Sutter Instrument Company, Novato, CA). Needle can be broken under the dissecting microscope using the highest magnification by gently scraping the tip of the needle with clean forceps immediately prior to use to produce a sharp pipette tip to penetrate the egg chorion during microinjection.
8. Microinjection system (PLI-100 microinjection system, Harvard Apparatus, Saint-Laurent, QC). This consists of a micropipette holder to hold micropipette, dissecting microscope, nitrogen gas cylinder connected to a microin-

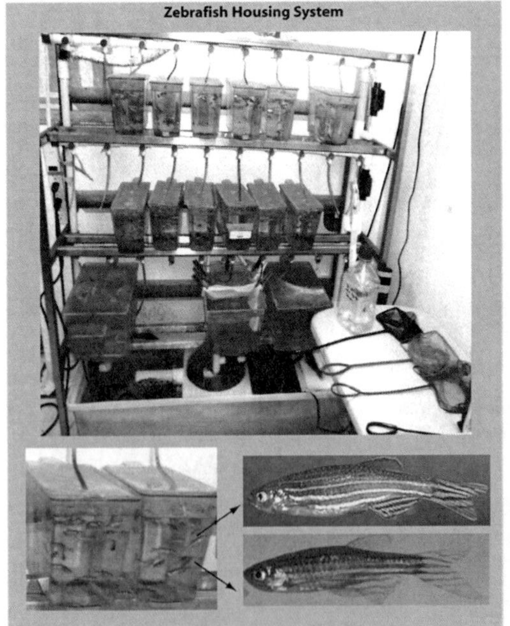

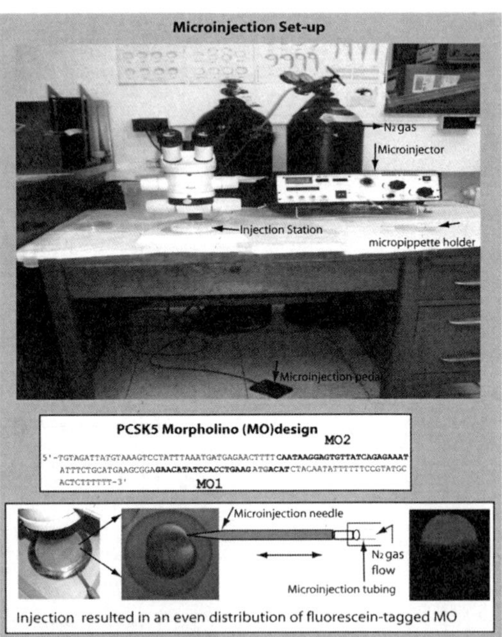

Fig. 15.2. Zebrafish housing system and knockdown technology: Microinjection system consists of a micropipette holder to hold micropipette, dissecting microscope, nitrogen gas cylinder connected to a microinjector (to apply pressure to deliver volumes from femtolitres to microlitres) and footswitch.

jector (to apply pressure to deliver volumes from femtolitres to microlitres) and footswitch (**Fig. 15.2**).

9. Incubator (28.5°C).
10. Dumont #5 watchmaker forceps.

3. Methods

3.1. Whole-Mount In Situ Hybridization

In order to gain greater insights into the relative abundance and specific location of the particular PCSK5 gene within the embryo during development, the analysis of the spatio-temporal expression was carried out by in situ hybridization. For all the stages examined the sense riboprobe of the corresponding *PCSK5* gene was used as a negative control (1, 12). Because of the discrete and well-documented expression pattern known for sonic hedgehog an antisense riboprobe for this gene was used as a positive control (2). Steps involved in whole-mount in situ hybridization are illustrated in **Fig. 15.3**.

3.1.1. Digoxigenin-Labelled RNA Probe Synthesis

PCR Amplification and Gene Cloning for Probe Template

1. Set up the following PCR reaction (50 μl) (**Table 15.1**) (*see* **Note 1**).
2. Verify the size of the PCR product by agarose gel electrophoresis (*see* **Note 2**).

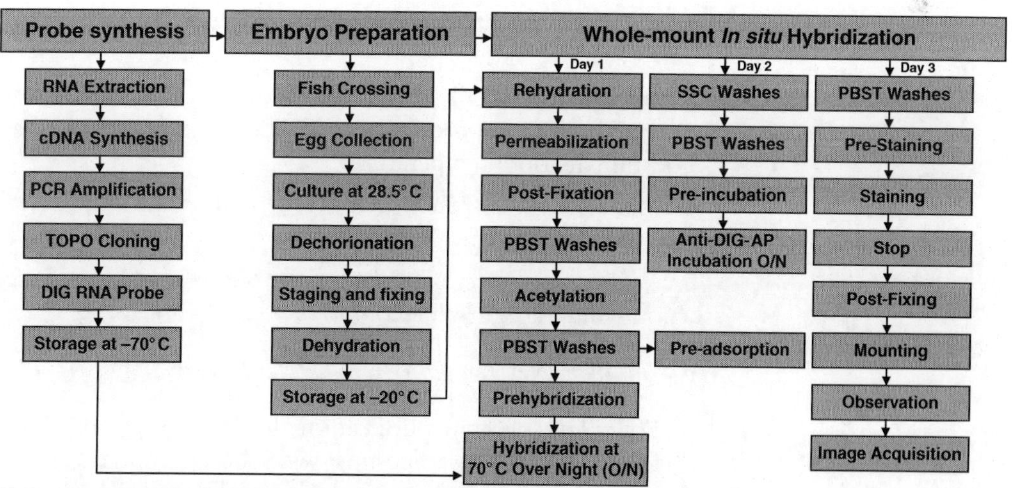

Fig. 15.3. Requirements for the whole-mount in situ hybridization procedure: (1) probe synthesis; (2) embryo preparation; and (3) in situ hybridization. Probe synthesis can be done in advance and probes stored at –80°C for long periods. Staged embryos can be stored at –20°C until used. Most whole-mount in situ hybridization experiments take 3 days to complete the procedure. For weakly expressed genes the staining reaction will take up to 16 h at room temperature or more if the staining reaction is carried out at 4°C.

Table 15.1
PCR mixture

Reagent	Volume
DNA template	10–100 ng
10X PCR buffer	5 μl
50 mM dNTPs	0.5 μl
Primers (100–200 ng each)	1 μM each
Water add to a final volume of	49 μl
Taq Polymerase (1 unit/μl)	1 μl
Total volume	50 μl

Table 15.2
TOPO cloning reaction

Reagent	Chemically competent *E. coli*	Electro competent *E. coli*
Fresh PCR product	0.5–4 μl	0.5–4 μl
Salt solution	1 μl	–
Dilute salt solution	–	1 μl
Water	Add to total volume of 5 μl	Add to total volume of 5 μl
TOPO vector	1 μl	1 μl
Final volume	6 μl	6 μl

3. Perform the TOPO cloning reaction (**Table 15.2**). Mix together PCR product and TOPO vector and incubate 5 min at room temperature.
4. Transform One Shot competent cells with the TOPO cloning product.
5. Analyse and select colonies for the presence of insert based on white or light blue colour.

Preparing the DNA Template from cDNA Clones

6. Following plasmid purification, linearize 5 μg of cDNA for each probe by digesting with the appropriate restriction enzyme that has a unique site located 5′ (for antisense probes) or 3′ (for sense probes) to the insert. Both PCSK5.1 and PCSK.2 antisense riboprobes were transcribed using SP6 polymerase with Not1-linearized templates and the corresponding sense riboprobes were

transcribed using T7 polymerase with *Hin*dIII-linearized templates.

7. Purify linearized DNA by phenol/chloroform extraction.
8. Precipitate DNA with sodium acetate and ethanol (*see* **Note 3**); centrifuge and wash with RNase-free 70% ethanol.
9. Resuspend DNA in DEPC-water.
10. Use an aliquot to verify purity and to recover by agarose gel electrophoresis.

RNA Labelling Reaction

11. Add 1–5 μg purified template DNA to a sterile, RNase-free reaction vial, and then, enough water (sterile, RNase-free, DMPC-treated, double distilled) to make the total sample volume of 13 μl.
12. Place the reaction vial on ice and then add the reagents listed in **Table 15.3**. Gently agitate the reaction mixture and centrifuge briefly.
13. Incubate the reaction mixture for 2 h at 37°C.
14. Add 2 μl DNase I, RNase free, to remove template DNA. Incubate the reaction mixture for 30 min at 37°C. Add 2 μl 0.2 M EDTA (pH 8.0) to stop the reaction.
15. Precipitate the reaction products with 2.5 μl 4 M LiCl and 75 μl 100% ethanol (*see* **Note 3**). Mix precipitating solution well and keep at –20°C overnight.
16. Spin suspension for 30 min in microfuge; wash pellet with 50 μl 70% ethanol; spun 3 min, dry and resuspend in 50 μl (30 min at 37°C).
17. Run 1–2 μl on an agarose gel to verify probe quality.

Table 15.3
RNA labelling reaction mixture

Reagent	Volume (μl)
10× NTP labelling mixture	2
10× transcription buffer	2
Protector RNase inhibitor	1
RNA polymerase SP6 or	
RNA polymerase T7	2
Final volume	20

This reaction yields about 10 μg of RNA (100–200 ng/μl). RNA transcripts are analysed by agarose gel electrophoresis.

3.1.2. Embryo Preparation

1. Collect embryos at the required developmental stages. Staging is performed according to Kimmel et al. (13).
2. Manually remove chorions with Dumont Watchmaker's Forceps no. 5.
3. Fix embryos in 4% paraformaldehyde (PFA) in 1X PBS for 2 h at room temperature or overnight at 4°C.
4. Rinse the embryos with PBS several times, transfer them into 100% methanol (MeOH) and store them at –20°C until used for whole-mount in situ hybridization or immunofluorescence analysis.
5. For RNA extraction, freeze staged embryos in liquid nitrogen and store at –80°C.
6. Extract RNA from staged embryos and synthesize cDNA for PCR amplification and subsequent cloning.

3.1.3. Whole-Mount In Situ Hybridization of Zebrafish Embryos

DAY 1

Rehydration

1. Prepare 6-well plates with successive dilutions of methanol in 1X PBS as 75% (v/v) methanol, 50% (vol/vol) methanol and 25% (v/v) methanol (*see* **Note 4**).
2. Place dehydrated embryos of the same developmental stage into baskets made of nylon mesh (**Fig. 15.4**) (*see* **Note 5**).
3. Rehydrate the embryos by moving the basket from one well to the next well of the plate for 5 min each.
4. Wash embryos three times, 5 min per wash, in 100% PBST.

Fig. 15.4. The 6-well sterile plates used for successive dilutions of methanol in 1X PBS. One plate includes six Eppendorf-nylon mesh baskets (*see* also **Fig. 15.1**).

Permeabilization

5. Permeabilize the rehydrated embryos by digestion with proteinase K (10 μg/ml) at room temperature for the indicated period (**Table 15.4**).

Post-fixation and PBST Washes

6. Incubate embryos for 20 min in 4% (wt/vol) paraformaldehyde in 1X PBS.
7. Wash embryos three times, 5 min per wash, in 1X PBST to remove residual paraformaldehyde.

Acetylation

8. Incubate embryos twice for 10 min each in acetylation mix (125 μl triethanolamine and 27 μl acetic anhydride in 10 ml ddH_2O) made immediately prior to use to minimize endogenous phosphatase activity.
9. Wash embryos twice, 10 min per wash in 1X PBST.

Pre-hybridization

10. Transfer embryos from their baskets to 1.5 ml sterile Eppendorf tubes and prehybridize with 200 μl hybridization mix for 2–4 h in a 70°C water bath.

Preadsorption of Antibody

11. Transfer about 30–50 embryos from their baskets to 1.5 ml sterile Eppendorf tubes.
12. Incubate embryos with 2 μl anti-DIG-AP (1:500) antibody in blocking buffer. Leave at room temperature for 2–4 h,

Table 15.4
Permeabilization time

Developmental stage (h)	Proteinase K digestion
0–6	15 s
6–12	30 s
12–18	3 min
24	12 min
48	25 min
72	40 min
96–120	50 min

and then stored at 4°C overnight. Take them out first thing in the morning to let them reach room temperature.

Hybridization

13. Add 100–200 ng of probe, mix the solution and hybridize overnight at 70°C in water bath (*see* **Note 6**).

DAY 2

SSC Washes

14. Prepare 6-well plates with successive dilutions of pre-warmed HM in 2X SSC as follows: 75% (v/v) HM, 50% (v/v) HM and 25% (v/v) HM (*see* **Note 7**).
15. Wash the hybridized embryos by moving the basket from one well to the next well of the plate for 15 min each in a 70°C water bath: twice, for 15 min per wash, in 2X SSC at 70°C and twice, for 30 min per wash, in 0.2X SSC at 65°C.

PBST Washes

16. Prepare 6-well plates with successive dilutions of 2X SSC in 1X PBST as follows: 75% (v/v) 2X SSC, 50% (v/v) 2X SSC and 25% (v/v) 2X SSC.
17. Wash hybridized embryos by moving the basket from one well to the next well of the plate for 5 min at room temperature on a rocker.

Preincubation and Incubation with Anti-DIG Antibody

18. The embryos are incubated for 3–4 h at room temperature in blocking buffer.
19. Incubate embryos with anti-DIG-AP antibody solution (1:5000) with blocking buffer overnight at 4°C with gentle agitation.

DAY 3

PBST Washes, Pre-staining, Staining

20. Briefly wash the embryos in PBST, then wash six times, 15 min per wash, in PBST at room temperature with gentle agitation.
21. Wash embryos two times for 15 min with alkaline Tris buffer (pre-staining buffer) at room temperature with gentle agitation (*see* **Note 8**).
22. Incubate embryos in staining solution at room temperature in the dark. The staining reaction is monitored with a stereomicroscope every 30 min. Reaction time varies from 15 min for highly expressed genes to 16 h for weakly expressed genes (*see* **Note 9**).

23. Stop the staining reaction when the desired staining intensity is reached by transferring the embryos containing baskets to the well containing stop solution.
24. Wash the embryos two times for 10 min at room temperature on a rocker with gentle agitation (*see* **Note 10**).

Post-fixation, Mounting, Observation and Image Acquisition

25. Incubate the embryos for 20 min in 4% (wt/vol) paraformaldehyde in 1X PBS.
26. Wash the embryos three times, 5 min per wash, in 1X PBST to remove residual paraformaldehyde (*see* **Note 11**).
27. Transfer the embryos in the minimum possible volume of PBS to a 6-well plate containing 100% glycerol. Place on a rocker and agitate gently overnight at room temperature in the dark.
28. The following day, mount the embryos in 100% glycerol and observe under microscope.
29. Capture and save images in a file format that allows improving the quality through image software such as Photoshop. Examples of these images are shown in **Fig. 15.5.**

Double In Situ Hybridization

The method described above is employed together with simultaneous incubation of both digoxigenin and fluorescein-labelled probes. The same procedure is followed for subsequent washes and blocking incubations but antibody and staining reactions are done sequentially. Briefly

30. Incubate embryos with AP-coupled-anti-digoxigenin Fab fragment, wash and react with BCIP-NBT. After detection, wash embryos with PBST twice for 20 min each at room temperature, then incubate for 30 min at 65°C in PBST with 1 mM EDTA to inactivate the antibody.
31. Incubate embryos in 100% methanol and rehydrate: 75, 50 and 25% methanol with PBST and back to PBST for 1 h.
32. Incubate embryos with AP-coupled fluorescein overnight and stain with fast red. Stop the staining reaction when the desired staining intensity is reached.
33. Follow the procedure mentioned above for the remaining steps.

3.2. Whole-Mount Immunofluorescence

To assess the spatio-temporal expression of PCSK5 protein levels, immunofluorescence was performed by monitoring antigen–antibody interactions using antigen-specific antibodies. The amino terminal sequences of mature PCSK5.1 and PCSK5.2 were

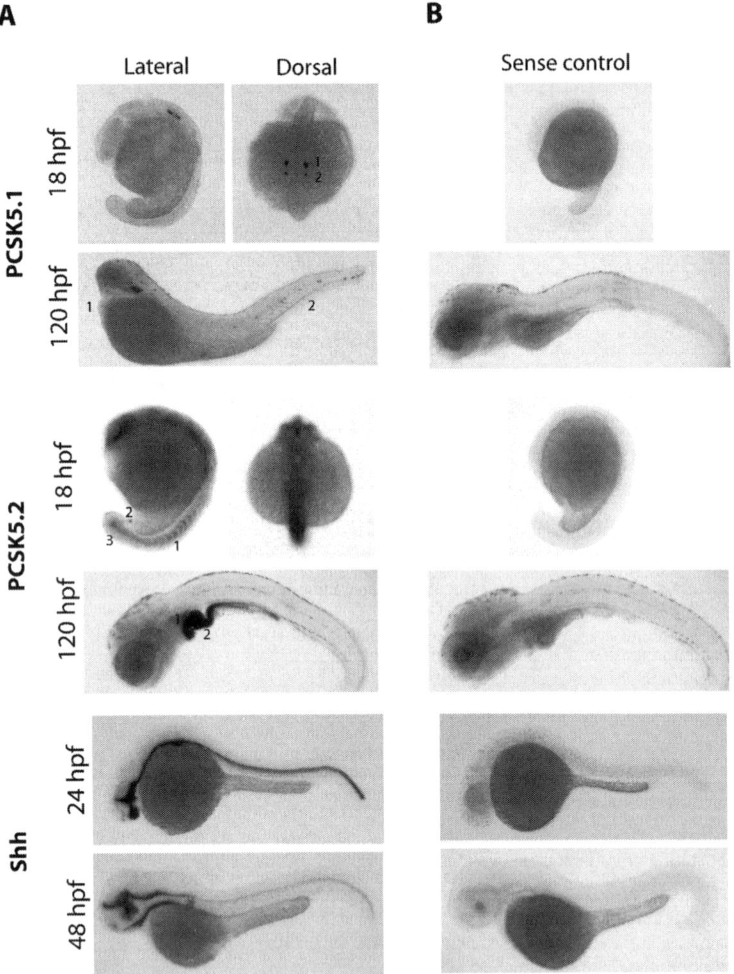

Fig. 15.5. Developmental expression of zebrafish PCSK5. (**a**) Whole-mount in situ hybridization analyses of PCSK5.1 and PCSK5.2 revealed by the corresponding antisense riboprobes. Lateral and dorsal view of 18hpf embryos show very discrete expression of PCSK5.1 within the anterior and posterior lateral line ganglia. At 120hpf PCSK5.1 is strongly localized within the anterior and posterior lateral line neuromasts. At 18hpf PCSK5.2 shows ubiquitous expression with distinct regionalization within the somites (1), Kupffer's vesicle (2) and in the tailbud (3). At 120 hpf PCSK5.2 is highly expressed within the liver (1) and intestine (2) in a manner similar to that found for mammalian PCSK5 (15). Results using Sonic Hedgehog (Shh) riboprobe showing well-documented neural expression (16) was used as positive control. (**b**) Absence of gene expression using respective sense riboprobes serve as negative controls.

used to raise site-specific antibodies. For this reason, while the in situ hybridization technique detects the message localization, immunofluorescence analysis labels only the mature processed forms of the processing enzyme. For the preadsorption negative control the respective antibody was preincubated for 2 h at room temperature with an excess of the purified peptide to which the antibody was raised. This peptide:antibody mixture was then diluted in blocking solution to replace the primary antibody step. Steps involved in whole-mount immunofluorescence are illustrated in **Fig. 15.6**:

1. Embryos are prepared as mentioned for in situ hybridization (**Section 3.1.2**).

DAY 1

2. Subject embryos to rehydration, permeabilization, post-fixation and PBST washes same as described for in situ hybridization technique (**Section 3.1.2,** steps 1–4).
3. Incubate embryos for 3–4 h at room temperature in blocking buffer.
4. Incubate embryos with anti-PCSK5.1 (1:500) or PCSK5.2 (1:1000) antibody solution or peptide: antibody mixture diluted with blocking buffer overnight at 4°C with gentle agitation.

DAY 2

Preadsorption of Antibody

5. Quickly rehydrate about 30–50 embryos in 1.5 ml sterile Eppendorf tubes.
6. Incubate embryos with 5 μl of goat anti-rabbit-Alexa Fluor 488 (1:200) antibody in blocking buffer. Leave at *room temperature* for 2–4 h.

PBST Washes, Incubation with Anti-rabbit-Alexa Fluor 488 and PBST Washes

7. Incubate embryos with primary antibodies.

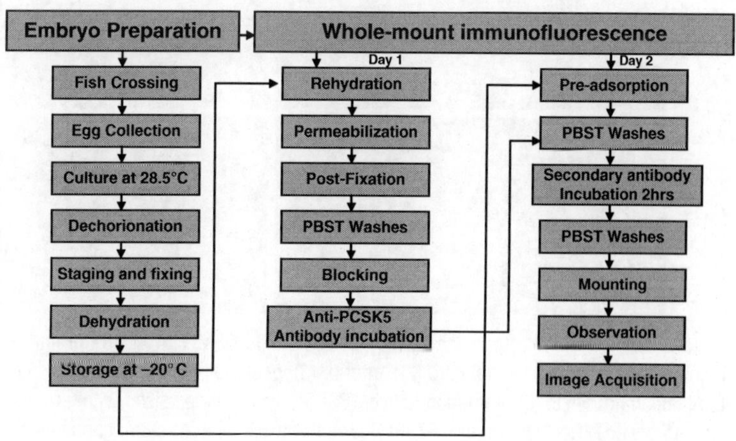

Fig. 15.6. Embryo preparation and whole-mount immunofluorescence. The technique takes 2–3 days using the staged and stored embryos to complete the procedure.

8. Wash briefly in PBST, then wash six times, 15 min per wash, in PBST at room temperature with gentle agitation.
9. Incubate embryos with goat anti-rabbit-Alexa Fluor 488 (1:200) secondary antibody in blocking buffer for 2 h.
10. Wash briefly in PBST, then wash four times, 15 min per wash, in PBST at room temperature with gentle agitation.

Mounting, Observation and Image Acquisition

11. Transfer embryos in the minimum possible volume of PBS to a 6-well plate containing 100% glycerol. Place on a rocker and agitate gently overnight at room temperature in the dark.
12. Next day, mount the embryos in 100% glycerol and observe under the microscope
13. Capture and save images in a file format that allows improvement of the image quality using software such as Adobe Photoshop 7.0. Examples of these images are shown in **Fig. 15.7.**

3.3. Microinjection

Microinjection has been widely used for generating transgenic fish (14) and analysing gene function (11). To uncover functional roles of PCSK5, we knocked down its protein level using MO oligonucleotides directed against PCSK5.1 starter methionine (MO-1) and splice-blocking morpholino at exon 4 and intron 4: exon–intron boundary (e4i4, MO-2). The MO-1 binds to target

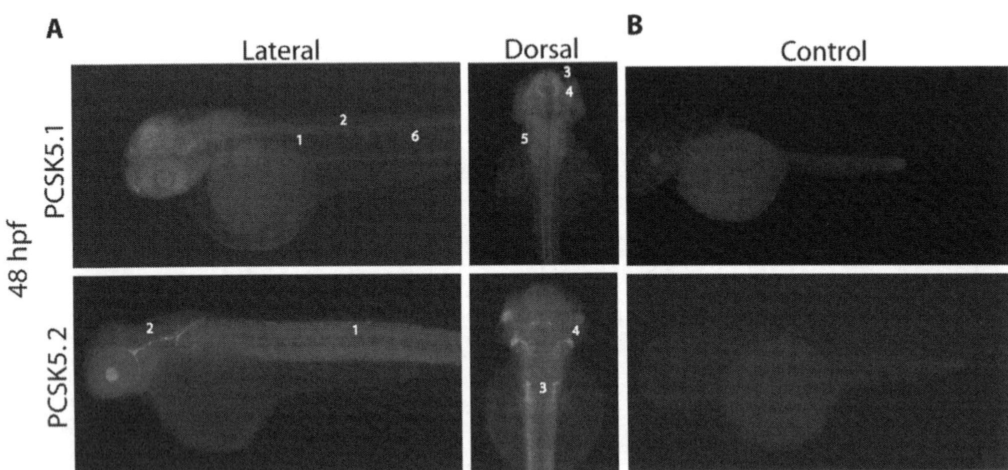

Fig. 15.7. Whole-mount immunolocalization of zebrafish PCSK5. (**a**) At 48 hpf immunoreactive PCSK5.1 was localized within motor neurons (1), spinal cord (2), anterior (3), postoptic (4) and hindbrain commissural neurons (5) in addition to lateral line neuromasts (6). Immunoreactive PCSK5.2 was also localized within motor neurons (1), cranial motor nerves which innervate branchial arches 1–7 (2), Mauthner axon (3) and retinal ganglial cell axons (4). (**b**) No immunoreactive signal was observed when control experiments employing antibodies pre-absorbed with peptides used as antigen were performed.

mRNA with high affinity specificity thereby blocking the initiation of PCSK5.1 translation. The MO-2 binds to the e4i4 splice site thereby blocking the pre-mRNA splicing and formation of the correct mRNA sequence. The mRNA product generated by a splice-blocking morpholino was analysed by RT-PCR using a single set of primers targeted to sequences outside (5′ and 3′) of the predicted deletion (Exon 4) and resulted a complete deletion of exon 4. Steps involved in microinjection procedure are illustrated in the following diagram (**Fig. 15.8**).

3.3.1. Embryo Preparation

1. In the late afternoon of the day before eggs are required, transfer fishes to a net positioned towards the top of a holding tank and covered. Alternatively transfer fishes in the ratio of one male per two female to the crossing tank with the separator.
2. In the morning, after the light cycle begins and 10–15 min after spawning has stopped, collect the eggs that have fallen through the net from the bottom of the tank, check under a dissecting microscope and select one- to two-cell stage embryos for microinjection. Remove the separator immediately after the light cycle begins if using the crossing tanks.

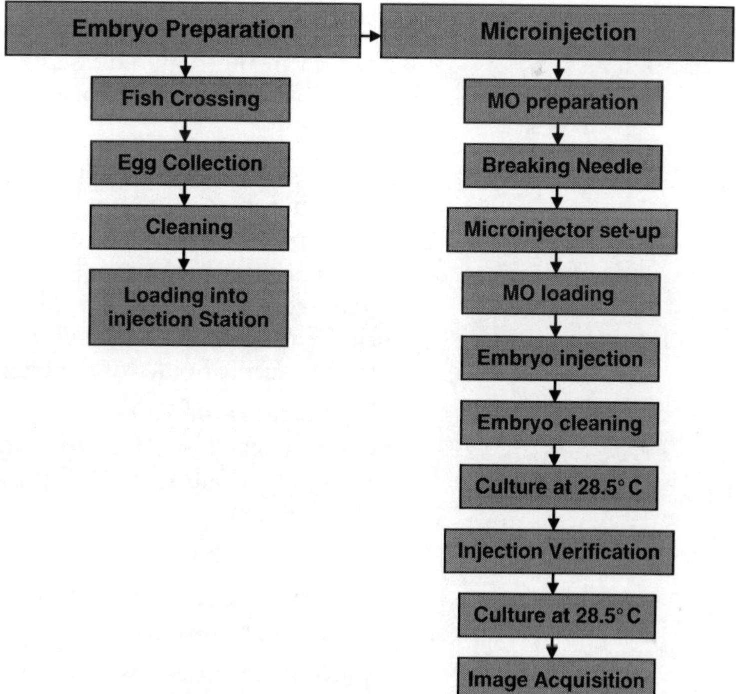

Fig. 15.8. Embryo preparation and microinjection. Injecting embryos should be performed immediately after collection by loading in the injection station and with the needle filled with morpholino solution.

Collect the eggs that have fallen to the bottom of the crossing tanks after 30 min.

3. Clean embryos by changing fresh egg water a few times. This is done quickly since the eggs remain in the one-cell stage only for 30 min after fertilization.
4. Collect the embryos with a wide-mouthed pipette and carefully arrange them in the furrows of the injection station. In order to facilitate microinjection, use only a little egg water to ensure that the embryos are held firmly within the furrow without floating away. Keep the rest of embryos in an 18°C incubator to slow down the cell division rate (not more than 1 h). This allows more time to inject embryos at the one- to two-cell stage.

3.3.2. Microinjection

1. Centrifuge the tube containing the morpholino (MO) solution before loading into the microinjection needle in order to avoid small particles from the MO solution clogging the needle (*see* **Notes 12** and **14**).
2. Using the highest magnification of the dissecting microscope, gently break the tip of the needle with clean forceps (*see* **Notes 13**).
3. Switch the microinjector on and open the pressure valve opened. Set P-clear at 100 psi and the injection pressure at 15 psi. Adjust the balance pressure to expel the MO solution in the required volume and to prevent medium from flowing back into the needle and diluting the MO solution.
4. Place the tip of the needle inside the tube that contains MO solution to allow the solution to migrate inside the needle by capillary force. When the MO has migrated to the tip, introduce the needle into the needle holder on the microinjector. Alternatively, MO can be aspirated into the needle using the fill button after introducing the needle into the needle holder of the injector.
5. Verify the flow of MO through the tip of the broken needle into injection chamber by pressing the foot pedal. For the best distribution of MO in the one-cell embryo, use a volume of 2 nl.
6. Set magnification of microscope to 1.6× for injection.
7. Bring the pipette tip close to the target embryo and focus on the cells to be injected.
8. Applying pressure using foot pedal ensures the flow of MO through the needle tip.
9. Adjust injection time and injection pressure balance for appropriate volume of MO discharge.

10. Pierce the tip of the pipette through the chorion directly into the cytoplasm of the —one- to two-cell stage embryo. Injection can also be done through the yolk and then into the cytoplasm of the dividing cell.
11. Deliver MO by applying pressure via the foot pedal. The injected MO is visible as red spot due to the presence of phenol red in the solution.
12. Slowly remove the pipette tip from the injected embryo by gently lifting the pipette.
13. Slowly move the needle to the adjacent embryo to be injected. Injection is performed for all of the embryos in the furrow.
14. Immediately remove from the Petri dishes and discard all dead and uninjected embryos.
15. Transfer the injected embryos to a dish with egg water and incubate at 28.5°C.
16. Check the embryos again a few hours (generally 3 h) after injection through GFP filter and remove dead or abnormal embryos.
17. Incubate the successfully injected embryos at 28.5°C until the required developmental stage is reached.
18. After day 1, manually dechorionate the embryos using Dumont #5 watchmaker's forceps. Morphological defects associated with the ATG-directed MO were photographed and are shown in **Fig. 15.9** as examples.
19. Fix the embryos for 2 h at room temperature and store at –20°C for staining.

4. Notes

1. Use less DNA if you are using plasmid DNA as a template and more DNA if you are using genomic DNA as a template for the PCR. Be sure to include a 10 min extension at 72°C after the last cycle to ensure that all PCR products are full length and 3′-adenylated to facilitate TOPO cloning.
2. There should be a single, discrete band on agarose gel from the PCR reaction. In the absence of a single, discrete band, gel-purify the fragment before using the TOPO TA Cloning Kit. Take special care to avoid sources of nuclease contamination. Alternatively, the PCR reaction can be optimized to eliminate multiple bands and smearing.

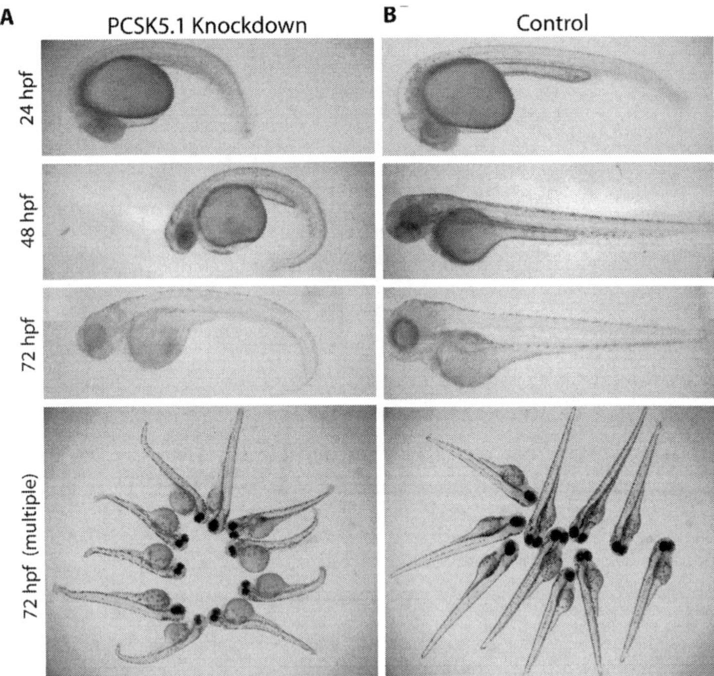

Fig. 15.9. Morpholino-based translation inhibition disrupts formation of the posterior lateral line. Phenotypic consequences of zfPCSK5.1 knockdown. MO-injected embryos (**a**) at 24 hpf–72 hpf displayed curved posterior body plan compared to control (**b**). PCSK5.1 morphants also display circular swimming pattern and lack a normal touch response.

3. We have found that precipitating DNA with sodium acetate/LiCl and ethanol overnight at –20°C works better than at –80°C for 40 min.
4. Unless otherwise stated, the in situ hybridization procedures are carried out with embryos in baskets as shown in **Fig. 15.2a, b** and RNase-free solutions in 6-well plates (4 ml/well) at room temperature under sterile conditions.
5. Baskets to place embryos are made in different colours with or without a rim on the top as shown in the figure. This permits easy orientation of the order of the baskets and the embryos within the basket to correspond to a particular probe.
6. Hybridization at 70°C with 50% deionized formamide reduces background staining and provides a clearer signal.
7. At day 2, pre-warm hybridization mix, 2X SSC and 0.2X SSC by placing them in a beaker with water and keep it in 70°C water bath for at least 20 min. Remove hybridization mix and probe from the tubes, replace with 1 ml of pre-warmed hybridization mix and incubate for 15 min at 70°C. During incubation, prepare 6-well plates with baskets for 2X SSC washes and place them at 70°C.

After 15 min transfer the embryos to basket and continue with 2X SSC washes. Maintain another water bath at 65°C. Place 6-well plate with 0.2X SSC for high-stringency washes to prevent non-specific hybridization of the probe.

8. After 1X PBST washes incubating embryos with alkaline Tris buffer twice for 15 min prior to incubation in the presence of BCIP and NBT will also provide clear signal.
9. If the gene of interest possesses antisense (AS) transcripts the sense probe will detect a signal for the AS transcript. For instance, we have found AS transcripts to zebrafish progranulin-1 and progranulin-2, members of the granulin family of secreted growth factors (8).
10. Placing embryos in methanol for 30 min or longer after the staining reaction is stopped converts the signal to a true purple colour and washes away non-specific staining.
11. Fixed embryos can be stored in the dark in stop solution at 4°C for several months.
12. In order to avoid generating non-specific abnormalities the MO injection solution and dilution should be prepared using pure water. Pilot experiments should be performed using various dilutions of MO to eliminate non-specific toxicity effects caused by high concentrations of MO.
13. To facilitate microinjection of embryos at the one-cell stage, prepare the injection needles and solutions in advance.

References

1. Zon, L. I. (1999) Zebrafish: A new model for human disease *Genome Res* **2**, 121–9.
2. Graham, J. L., and Peter, D. C. (2007) Animal models of human disease: Zebrafish swim into view *Nature Reviews Genetics* **8**, 353–67.
3. Kari, G., Rodeck, U., and Dicker, A. P. (2007) Zebrafish: An emerging model system for human disease and drug discovery *Clin Pharmacol Ther* **82**, 70–80.
4. Love, D. R., Pichler, F. B., Dodd, A., Copp, B. R., and Greenwood, D. R. (2004) Technology for high- throughput screens: The present and future using zebrafish *Curr Opin Biotechnol* **15**, 564–71.
5. Guo, S. (2009) Using zebrafish to assess the impact of drugs on neural development and function *Expert Opin Drug Discov* **7**, 715–26.
6. Ausoni, S., and Sartore, S. (2009) From fish to amphibians to mammals: In search of novel strategies to optimize cardiac regeneration *J Cell Biol* **184**, 357–64.
7. Major, R. J., and Poss, K. D. (2007) Zebrafish heart regeneration as a model for cardiac tissue repair *Drug Discov Today Dis Models* **4**, 219–25.
8. Cadieux, B., Chitramuthu, B. P., Baranowski, D., and Bennett, H. P. (2005) The zebrafish progranulin gene family and antisense transcripts *BMC Genomics* **6**, 156.
9. Poirier, S., Prat, A., Marcinkiewicz, E., Paquin, J., Chitramuthu, B. P., Baranowski, D., Cadieux, B., Bennett, H. P. J., and Seidah, N. G. (2006) Implication of the proprotein convertase NARC-1/PCSK9 in the development of the nervous system *J Neurochem* **98**, 838–50.
10. Westerfield, M. (2007) *The Zebrafish Book*, 5th Edition. A guide for the laboratory use of zebrafish (Danio rerio). Eugene, OR: University of Oregon Press (4th Edition available online).

11. Nasevicius, A., and Ekker, S. C. (2000) Effective targeted gene 'knockdown' in zebrafish *Nature Genet* **26**, 216–20.
12. Chitramuthu, B. P., Baranowski, D. C., Cadieux, B., Rousselet, E., Seidah, N. G., and Bennett, H. P. (2010) Molecular cloning and embryonic experssion of zebrafish PCSK5 co-orthologues: Functional assessment during lateral line development. *Dev Dyn* **239**, 2933–46.
13. Kimmel, C. B., Ballard, W. W., Kimmel, S. R., Ullmann, B., and Schilling, T. F. (1995) Stages of embryonic development of the zebrafish *Dev Dyn* **203**, 253–310.
14. Culp, P., Nusslein-Volhard, C., and Hopkins, N. (1991) High frequency germline transmission of plasmid DNA sequences injected into fertilised zebrafish eggs *Proc Natl Acad Sci USA* **88**, 7953–7.
15. Lusson, J., Vieau, D., Hamelin, J., Day, R., Chrétien, M., and Seidah, N. G. (1993) cDNA structure of the mouse and rat subtilisin/kexin-like PC5: A candidate proprotein convertase expressed in endocrine and nonendocrine cells *Proc Natl Acad Sci USA* **90**, 6691–5.
16. Krauss, S., Concordet, J. P., and Ingham, P. W. (1993) A functionally conserved homolog of the Drosophila segment polarity gene hh is expressed in tissues with polarizing activity in zebrafish embryos *Cell* **75**, 1431–44.

Chapter 16

Characterization of Impaired Processing of Neuropeptides in the Brains of Endoprotease Knockout Mice

Margery C. Beinfeld

Abstract

With the development of mice in which individual proteolytic enzymes have been inactivated, it has been of great interest to see how loss of these enzymes alters the processing of neuropeptides. In the course of studying changes in the peptide cholecystokinin (CCK) and other neuropeptides in several of these knockout mice, it has become clear that neuropeptide processing is complex and regionally specific. The enzyme responsible for processing in one part of the brain may not be involved in other parts of the brain. It is essential to do a detailed dissection of the brain and analyze peptide levels in many brain regions to fully understand the role of the enzymes. Because loss of these proteases may trigger compensatory mechanisms which involve expression of the neuropeptides being studied or other proteases or accessory proteins, it is also important to examine how loss of an enzyme alters expression of the neuropeptides being studied as well as other proteins thought to be involved in neuropeptide processing. By determining how loss of an enzyme alters the molecular form(s) of the peptide that are made, additional mechanistic information can be obtained. This review will describe established methods to achieve these research goals.

Key words: Mouse brain dissection, prohormone convertase, RIA, HPLC, mRNA, qPCR.

1. Introduction

The development of mice in which individual proteolytic enzymes have been inactivated is a major scientific breakthrough that has allowed examination of what role these enzymes might play in neuropeptide processing. A number of neuropeptides have been studied in these mice. The assumption in these studies is that if levels of the neuropeptide are lower in the knockout mice than

M. Mbikay, N.G. Seidah (eds.), *Proprotein Convertases*, Methods in Molecular Biology 768,
DOI 10.1007/978-1-61779-204-5_16,

in the wild-type controls, the enzyme that has been inactivated is involved in processing of that neuropeptide.

Detailed examinations of changes in the level of a number of neuropeptides in the brains of these mice have demonstrated that neuropeptide processing is regionally specific and complex. In the case of CCK, loss of PC1 (1) causes a specific pattern of decrease in CCK that differs from that of loss of PC2 (2), PC7 (unpublished observations), or cathepsin L (3). Substantial differences in peptide levels and peptide and enzyme expression have been observed between male and female mice, so both genders should be examined. As most peptides develop postnatally and may decline in aged mice, it is important to compare adult wild-type and knockout mice of similar ages.

It is rare that loss of a single enzyme causes peptide levels to be undetectable. In the case of CCK, there is no instance where loss of one prohormone convertase, cathepsin L, or carboxypeptidase E (4) results in complete elimination of CCK. It has even been observed that loss of PC2 increases proCCK and CCK levels in cerebral cortex and related structures. This is actually accompanied by a decrease in CCK mRNA levels (2). The mechanism for this is not clear.

Given this information, it is essential to examine how loss of an enzyme affects peptide levels in specific brain regions. As loss of an enzyme in utero could cause compensatory changes in the expression of the peptide(s) being measured as well as other enzymes involved in the processing, it is also important to examine expression of the peptide mRNA and other proteins involved in the processing of the peptide.

An examination of how loss of an enzyme alters the molecular forms produced provides validation of the RIA results and may provide some mechanistic information about the role that a protease may be playing in peptide processing.

This review provides an overview of the techniques required to perform these experiments along with specific methods that have been used in the past.

2. Materials

2.1. Mouse Brain Dissection

1. Ketamine (100 mg/ml) and xylazine (100 mg/ml).
2. Mouse Brain Mold with 1 mm spacing (Zivic-Miller, Pittsburgh, PA).
3. Double-edge razor blades.
4. Dissecting tools, glass Petri dish.
5. Trizol Reagent (Life Technologies, Carlsburg, CA).

2.2. Peptide Extraction, RIA, and Protein Assay

1. Microsonicator (Misonix, Farmingdale, NY).
2. Savant SpeedVac concentrator (Thermo-Fisher).
3. Peptide RIA.
4. Micro BAC Kit (Pierce Chemical, Rockford, IL).

2.3. Peptide HPLC

1. HPLC system with column, UV detector, and fraction collector.
2. C18 Sep-Pack cartridges (Waters Co, Milford, MA).
3. Solvents (HPLC-grade): water, acetonitrile, and trifluoroacetic acid (TFA).
4. Peptide standards.

2.4. RNA Extraction, Reverse Transcription, and Real-Time qPCR

1. Chloroform, RNAeasy columns (Qiagen, Valencia, CA).
2. Nanodrop Spectrometer (Molecular Devices).
3. RETROscript Kit (Ambion, Austin, TX) or Protoscript Kit (New England Biolabs, Beverly, MA).
4. SYBR Green mix (Applied Biosystems, Foster City, CA).
5. Gene-specific primers.
6. Real-time qPCR instrument.

3. Methods

3.1. Mouse Brain Dissection (See Notes 1–6)

1. Mice are euthanized with a method approved by the Institutional Animal Care Committee. We have used an overdose of ketamine/xylaxine. Each mouse receives an ip injection of 2 ml/kg of a 1:10 mixture of 100 mg/ml xylazine and 100 mg/ml ketamine.
2. After the mouse is anesthetized (does not respond to a toe pinch), the neck is cut with scissors and the body given to the person who is removing the peripheral organs.
3. The skin on the head is cut from the rear (spinal cord end) with scissors and removed. Using a small scissor and working from the rear (inserting one blade in where the spinal cord is attached), cut the scull open at the midline while trying to damage the brain as little as possible. Use a forceps or a small pliers, gently remove the bone flaps on both sides to free the forebrain, cerebellum, and whatever portion of the spinal cord that remained with the head. Carefully remove the bone covering the olfactory bulbs, remove and save them. Insert a small spatula or scissors underneath the front of the brain to cut the optic nerves where they attach to the hypothalamus. Continue lifting the forebrain with a

small spatula and it should be possible to remove it completely with the cerebellum and spinal cord still attached. This process is described in detail (with pictures) for the rat brain (5). The subsequent dissection should take place on a cold surface; a large glass Petri dish inverted on wet ice works well.

4. The pituitary should have remained in the skull enclosed in a membrane. It should be possible to remove it with a small forceps.
5. The cerebellum can be dissected off with a small spatula leaving the pons, medulla, and spinal cord attached. The cerebellum is usually discarded because it does not have many peptides. That is an individual choice as more peptides have been discovered in the cerebellum recently. The pons, medulla, and spinal cord (if present) are dissected with a scalpel or razor blade as in **Fig. 16.1**.
6. The forebrain is then inverted so the hypothalamus is up. The entire hypothalamus can easily be removed in one piece by making small scissor cuts (about 1.5 mm deep) in the front (optic chiasm), the rear (mammillary bodies), and on the sulci on the sides. After the cuts are made, press down from the rear on the two side cuts with a small scissor and cut out the entire structure.
7. Place the forebrain in a mouse brain mold with the ventral (hypothalamic) surface up. Insert double-edge razor blades cut to make single blades in the mold starting at the front of the block of hypothalamic tissue removed and fill up all the slots with blades. To insure that the brain is completely

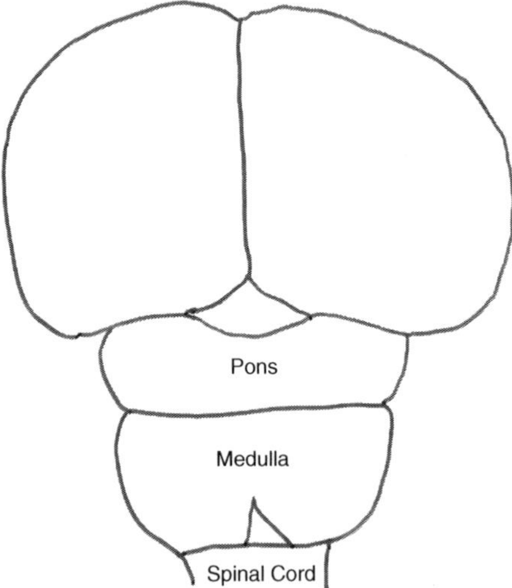

Fig. 16.1. Drawing of the ventral surface of the mouse brain with the cerebellum and olfactory bulbs removed to show the location of the pons, medulla, and spinal cord.

cut hold onto the blades on the outside and gently move them back and forth laterally as a group. Remove individual blades and transfer the sections to glass slides on top of the cold surface. Orient the sections anterior to posterior on the cold surface. Dissect the areas of interest in reference to the brain atlas with a scalpel or razor blade and place them in Eppendorf tubes on dry ice. Keep them moist with cold saline. Sometimes it is useful to use a tiny paint brush to move the sections and pick up small pieces that have been dissected.

8. The brain areas that are the easiest to obtain are cerebral cortex at multiple levels, caudate putamen + globus pallidus, basal forebrain (nucleus accumbens, adjacent pyriform cerebral cortex, and olfactory tubercle), septum, thalamus, hippocampus, amygdala (amygdala + adjacent pyriform cerebral cortex), substantia nigra plus the ventral tegmental area. Sometimes these structures extend for several sections and are pooled in one tube. Sometimes it is necessary to turn the sections over because the structure of interest is on the other side.

9. It is much better to dissect the live animals at the site where they are to be studied; that way the tissues are only frozen once. This is not always feasible and we have successfully worked with brains that were removed at other facilities, frozen and shipped to us. In this case, the brains can be partially thawed on a cold surface, sectioned as described. They section very well when they are partially frozen. The difference is that each brain sample for RIA should be extracted immediately after dissection, while the RNA samples taken into Trizol can be frozen and worked on later. We have succeeded in getting reasonable RIA values and have obtained RNA that was of high enough quality to do qPCR.

3.2. Peptide Extraction and RIA

1. The choice of extraction method depends on the peptide(s) being measured. In the late 1970s, we started using sonication in 0.1 N HCl as an extraction that allowed us to measure multiple peptides in the same sample (7). A microsonicator is very useful for these small samples. The tip should be washed between samples. Dilute HCl may not be optimal for all peptides; probably some extraction that involves heating the sample in dilute acetic acid to inactivate proteases would be better for long-term storage of samples. Acetic acid has the advantage that it is volatile, while dilute HCl is neutralized with an equal volume of 0.1 NaOH prior to RIA.

2. If acetic acid is used, the best way to remove it is to aliquot the samples into the tubes to be used for the RIAs and dry them in a SpeedVac concentrator.

3. The volume of extraction media would need to be customized to the sample size, peptide abundance, etc. For CCK, we extract cerebral cortex, hippocampus, thalamus samples in 0.5 ml, and smaller samples in 0.1 or 0.2 ml. CCK is a very abundant peptide, so this needs to be considered.

3.3. Protein Assay

1. In order to normalize the RIA data to tissue size, a small aliquot (5 μl) should be taken of the sonicated sample before it is clarified for RIA to use for the protein assay. We use a micro-BCA protein method in 96-well plates (with a volume of 0.2 ml/well) that can be read in a plate reader.
2. Sample, protein standard (BSA, from 1 to 50 μg) or sample is added to a maximum volume of 0.1 ml.
3. 0.1 ml of the assay mix is added.
4. After 30 min at room temperature, the color is read in a plate reader.

3.4. Peptide HPLC

1. The RIA is performed first and changes in peptide levels can be further investigated by HPLC with RIA detection of molecular forms using sample leftover from the RIA. It may be necessary to pool extracts from similar individuals to obtain enough material for HPLC/RIA. This requires an HPLC with a UV detector (set preferably at the peptide bond absorption, 220 nm), appropriate synthetic peptide standards, a fraction collector, and a peptide RIA to detect where peptides in brain extracts elute. Gradients of acetonitrile in water with 0.1% trifluoroacetic acid (TFA) as an ion pairing reagent are fairly standard. This solvent has the advantage that it is completely volatile. The best way to remove it is to aliquot the samples (in this case the HPLC factions) into the tubes to be used for the RIAs and dry them in a SpeedVac concentrator.
2. It may be possible to evaluate molecular forms with ion exchange chromatography, size exclusion chromatography, specific antisera, mass spectrometry, or Western blotting. These techniques will not be described in this review.
3. To extend the life of the HPLC column, it is advisable to purify the samples before injecting them. The classical method is to use C18 Sep-Pack cartridges. The cartridges are prepared by passing 3 ml of methanol through them with a syringe, followed by passing air with the syringe to completely remove the solvent. Then the cartridge is equilibrated with 0.1%TFA by passing 20 ml of the solvent followed by air. The biological sample is passed three times through the cartridge (and retained to calculate recovery), and the cartridge is rinsed with 20 ml of 0.1%TFA. The purified sample

is eluted with 3 ml of 80% acetonitrile/20% water containing 0.1% TFA. This 3 ml sample is concentrated with a SpeedVac concentrator to an appropriate volume for the HPLC system. It is usually advisable to centrifuge the concentrated sample to remove any particulate matter before separation. A small aliquot of the final sample should be retained for recovery measurements.

4. An HPLC system needs to be developed to separate the known forms of the peptide. This usually involves an increasing gradient of acetonitrile in TFA. There are many good HPLC columns: a Waters 4.6 × 250 mm Symmetry Shield RP 18 column has been good for separating CCK peptides.
5. Great care needs to be taken not to contaminate biological samples with the peptide standards. After the HPLC system is set up, it is important to make a number of blank water injections and right before the actual samples to do a blank run with water and collect samples and assay them as if it were a biological sample to ensure that the column and injector were completely clean. After all the biological samples are run, the standards can be re-injected to insure that they still elute in the same position.
6. The fraction collector attached to the HPLC system is usually set to collect 1 min (and usually 1 ml) samples following blank or sample injection. This could be different with microbore instruments which operate at different flow rates. An aliquot of each sample is removed into the test tube where the RIA will be performed and the aliquot is dried down with a SpeedVac concentrator.

3.5. RNA Extraction, Reverse Transcription, and Real-Time qPCR

1. Tissues to be used for qPCR are placed in 100 μl of Trizol reagent on dry ice and stored at –80°C until extraction.
2. For the extraction, the samples are defrosted, sonicated, and further extracted with the standard Trizol protocol, scaled down to the small size of the sample.
3. They are further purified individually with RNAeasy columns by Qiagen, using the method described in the kit.
4. The quality and quantity of the RNA is determined with a Nanodrop spectrometer.
5. For qPCR, first-strand cDNA was synthesized using 2 μg (or less) of total RNA (previously treated for 10 min at room temperature with DNase) in a 20 μl reverse transcriptase reaction mixture using random decamers following either the RETROscript Kit or the Protoscript Kit.
6. Real-time PCR reactions were performed in a 25 μl reaction mixture containing 1/20th volume of cDNA preparation,

1X SYBR Green mix with 5 pmol of each primer. The expression of the genes of interest was compared to expression of 18S mRNA. The cycle threshold (Ct) of the gene of interest subtracted from the Ct of 18S giving the dCt. The value of the control dCt (wild type) was set to 1 and the control dCt was subtracted from the experimental (knockout) dCt yielding the ddCt.

7. The difference in expression between control and experimental is calculated as 2^{-ddCt}, as described (8).

4. Notes

1. The dissection being described is a freehand dissection performed with unfixed tissue. Even greater anatomical detail could be obtained with frozen sections and micropunches as has been done with rat brain (5).
2. Before beginning this process, it is important to establish what tissues are being harvested, set up protocols, number tubes, etc. It is essential to compare wild-type and knockout mice that are the same age, including both males and females. The number of individuals in each group required to see significant changes depends on the magnitude of these changes, but having at least six individuals per group is usually sufficient. As it is often difficult to obtain large numbers of mice suitable for these studies at the same time, it is acceptable to dissect fresh tissue and freeze it for later extraction once a complete group of individuals has been obtained.
3. It is essential to work quickly once the mouse has been sacrificed. Additional task lighting is very helpful. The mouse tissues are small enough that they can usually be put in Eppendorf tubes. I prefer to use metal tube racks placed on top of dry ice to quick freeze tissues.
4. As knock-out mice are a valuable research resource, we have always harvested all tissues known to contain peptides, including pituitary, adrenals, stomach, pancreas, intestines, and colon. Even if we are not set up to investigate these other tissues, we always remove them and quick freeze then and provide them to other investigators. In this case, it is essential to have two people to do the dissection at the same time so that the tissues can be removed as quickly as possible.
5. For the fine dissection of the mouse brain, it is essential to become familiar with the anatomy of the mouse brain. It is actually very similar to the rat brain, only smaller. There

is a good atlas of the mouse brain (6). It is a good idea to practice the dissection with wild-type or control mice to become comfortable with it before killing valuable knock-out mice. Reproducible dissection is the goal.

6. When mice are killed in-house and tissues are to be used for both peptide RIA and qPCR, we have always divided the tissue, half frozen for later peptide extraction and the other half placed in Trizol for future RNA extraction. It is theoretically possible to recover protein from these Trizol samples for peptide RIA when the RNA is processed. The author has no experience with this protocol which appears to involve precipitating the protein with ethanol. This may not be effective for recovering small peptides for RIA, some of which like CCK are soluble in ethanol or methanol.

References

1. Cain, B. M., Connolly, K., Blum, A. C. et al. (2004) Genetic inactivation of prohormone convertase (PC1) causes a reduction in cholecystokinin (CCK) levels in the hippocampus, amygdala, pons and medulla in mouse brain that correlates with the degree of colocalization of PC1 and CCK mRNA in these structures in rat brain *J Neurochem* **89**, 307–13.
2. Beinfeld, M. C., Blum, A., Vishnuvardhan, D., Fanous, S., and Marchand, J. E. (2005) Cholecystokinin levels in prohormone convertase 2 knock-out mouse brain regions reveal a complex phenotype of region-specific alterations *J Biol Chem* **280**, 38410–15.
3. Beinfeld, M. C., Funkelstein, L., Foulon, T. et al. (2009) Cathepsin L plays a major role in cholecystokinin production in mouse brain cortex and in pituitary AtT-20 cells: Protease gene knockout and inhibitor studies *Peptides* **30**, 1882–91.
4. Cain, B. M., Wang, W. G., and Beinfeld, M. C. (1997) Cholecystokinin (CCK) levels are greatly reduced in the brains but not the duodenums of Cpe^{fat}/Cpe^{fat} mice: A regional difference in the involvement of carboxypeptidase E (Cpe) in pro-CCK processing *Endocrinology* **138**, 4034–7.
5. Palkovits, M., and Brownstein, M. J. (1988) *Maps and Guide to Microdissection of the Rat Brain.* Elsevier Press, New York, NY.
6. Franklin, K. B. J., and Paxinos, G. (1997) *The Mouse Brain in Stereotaxic Coordinates.* Academic Press, San Diego, CA.
7. Eiden, L. E., Mezey, E., Eskay, R. L., Beinfeld, M. C., and Palkovits, M. (1990) Neuropeptide content and connectivity of the rat claustrum *Brain Res* **523**, 245–50.
8. Livak, K. J., and Schmittgen, T. D. (2001) Analysis of relative gene expression data using real-time quantitative PCR and the 2(-Delta Delta C(T)) Method *Methods* **25**, 402–8.

Chapter 17

Quantitative Peptidomics of Mice Lacking Peptide-Processing Enzymes

Jonathan Wardman and Lloyd D. Fricker

Abstract

Peptidomics is defined as the analysis of peptides present in a tissue extract, usually using mass spectrometry-based approaches. Unlike radioimmunoassay-based detection techniques, peptidomics measures the precise form of each peptide, including post-translational modifications, and can readily distinguish between longer and shorter forms of the same peptide. Also, peptidomics is not limited to known peptides and can detect hundreds of peptides in a single experiment. Quantitative peptidomics enables comparisons between two or more groups of samples and is perfect for studies examining the effect of gene knockouts on tissue levels of peptides. We describe the method for quantitative peptidomics using isotopic labels based on trimethylammonium butyrate, which can be synthesized in five different isotopic forms; this permits multivariate analysis of five different groups of tissue extracts in a single liquid chromatography/mass spectrometry run.

Key words: Prohormone convertase, proprotein convertase, carboxypeptidase, peptidomics, proteomics, peptidase, protease.

1. Introduction

Prohormone convertases PC1/3 and PC2 function as endopeptidases in the neuroendocrine system, contributing to the processing of peptide hormones and neuropeptides (1, 2). In addition to the endopeptidase step, most bioactive peptides require an additional step mediated by carboxypeptidase E and/or D; these enzymes remove C-terminal basic residues from the prohormone convertase reaction products (3). For some peptides, additional modifications are required such as C-terminal amidation (4). To study the peptide-processing enzymes, it is useful to have

M. Mbikay, N.G. Seidah (eds.), *Proprotein Convertases*, Methods in Molecular Biology 768,
DOI 10.1007/978-1-61779-204-5_17, © Springer Science+Business Media, LLC 2011

a method of measuring the precise form of each peptide present in a tissue. Mass spectrometry-based quantitative peptidomics techniques can be used to distinguish peptide-processing profiles in various model biological systems (5). This technique involves extraction and purification of peptides from mouse brain regions followed by labeling with isotopic tags and analysis using liquid chromatography and electrospray ionization mass spectrometry. The analysis provides a detailed view of the precise molecular forms of each peptide, including any post-translational modifications. By examining how the form and the relative amount of a peptide vary between wild-type mice and mice lacking a peptide-processing enzyme, one can understand the contribution of that enzyme in a given tissue (6–11). Peptides which increase in null mice relative to wild type are putative substrates of the knocked-out peptidase, whereas peptides that decrease in null mice relative to wild-type mice are putative products of the enzyme. Peptides that do not change between null mice and wild-type mice are presumably processed redundantly by multiple peptidases such that deletion of any one enzyme does not have an effect on peptide levels. By comparing the amino acid sequences of the peptides that do change with those that do not change in the absence of an enzyme, one can better understand the specificity of each enzyme. The chemical character of the amino acids surrounding the cleavage site, such as presence of hydrophobic, charged, or bulky side chains, in turn renders information about the physical aspects of the catalytic site. By comparing quantitative peptidomics data from several different brain regions in multiple groups of null mice, a global view of peptidase processing of neuroendocrine peptides is attained.

Most of the previous in vivo studies of PC1/3 and PC2 null mice have used radioimmunoassays to quantify changes in peptide processing (12). If antibodies are available that recognize a specific form of a peptide, this method provides useful information. However, this technique is limited by the requirement for specific antisera to known peptides. Despite the sensitivity of these antibody-based experiments, detection can be unspecific, as antibodies will often recognize several different forms of a peptide, including N- or C-terminally extended peptides or peptides with post-translational modifications such as amidation or acetylation. This makes determining the precise molecular form of the peptide difficult. Using a mass spectrometry-based quantitative peptidomics approach, peptide levels in a variety of tissues from both prohormone convertase knockout mice and their wild-type counterparts can be analyzed in a systematic manner (7, 8, 10, 11). This method allows for the relative quantification of both known peptides and previously unknown peptides; these can often be identified from tandem mass spectrometry analysis. The following protocol describes the basic method of peptide extraction, isotopic labeling, and mass spectrometry analysis.

2. Materials

2.1. Extraction of Peptides

1. Distilled water Milli-Q system (Millipore, Bedford, MS, USA).
2. 0.1 M Hydrochloric Acid (6N, sequanal grade, constant boiling; Pierce, Rockford, IL, USA).
3. 0.4 M NaH_2PO_4 (Sigma-Aldrich, Inc., St Louis, MO, USA).
4. Ultrasonic processor (W-380; Ultrasonic, Inc., Farmingdale, NY, USA).
5. Low-retention microcentrifuge tubes (Eppendorf).

2.2. Peptide Labeling

Several isotopic tags can be used for quantitative peptidomics, some of which are commercially available (such as succinic anhydride with four hydrogens or deuteriums). Reagents that react with primary amines are usually used for peptidomics because a free amine is frequently found on peptides on either the N terminus and/or Lys side chains (13). However, some peptides do not contain a free amine due to N-terminal acetylation, pyroglutamylation, or another modification, and the absence of Lys residues; these peptides will not be quantifiable using the following approach. In our experience, the optimal isotopic tags for labeling amines are based on 3-(2,5-dioxopyrrolidin-1-yloxycarbonyl)propyl trimethylammonium chloride (14), which is the *N*-hydroxysuccinimide (NHS) ester of trimethylammonium butyrate (TMAB). One advantage of these reagents over succinic anhydride is that the TMAB reagent contains a positive charge (a quaternary amine) and so the positive charge(s) of the peptide's N terminus and any Lys residues is/are maintained after labeling with TMAB-NHS. In contrast, peptides labeled with succinic anhydride will have the positive charges of the free amines converted into negative charges, and unless there is another positive charge on the peptide, it will not be observed in mass spectrometry performed in positive ion mode (which is the most common procedure). Furthermore, succinic anhydride-labeled peptides tend to show weaker signals than do TMAB-labeled peptides (15). In addition, peptides labeled with TMAB reagents containing nine deuteriums co-elute from HPLC with peptides labeled with the nine-hydrogen form of the reagent; this is important for quantification. In contrast, peptides labeled with succinic acid containing four deuteriums do not precisely co-elute with peptides labeled with the hydrogen form of succinic acid (15). Finally, the TMAB reagents can be produced in five different isotopic forms containing either all hydrogen (referred to as D0-TMAB-NHS), three deuteriums (D3-TMAB-NHS), six deuteriums (D6-TMAB-NHS), nine deuteriums (D9-TMAB-NHS), or nine deuteriums and three atoms of ^{13}C (referred to as D12-TMAB-NHS because the mass difference is 12 Da greater than the D0 form) (16). Structures of the reagents are shown in **Fig. 17.1** along with the masses added to the peptide by the addition of the TMAB group to an amine and the loss of one hydrogen atom. The advantage of multiple isotopic forms of a label is that it permits multivariate analysis of different tissue extracts, from either completely different groups of mice or replicates of the same genotype, as shown in **Fig. 17.2**:

1. TMAB-NHS compounds. The synthesis of these compounds has been previously described (16).
2. NaOH (Sigma-Aldrich, Inc., St Louis, MO, USA).
3. Dimethyl sulfoxide (Sigma-Aldrich, Inc., St Louis, MO, USA).
4. $NH_2OH \cdot HCl$ (Sigma-Aldrich, Inc., St Louis, MO, USA).
5. Glycine (Sigma-Aldrich, Inc., St Louis, MO, USA).

Reagent	Structure
D0-TMAB-NHS (+ 129 Da)	$(CH_3)_3N^+$–CH₂CH₂CH₂–C(=O)–O-NHS
D3-TMAB-NHS (+ 132 Da)	$(CH_2d)_3N^+$–CH₂CH₂CH₂–C(=O)–O-NHS
D6-TMAB-NHS (+ 135 Da)	$(CHd_2)_3N^+$–CH₂CH₂CH₂–C(=O)–O-NHS
D9-TMAB-NHS (+ 138 Da)	$(Cd_3)_3N^+$–CH₂CH₂CH₂–C(=O)–O-NHS
D12-TMAB-NHS (+ 141 Da)	$(^{13}Cd_3)_3N^+$–CH₂CH₂CH₂–C(=O)–O-NHS

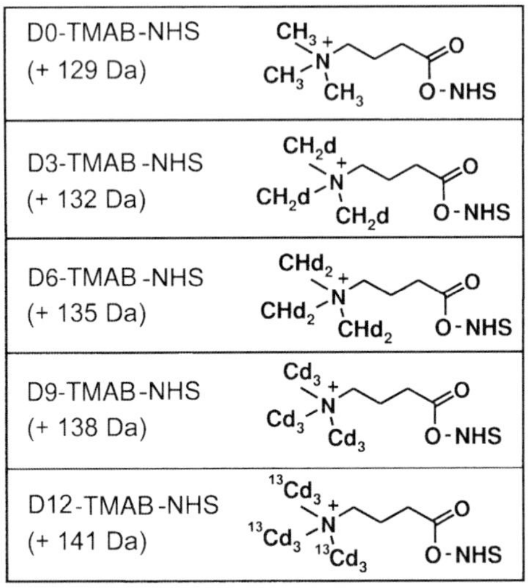

Fig. 17.1. TMAB labels used for isotopic tagging. The form of peptide shown includes the *N*-hydroxysuccinimide (NHS) moiety, which is replaced by the amine (N terminus or Lys side chain) in the reaction with the peptide. The indicated mass is the net additional mass after one tag is incorporated into the peptide, replacing a hydrogen on the amine. Note the quaternary amine group which provides a positive charge after labeling, maintaining the charge of the primary amine on the peptide. The TMAB-NHS labeling reagents are synthesized as the chloride salt (not shown).

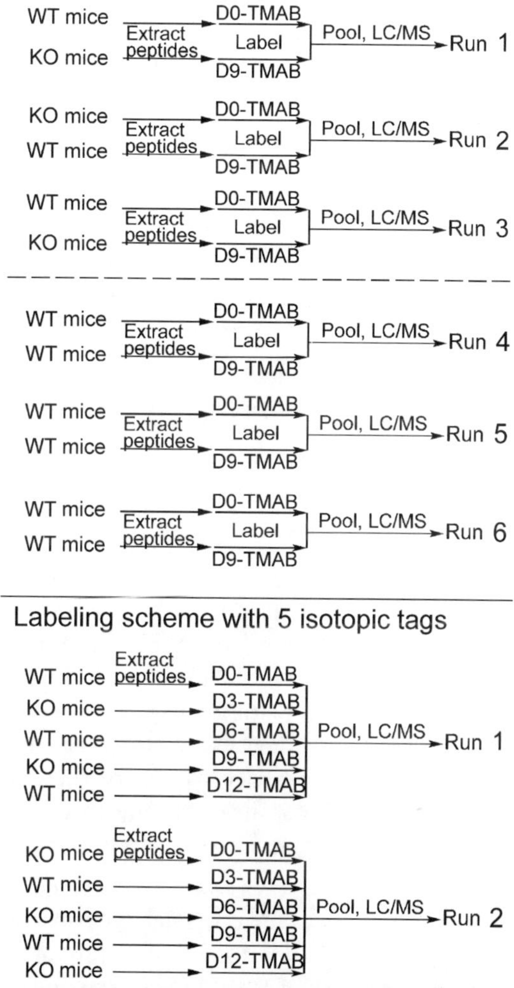

Fig. 17.2. Scheme for isotopic labeling of peptides extracted from tissues. *Top*: If two experimental conditions are to be compared (such as WT and KO), only two isotopic tags are needed for each replicate. Peptides are extracted from the appropriate brain region or other tissue, labeled with isotopic tags, and pooled as indicated. Note that the samples shown for run 2 are labeled with tags in the reverse scheme as those in runs 1 and 3; this is done to control for possible artifacts that could result from differential reactivity of the isotopic tags (not previously found to be a problem, but if the synthesis or the storage of one of the labels is not optimal, this could result in stronger labeling with one of the reagents). To obtain data for statistical tests, it is necessary to repeat the analysis with comparable groups of WT animals, as shown for runs 4–6. This provides a control for peptide variability among groups of animals and allows for statistical comparison of the results from runs 1–3. *Bottom*: The use of five isotopic labels to compare two different groups of animals allows for replicates to be performed within each LC/MS run, as shown in the scheme. Note the reversal of the labels in the scheme for run 2, as compared to run 1.

6. Hydrion pH papers (8.0–9.5) (Micro Essential Laboratory, Brooklyn, NY, USA).

2.3. Peptide Purification

1. Amicon Ultra 4-ml Ultracel 10,000 molecular weight cutoff centrifugal filter devices (Millipore, Bedford, MS, USA).
2. PepClean™ C-18 spin column (Pierce, Rockford, IL, USA).
3. Acetonitrile (HPLC grade; Fisher Scientific, Fair Lawn, NJ).
4. Trifluoroacetic acid (Pierce, Rockford, IL, USA).

3. Methods

3.1. Peptide Extraction

1. *Sacrifice and microwaving.* Mice are sacrificed by cervical dislocation. Mice are decapitated and heads are immediately placed in a conventional microwave oven at full power for 8 s, until the internal temperature of the brain reaches 80°C (17). Alternatively, mice can be sacrificed by microwave irradiation (18). Because the strength of different microwave devices varies considerably, it is essential to calibrate the device to ensure that the brain is heated to a temperature sufficient to inactivate proteases without damaging the brain structure by excessive heating.
2. *Dissection.* Brain is allowed to cool, then removed and cut into coronal sections using a razor blade for further dissection. Prefrontal cortex is removed by cutting at Bregma 1.94; all measurement references are taken from (19). Additional coronal cuts are made on either side of the hypothalamus at Bregma 0.00 and −3.00. The striatum is dissected from the 1.94 to 0.00 section by removing the cortex. This "striatum" sample includes the caudate putamen, the nucleus accumbens, the septum, and the ventral palladium. The coronal section of Bregma 0.00 to −3.00 is dissected into the hippocampus, the thalamus, the amygdala, and the hypothalamus. Cortex can be dissected from this section as well. Cerebellum is dissected from the remaining brain by removing the forebrain and brainstem sections from the cerebellum.
3. *Tissue storage.* At this point, tissues from multiple animals can be pooled into single tubes if multiple animals are to be used for each group. It is imperative that low-retention tubes are used for this, as loss of peptides can occur if normal tubes are used. Washing the tubes with double distilled water

beforehand is recommended. Tissues should be stored at –70°C until extraction.

4. *Sonication*. Dissected tissue pools are sonicated two times for 20 s at 1 pulse/s at duty cycle 3, 50% output, in an appropriate volume of ice-cold water using an ultrasonic processor. Tissues are sonicated in low-retention, 2-ml-volume microfuge tubes, using 5 μl of water per μg of tissue, with a minimum of 200 μl water to prevent splatter. A lower additional volume of water is used to rinse the tip of the sonicator into the tube containing the brain extracts, ensuring all peptides are collected. The sonicator should be rinsed between tissue extractions.
5. *Extraction of* peptides. The homogenates are incubated in a 70°C water bath for 20 min and then cooled on ice for 15 min. Extracts are acidified with ice-cold 0.1 M HCl to a final concentration of 10 mM HCl followed by vortex mixing. It is important to be sure that extracts are ice cold before adding acid to prevent acid-labile bonds from breaking. After incubation of extracts with acid for 15 min at 4°C, homogenates are centrifuged at 13,000×*g* for 40 min at 4°C and the supernatant is transferred to a new low-retention tube. These extracts can be frozen and concentrated to a lower volume in a vacuum centrifuge. This may be necessary if large volumes are to be combined during the filtration step (it is optimal if the total volume of all labeled extracts is under 4.0 ml to fit in filters used in a subsequent step).
6. *Buffering*. The pH of the peptide extracts is then adjusted to 9.5 by the addition of 0.4 M phosphate buffer (pH 9.5). These extracts can be stored at –70°C until labeling.

3.2. Labeling

The number of labels used and the number of replicates depend on the experimental questions being addressed. If comparing two conditions, such as wild-type (WT) and knockout (KO) mouse brain extracts, only two isotopic tags need to be used. However, it is essential to perform replicates, and at least one of the replicates should have the tags switched so that WT and KO are labeled with the opposite tags used for the other groups, as shown for the example indicated in run 2 as compared to runs 1 and 3 (**Fig. 17.2**, top). Because the resulting analysis provides relative values, and not absolute values, in order to perform statistical analysis of the data, it is important to also compare the relative level of peptides in WT vs WT groups (**Fig. 17.2**, middle). This allows ratios from the KO/WT experiments to be directly compared to the ratios from WT/WT experiments and statistical testing to be performed. An alternative approach is to use additional isotopic tags, such as the five forms of the TMAB-NHS labels; this allows for replicates to be included in a single run, and with two LC/MS runs, it is possible to compare five KO and five WT groups (**Fig. 17.2**, bottom):

1. *Labels.* Depending on how many samples are to be compared, two to five labels may be used to label sample extracts. The TMAB-NHS labeling reagents are dissolved in DMSO at 350 μg/μl and used to label each sample. Typically, a total of 5 mg of TMAB-NHS reagent is used for labeling of each mouse brain region (i.e., 5 mg for each hypothalamus, striatum, etc. present in the tube).
2. *Label volume calculation.* To determine the volume of labeling solution added per round of labeling, divide the total volume of label/DMSO solution by seven, as seven rounds of labeling will be carried out.
3. *Labeling.* Each round of labeling consists of adding one-seventh of the label volume to the appropriate sample individually. Samples are allowed to incubate at room temperature for 10 min before the pH is adjusted to 9.5 using 1.0 M NaOH, with pH paper used to test the pH of each sample by blotting <1 μl of sample onto the paper. Samples are incubated for another 10 min at room temperature before the next round of label is added. This process is repeated seven times to ensure all peptides are properly labeled. After the final addition of TMAB-NHS reagent, the samples should be incubated at room temperature for another 10–30 min. The entire labeling procedure requires approximately 4 h.
4. *Quenching.* After the labeling and prior to combining the samples for subsequent steps, it is essential to quench any unreacted TMAB-NHS reagent; this is accomplished by the addition of 2.5 M glycine in water. For this, 10 μl of the glycine solution is added per 5 mg of TMAB-NHS reagent and the mixture is incubated at room temperature for 40 min.

3.3. Peptide Purification

1. *Pool and filter.* Labeled samples that are to be compared are pooled together; typical schemes are shown in **Fig. 17.2**. After pooling, samples are applied to Amicon Ultra 4-ml Ultracel filters to remove proteins >10 kDa. Before using the filters, they should be washed with 2 ml water to remove any glycerol that remains on the filter from the manufacturer. Once rinsed, labeled samples are pooled and filtered according to the filter instructions. The flow-through is retained; this is the filtrate which will be analyzed.
2. *Hydroxylamine step.* In order to ensure that all labeled peptides are labeled only on free amino termini and lysines, and not on tyrosines, the filtrate must be treated with hydroxylamine (NH_2OH). First the pH of the filtrate is adjusted to 9.0 using 1.0 M NaOH. Then, 2.0 M hydroxylamine in

DMSO is added to the filtrate at a ratio of 3 μl of hydroxylamine solution for every 10 mg of TMAB label in the filtrate (if 5 mg of D0- and 5 mg of D9-TMAB-NHS are used, the total amount after pooling is 10 mg; more TMAB-NHS reagent will be present if additional isotopic tags are used, and so additional hydroxylamine should also be used). The reaction is carried out in three rounds, with one-third of the 2.0 M hydroxylamine added per round. After the first aliquot of hydroxylamine is added, the filtrate is incubated at room temperature for 10 min. The pH of the filtrate is adjusted back to 9.0 using 1.0 M NaOH. The addition of hydroxylamine followed by incubation and pH adjustment is then carried out twice more, for a total of three rounds. These filtrates can now be stored at –70°C until desalting.

3. *Desalting*. Desalting is carried out with a PepClean™ C-18 spin column (Pierce). There are many peptides in these filtered brain extracts, so for each filtered pool of samples, the resin from two C-18 columns is combined into one column by pouring the resin from one column to the other. Other than this step, desalting of the sample is carried out according to the manufacturer's instructions using solutions made with acetonitrile and trifluoroacetic acid. Peptides are eluted with 80 μl of 70% acetonitrile and 0.1% trifluoroacetic acid in water. These eluents are frozen and concentrated to 10–20 μl in a vacuum centrifuge. Aliquots of the peptide samples are stored at −70°C until mass spectrometry (MS) analysis.

3.4. Liquid Chromatography and Mass Spectrometry (LC/MS)

In order to obtain quantitative information on the relative levels of peptides, it is necessary to analyze the samples on MS. In our experience, best results are obtained by chromatography on a reversed-phase column with direct electrospray ionization mass spectrometry on a quadrupole time-of-flight (q-TOF) instrument. Other types of instruments, such as ion traps, provide less information, and some (such as the LCQ ion trap) did not yield any data for test samples. Matrix-assisted laser desorption ionization time-of-flight (MALDI-TOF) instruments are also less effective than q-TOF instruments because the MALDI-TOF laser causes decomposition of the TMAB tag and the loss of the isotopic label (i.e., the loss of trimethylamine). The extent of the decomposition depends on the intensity of the laser in the MALDI instrument and can vary from a small percentage to the majority of the signal. The loss of trimethylamine from the TMAB removes the isotopic tag, and therefore the resulting signal cannot be quantified. The q-TOF instruments also show loss of the trimethylamine label, but only after collision-induced dissociation, which is actually an advantage; the MS mode has the trimethyl group intact and provides quantification, while the MS/MS mode shows nearly complete dissociation of the trimethylamine tag which makes interpretation of the data easier (i.e., all of the five isotopic forms of the TMAB tags produce identical MS/MS spectra because they lose the trimethylamine moiety).

A variety of LC systems have been used with success. When limiting amounts of sample are available, such as when analyzing individual mouse hypothalami (or other brain regions), nanospray is optimal. If larger sample sizes are available, then nanospray is not critical. A typical protocol for LC/MS analysis on a quadrupole time-of-flight mass spectrometer (Ultima Q-TOF Waters/Micromass, Manchester, UK) is described below:

1. Frozen samples are thawed and briefly centrifuged in a microfuge to remove particulates.
2. An aliquot (typically 2–5 μl) is injected onto a Symmetry C-18 trapping column (5 μm particles, 180 μm i.d. × 20 mm; Waters, USA).
3. The material is desalted online for 15 min.
4. The trapped peptides are separated by elution with a water/acetonitrile/0.1% formic acid gradient through a BEH 130 C-18 column (1.7 μm particles, 100 μm i.d. × 100 mm; Waters, USA), at a flow rate of 600 nl/min.
5. Data are acquired in data-dependent mode and selected peptides dissociated by collisions with argon, using standard procedures. Note that for optimal MS/MS analysis, it may be helpful to use higher collision energy values for the TMAB-labeled peptides than typically used for non-labeled peptides.

3.5. Data Analysis

Relative levels of peptides in treatment groups compared to controls are determined by measuring peak intensities of mass spectra in the appropriate software program (for example, Masslynx is used for viewing data obtained on a Waters instrument). Spectra are scanned in this program and peaks sets are recognized by number and distance from one another. Peptides usually show mass-to-charge ratios in the 300–1,400 range (**Fig. 17.3**, top) and elute over a 20–30 min period, with individual peptides eluting within 5–30 s in a typical experiment. A typical spectrum is shown in **Fig. 17.3**, with the 300–1,400 range shown in the top panel for a single elute time. In this spectrum, a number of peak pairs can be observed, most of which show comparable peak intensities for the light and heavy forms of each peptide. An expanded view of the mass/charge range from 480 to 540 is shown in the middle panel, revealing a set of peaks that are not at equal intensities. The *m/z* difference between the two sets of peaks is ~4.5 (507.183–502.662). In this example, the charge state is 2; this is determined by comparing the ^{13}C-containing peaks to the monoisotopic peak (507.183 for the monoisotopic peak, 507.682 for the peak containing one atom of ^{13}C; the difference is 0.5 or 1/2, meaning the charge is 2). Therefore, the m/z difference of 4.5 for $z=2$ means that the mass difference is 9 Da, which indicates that a single isotopic tag has been incorporated. The mass of this peptide without isotopic tags or protons can be calculated from the following formula:

$$\text{Mass of unmodified peptide} = (m/z \cdot z) - (c \cdot T) - (1.008 \cdot (z - T)$$

where m/z is the observed mass-to-charge value for the monoisotopic peak; z is the charge state; c is the mass of the TMAB tag (128.118 for the D0 TMAB, 137.170 for the D9 TMAB); T is the number of tags incorporated; 1.008 is the mass of a proton, and $(z-T)$ is the calculation of the number of protons (i.e., the difference between the charge and the number of tags).

This last part of the equation is essential because the TMAB tags add a positive charge due to the quaternary amine group and therefore the charge state is not equal to the number of protons.

Typically, the mass of the unmodified peptide is taken as the average of the values determined from all of the isotopic tags; in the example shown in **Fig. 17.3**, this would be the average of the D0-TMAB and D9-TMAB groups. For samples using the five TMAB tags, the average of all five monoisotopic peaks would be used for this calculation.

The value of determining the mass of the unmodified peptide is for comparison to databases; because the tags were added as a means of quantifying the peptide, the mass of this tag needs to be subtracted in order to represent the endogenous form of the peptide. If additional modifications are detected (such as phosphorylation or acetylation), the extra mass of the modification is not subtracted because these modifications represent the endogenous form in the sample.

Quantification of the results is performed by measuring peak intensity. This can be done using the mass spectrometry program, although it is often faster to use a ruler and simply calculate the relative peak height. Typically, the monoisotopic peak and the peak containing one atom of ^{13}C are averaged so that the peak intensity is based on multiple points and not a single peak (which can show spikes that are not reflective of the overall level). In the example shown (**Fig. 17.3**, middle panel inset), the average peak intensity for the sample from the KO mouse striatum is 18% of the value from the WT mouse striatum, and the ratio is therefore 0.18.

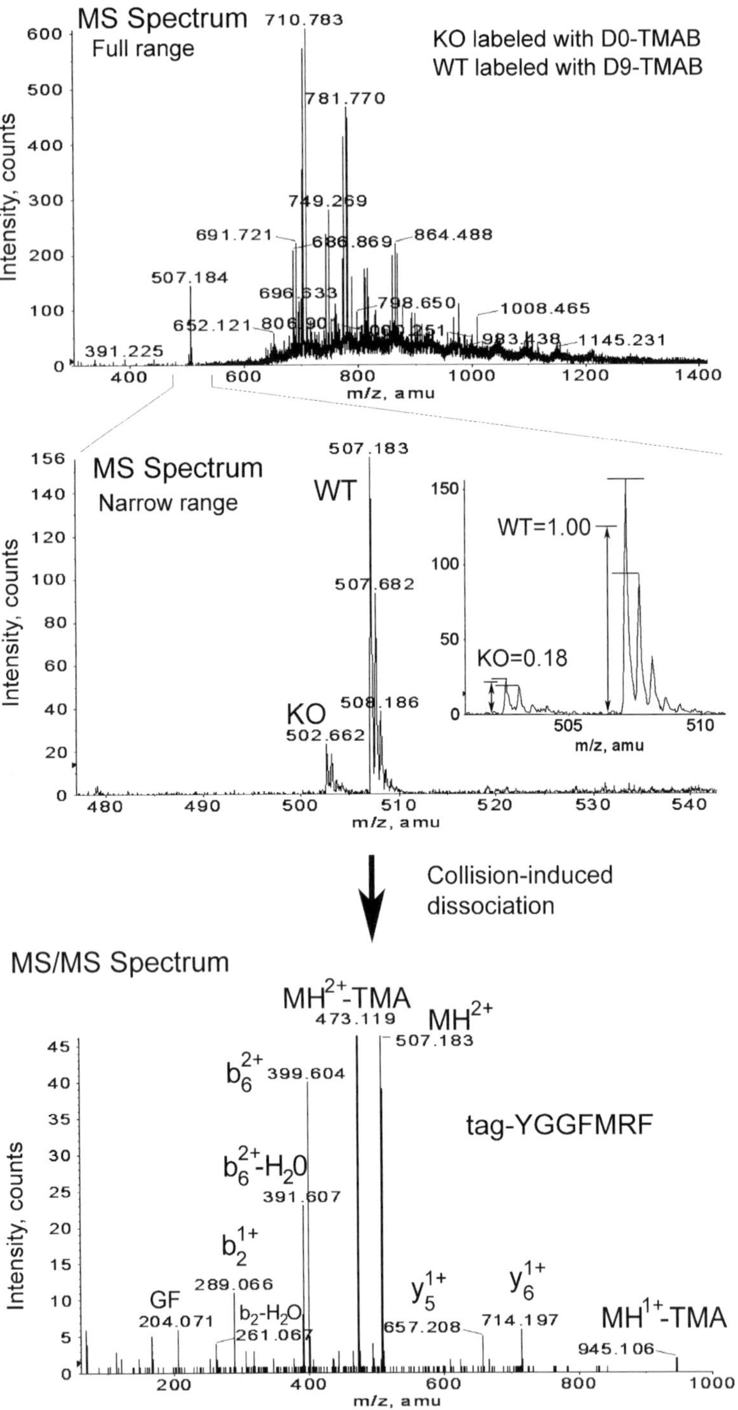

Fig. 17.3. Representative MS and MS/MS data. *Top*: A spectrum showing MS over the mass/charge (*m/z*) range 300–1,400 at 58.2 min from an experiment in which PC2 KO mouse striatal extract was labeled with D0-TMAB and WT striatal extract was labeled with D9-TMAB. Note the appearance of several peak pairs that appear roughly equal in intensity: these include 686.869 and 691.721; 704.216 and 710.783; 744.016 and 749.269; 774.553 and 781.770 (some of these are not labeled in the figure). The

If five labels are used, the analysis is generally similar except that the final value is not a simple ratio but the relative peak height of each of the five peaks. For an experiment setup as in **Fig. 17.2**, bottom, the controls would be averaged and the ratio of each KO peak group to the average control peak would be determined. When using five labels, spectra often contain overlapping peaks in which the signals with multiple ^{13}C atoms extend into the range of the next TMAB tag. In these cases, the isotopic distribution of the peptide must be calculated and the contribution from the ^{13}C-containing peaks subtracted as described (16). This is not a problem when using just the D0-TMAB and D9-TMAB labels because the 9 Da mass difference of these two labels is relatively much larger than the natural isotopic abundance.

Statistical significance is calculated using Student's two-tailed *t*-test to compare the ratio of each peptide determined from treatment vs control LC/MS runs to the ratio of the peptide in the control vs control LC/MS runs. If using two labels, as per the scheme in **Fig. 17.2**, top, then this requires additional runs of WT vs WT samples. If using five labels, as in **Fig. 17.2**, bottom panel, then the variation among replicates of WT samples can be determined and included in the statistical calculations.

3.6. Identification of Peptides by MS/MS Sequencing

Interpretation of MS/MS data is usually performed by computer-assisted searching of databases consisting of proteins or translated cDNA. A number of programs are available for database search-

Fig. 17.3. (Continued) 704/710 ions were subsequently identified by MS/MS as Purkinje cell protein 4 with eight TMAB tags and three protons (11+ charge state), while the 774/781 ions were identified as this protein with eight TMAB tags and two protons (10+ charge state). The 686/691 and 774/749 ions represent an unidentified peptide with seven TMAB tags and charge states of 13+ and 12+, respectively. *Middle*: The spectrum in the *top panel* was analyzed over the 480–540 mass/charge range, showing ions of 502.662 and 507.183. These ions co-elute (not shown), have a charge of 2+ (based on analysis of the difference between *m/z* of the monoisotopic and ^{13}C-containing peaks), and have a mass difference of 9.0 Da, indicating that a single TMAB tag was incorporated. Quantification is performed by measuring peak intensity of the monoisotopic peak and the peak containing one atom of ^{13}C, taking an average of these two values, and then comparing the ratio of the experimental group (KO) to the control group (WT). In this example, the ratio of KO to WT is 0.18. *Bottom*: The 507.183 ion was selected by the mass spectrometer for collision-induced dissociation, resulting in the MS/MS spectrum shown. From this spectrum, and using computer search programs, it is possible to identify this peptide as the proenkephalin heptapeptide (sequence YGGFMRF). Fragment ions labeled "b" represent N-terminal pieces of the peptide, while those labeled "y" represent C-terminal pieces. Note that the trimethylamine (TMA) moiety of the TMAB tags is removed from the labeled peptide upon collision-induced dissociation, resulting in b ions that contain only the butyryl tag (which adds 69 Da to the mass of the peptide). Strong peaks for the precursor ion (507.183, 2+) and precursor ion after loss of TMA (473.119, 2+, and 945.106, 1+) are visible in the spectrum.

ing. For peptides labeled with the TMAB reagents, we have found the optimal program to be Mascot. This program currently has four of the five TMAB labels included as options (the missing one is the D12-TMAB). Most importantly, the Mascot program considers the neutral loss of TMA from the peptides during collision-induced dissociation; this causes the loss of 59 Da from peptides labeled with one D0-TMAB tag, 62 Da from peptides labeled with one D3-TMAB tag, 65 Da from peptides labeled with one D6-TMAB tag, and 68 Da from peptides labeled with one D9-TMAB. For example, the proenkephalin heptapeptide (YGGMFRF) incorporates one TMAB tag and when the D9-TMAB form is fragmented by collision-induced dissociation, both the precursor ion (507.183, 2+) and a 68-Da smaller form (473.119, 2+) are observed (**Fig. 17.3**). The fragments resulting from cleavage of amide bonds (the b and y ions) usual lack the TMA group, as observed for the heptapeptide (**Fig. 17.3**).

Mascot searches need to be followed by manual interpretation to eliminate false positives. Several criteria are used to accept or decline the peptides identified by Mascot:

1. The isotopic form of TMAB matched by Mascot is the correct one based on the analysis of the peak set. While this may seem obvious, Mascot does not consider the peak set and know which of the individual peaks correspond to the D0 form, the D9 form, or any other form used. Therefore, if using the five isotopic forms of TMAB, there is a 1 in 5 chance that a false-positive labeled with one tag is correct. If using two tags, there is a 1 in 25 chance that a false positive has the correct tags. Therefore, a correlation of the isotopic TMAB form in the observed peak set with the predicted Mascot match is a simple and necessary step.

2. The number of tags incorporated into the peptide matches the number of free amines (N terminus and side chains of Lys). If using multiple tags, all should be the same isotopic form on a particular peptide (i.e., all D0-TMAB, or D9-TMAB, and not one D0-TMAB and one D9-TMAB on the same form of a peptide).

3. The Mascot score is either the top score of all potential peptides or the other peptides with comparable scores can be excluded by criteria 1, 2, and 4–7, leaving only one peptide that matches all criteria.

4. The majority (>80%) of the major MS/MS fragment ions match predicted a, b, or y ions, or precursor ions with loss of trimethylamine.

5. The mass accuracy of the fragment ions is within the accepted specification for the q-TOF instrument used for the analysis.

6. A minimum of five fragment ions match b or y ions. For small peptides, this can be a problem.
7. The charge state should be reasonable based on the peptide sequence.

4. Notes

1. While the quantitative peptidomics technique is able to identify hundreds of peptides and their specific processing forms from a single LC/MS run, the method does not detect every peptide in the sample. For example, peptides lacking an N-terminal free amine (due to post-translational modification such as acetylation or pyroglutamylation) that also lack an internal lysine residue are not labeled by the TMAB reagent. These peptides are therefore present as single peaks, and no information on relative levels can be obtained using this method. In other cases, intrinsic factors can cause low ionization efficiency of certain peptides during mass spectrometry. The mass/charge ratios used in these studies (300–1,800 a.m.u.) exclude some very small or very large peptides from detection. Finally, the dynamic range of peptide levels in biological samples varies beyond the detection limit of the mass spectrometry equipment and low-abundance peptides are not detectable above the background.
2. Avoiding contaminants is important. Small molecules and polymers can substantially interfere with the MS analysis. Clean water is essential. Some brands of microfuge tubes and filtration devices have polymeric contaminants that appear as polyethylene glycol-related compounds on MS; these contaminant signals completely overwhelm the signals from the tissue-derived peptides.
3. Using low-retention microfuge tubes and pipette tips is important, as peptides can bind regular tubes and tips, and many peptides can therein be lost during transfers and labeling steps. Whenever possible, rinse and dry tubes and tips with double-deionized water before use.
4. Prepare labeling solutions as well as glycine and hydroxylamine solutions fresh for each set of labeling reactions. It is extremely important to prepare all solutions (HCl, NaOH, glycine, phosphate buffer, and desalting solutions) with double-deionized water to avoid contamination from small organic molecules that can interfere with the mass spectrometry.

5. It is extremely important to clean Centricon filters before filtering peptides. These filters often contain glycerol which needs to be washed off before filtering to avoid sample contamination which will substantially interfere with mass spectrometry. Make sure that filters are covered with double-deionized water and then run the water through the filters at the designated centrifugal speed. Be sure to empty any remaining water before loading extracts for filtering.

Acknowledgments

The development of the techniques described in this chapter was supported by National Institutes of Health grant DA-04494 (L.D.F.).

References

1. Zhou, A., Webb, G., Zhu, X., and Steiner, D. F. (1999) Proteolytic processing in the secretory pathway *J Biol Chem* **274**, 20745–8.
2. Seidah, N. G., and Chretien, M. (1994) Proprotein convertases of subtilisin/kexin family *Meth Enzymol* **244**, 175–88.
3. Arolas, J. L., Vendrell, J., Aviles, F. X., and Fricker, L. D. (2007) Metallocarboxypeptidases: Emerging drug targets in biomedicine *Curr Pharm Des* **13**, 349–66.
4. Prigge, S. T., Mains, R. E., Eipper, B. A., and Amzel, L. M. (2000) New insights into copper monooxygenases and peptide amidation: Structure, mechanism and function *Cell Mol Life Sci* **57**, 1236–59.
5. Fricker, L. D., Lim, J., Pan, H., and Che, F. -Y. (2006) Peptidomics: Identification and quantification of endogenous peptides in neuroendocrine tissues *Mass Spectrom Rev* **25**, 327–44.
6. Che, F. -Y., Biswas, R., and Fricker, L. D. (2005) Relative quantitation of peptides in wild type and $Cpe^{fat/fat}$ mouse pituitary using stable isotopic tags and mass spectrometry *J Mass Spectrom* **40**, 227–37.
7. Pan, H., Nanno, D., Che, F. Y., Zhu, X., Salton, S. R., Steiner, D. F., Fricker, L. D., and Devi, L. A. (2005) Neuropeptide processing profile in mice lacking prohormone convertase-1 *Biochemistry* **44**, 4939–48.
8. Pan, H., Che, F. Y., Peng, B., Steiner, D. F., Pintar, J. E., and Fricker, L. D. (2006) The role of prohormone convertase-2 in hypothalamic neuropeptide processing: A quantitative neuropeptidomic study *J Neurochem* **98**, 1763–77.
9. Zhang, X., Che, F. Y., Berezniuk, I., Sonmez, K., Toll, L., and Fricker, L. D. (2008) Peptidomics of Cpe(fat/fat) mouse brain regions: Implications for neuropeptide processing *J Neurochem* **107**, 1596–613.
10. Zhang, X., Pan, H., Peng, B., Steiner, D. F., Pintar, J. E., and Fricker, L. D. (2010) Neuropeptidomic analysis establishes a major role for prohormone convertase-2 in neuropeptide biosynthesis *J Neurochem* **112**, 1168–79.
11. Wardman, J. H., Zhang, X., Gagnon, S., Castro, L. M., Zhu, X., Steiner, D. F., Day, R., and Fricker, L. D. (2010) Analysis of peptides in prohormone convertase 1/3 null mouse brain using quantitative peptidomics *J Neurochem* **114**, 215–25.
12. Scamuffa, N., Calvo, F., Chretien, M., Seidah, N. G., and Khatib, A. M. (2006) Proprotein convertases: Lessons from knockouts *FASEB J* **20**, 1954–63.
13. Julka, S., and Regnier, F. E. (2004) Quantification in proteomics through stable isotope coding: A review *J Proteome Res* **3**, 350–63.
14. Zhang, R., Sioma, C. S., Thompson, R. A., Xiong, L., and Regnier, F. E. (2002) Controlling deuterium isotope effects in comparative proteomics *Anal Chem* **74**, 3662–9.

15. Che, F. -Y., and Fricker, L. D. (2005) Quantitative peptidomics of mouse pituitary: Comparison of different stable isotopic tags *J Mass Spectrom* **40**, 238–49.
16. Morano, C., Zhang, X., and Fricker, L. D. (2008) Multiple isotopic labels for quantitative mass spectrometry *Anal Chem* **80**, 9298–309.
17. Che, F. -Y., Lim, J., Biswas, R., Pan, H., and Fricker, L. D. (2005) Quantitative neuropeptidomics of microwave-irradiated mouse brain and pituitary *Mol Cell Proteomics* **4**, 1391–405.
18. Svensson, M., Skold, K., Svenningsson, P., and Andren, P. E. (2003) Peptidomics-based discovery of novel neuropeptides *J Proteome Res* **2**, 213–19.
19. Paxinos, G., and Franklin, K. B. J. (2001) *The Mouse Brain in Stereotaxic Coordinates*, Academic Press, San Diego, CA.

Chapter 18

A Proteomic Protocol to Identify Physiological Substrates of Pro-protein Convertases

Guiying Nie and Andrew N. Stephens

Abstract

Proprotein convertases (PCs) convert pro-proteins into their bioactive forms through limited proteolytic cleavage, thereby regulating the temporal and spatial activation of a large number of functionally important proteins. This "converting" process is involved in a wide range of essential physiological and pathological processes, making PCs valuable therapeutic targets. One of the challenges in the field of PC research has been to identify the physiological substrates of a particular PC in a specific tissue or cellular process. Proteomics provides an unprecedented opportunity to identify novel PC substrates in a physiological context. Here we provide a detailed practical procedure utilizing two-dimensional fluorescent differential gel electrophoresis (2D-DiGE) and tandem mass spectrometry techniques, in combination with other standard molecular and biochemical methods, to identify and subsequently validate novel PC6 substrates in a critical uterine event called decidualization. This method is applicable to the study of any PC members and their relevant cellular processes.

Key words: Proprotein convertase, PC, PC6, proteomics, 2D-DiGE, substrate.

1. Introduction

Proprotein convertases (PCs) play a central role in the processing and/or activation of a large number of pro-proteins in many physiological and pathological processes (1–3). PCs are therefore important regulatory molecules that control tissue-specific and functionally essential bioactive proteins through temporal and spatial activation of their pro-protein substrates. It is well established that the PCs cleave their target proteins at pairs and single basic amino acids in the general consensus sequence (K/R) (X)

M. Mbikay, N.G. Seidah (eds.), *Proprotein Convertases*, Methods in Molecular Biology 768,
DOI 10.1007/978-1-61779-204-5_18, © Springer Science+Business Media, LLC 2011

n– (K/R)↓P_i', where $n = 0, 2, 4$ or 6 and X is any amino acid (1, 4–6). The X and P_i' residues are believed to further define the fine specificity of each PC (1). To date, the proteins identified or predicted to be processed by PCs include growth factors, peptide hormones, neuropeptides, extracellular matrix proteins, adhesion molecules, proteolytic enzymes, and integral membrane proteins (6, 7).

One of the challenges in the field of PC research has been to identify and assign the physiological substrates of a particular PC in a specific tissue or cellular process. Studies using techniques such as gene silencing, gene knockdown, co-transfection, and co-expression have greatly advanced this front. Recently developed proteomics approaches have also proven to be important and novel tools for the identification of native substrates of a specific protease (8, 9).

Here we describe a detailed procedure using proteomics in combination with other standard molecular and biochemical methods to identify novel PC6 substrates in a uterine event called decidualization. This method is applicable to the study of any of the PC members and their relevant cellular processes. The overall experimental procedures are schematically illustrated in **Fig. 18.1**.

We have previously shown that PC6 is the sole PC member upregulated during, and critical for, decidualization in the mouse and human (10, 11). Decidualization is a uterine remodelling event essential for embryo implantation. Knockdown of PC6 during early pregnancy in vivo in the mouse uterus results in complete failure of implantation due to decidualization arrest (10). Blocking PC6 production in human endometrial stromal cells (HESCs) significantly inhibits decidualization (11). We take advantage of the fact that HESCs can be isolated from the human uterus and decidualized in culture, and that PC6 is the sole PC member associated with decidualization. We reason that when decidual cellular proteins are cleaved by purified active recombinant human PC6, the abundance of PC6 substrates will change, the precursor forms [higher molecular weight (MW)] will decrease, whereas the processed forms (lower MW) will increase. These can be identified by mass spectrometry following analysis by CyDye labelling and two-dimensional fluorescent differential gel electrophoresis (2D-DiGE). Three validation steps are taken to confirm the authenticity of the identified proteins as physiological substrates: bioinformatics analysis of the primary amino acid sequence to confirm that they contain the consensus PC cleavage sites (12), Western blotting of the cellular lysates to validate that they change MW and abundance following PC6 cleavage, and immunohistochemical localization in human uterine tissues to confirm that they are co-expressed with PC6 in the same cell in vivo.

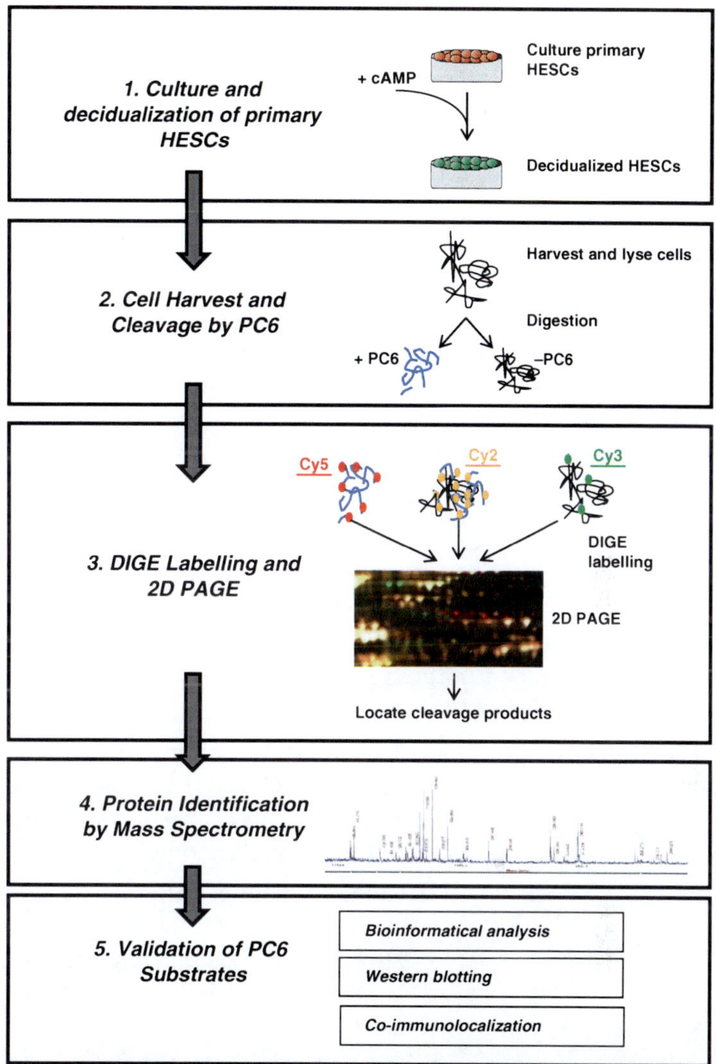

Fig. 18.1. Overview of the proteomic strategy applied to identify PC6 substrates in human endometrial stromal cells (HESCs). HESCs are treated with cAMP to induce decidualization. The decidualized cells are harvested and lysed, and equal amounts of protein incubated with or without PC6. Digested and undigested preparations are compared by 2D-DiGE, followed by identification by mass spectrometry. The status as *bona fide* PC substrates is confirmed by subsequent bioinformatics analysis, Western blotting of cell lysates, and co-immunolocalization in vivo.

2. Materials

2.1. Cell Culture and Decidualization of Primary Human Endometrial Stromal Cells

1. Phosphate buffered saline (PBS): 137 mM NaCl, 2.7 mM KCl, 10 mM sodium phosphate dibasic, 2 mM potassium phosphate monobasic, pH 7.4.
2. Dulbecco's modified Eagle's medium (DMEM)/Ham's F12 (Sigma, MO).
3. Charcoal-stripped foetal calf serum (CS-FCS) (Thermo Electron Corporation).
4. DNase I (Roche), stock solution made in 1 mg/ml with PBS, aliquots stored at –20°C.
5. Collagenase type 3 (Worthington), stock solution made in 10 mg/ml with PBS, aliquots stored at –20°C.
6. Streptomycin and penicillin (Gibco), L-glutamine (Sigma).
7. 1 M cAMP (Sigma), stock solution made in PBS, aliquots stored at –20°C.
8. Scissors and forceps (autoclaved), 11- and 44-μm nylon mesh (autoclaved), 50-ml falcon tubes (BD Bioscience).

2.2. PC6 Cleavage Assay

1. Solubilization buffer: 7 M urea, 2 M thiourea, 40 mM Tris, 1% (w/v) C7BzO. Dissolve completely and store in 1 and 20 ml aliquots at –80°C for up to 3 months. Thaw well prior to use. Single use only.
2. CyDye labelling buffer: 7 M urea, 2 M thiourea, 30 mM Tris, 1% (w/v) C7BzO, pH 8.1, at room temperature with HCl. Dissolve completely and store in 1 and 20 ml aliquots at –80°C for up to 3 months. Thaw well prior to use. Single use only.
3. DMEM/Ham's F12 medium.
4. pERTKR-AMC fluorogenic peptide substrate (Bachem, PA).
5. Recombinant human PC6-A (rhPC6A) (PhenoSwitch Bio-Sciences, Canada).
6. Tributylphosphine (TBP) 200 mM: Purchased as single-use vials and stored under N_2 at 4°C for no more than 1 week after opening (Sigma).
7. Wallac Victor 2 spectrophotometer (Perkin Elmer, MA).
8. B212 hand-held pH meter (Australian Scientific).

2.3. Labelling and 2D-PAGE

1. CyDyes^{TM} minimal labelling dyes (GE Healthcare), resuspended in dimethylformamide (DMF; Sigma) that has been opened no more than 3 months prior.

2. IEF buffer: 7 M urea, 2 M thiourea, 1% (w/v) C7BzO. Dissolve completely and store in 1 and 20 ml aliquots at –80°C for up to 3 months. Thaw well prior to use. Single use only.
3. Ampholytes appropriate to the pH gradient to be used for separation.
4. Bradford microassay (Bio-Rad).
5. Hand-held conductivity meter (Horiba).
6. IPGPhor II instrument (GE Healthcare).
7. Low-fluorescence glass plates (GE Healthcare).
8. Dodeca gel apparatus (Bio-Rad).
9. FLA 5100 multiwavelength fluorimager and specialty dual-wavelength filter sets (FujiFilm, Japan).
10. Progenesis PG240 SameSpots software (Nonlinear Dynamics, Newcastle Upon Tyne).
11. 96-Well V-bottom polypropylene microplates (Greiner Bio-One).
12. ProPicII robotic spot picker (Genomic Solutions, MI).

2.4. Mass Spectrometry

1. Recombinant porcine trypsin (Promega): Store lyophilized vials at –20°C. Resuspend immediately prior to use in freshly prepared 25 mM ammonium carbonate (final trypsin concentration 20 ng/μl). Keep on ice; discard after use (*see* **Note 1**).
2. MALDI-TOF matrix solution: Recrystallize α-cyano-4-hydroxycinnamic acid (CHCA) MALDI matrix (Laser-BioLabs). Dissolve fresh in 50% acetonitrile and 0.1% trifluoroacetic acid at a concentration of 5 mg/ml. Other concentrations can be used as required for best crystallization; we typically find that concentrations between 2–10 mg/ml are suitable.
3. C18 ZipTips: Standard bed volume of 0.6 μl resin (Millipore).
4. ProP program: Hosted by the Center for Biological Sequence Analysis (CBS) prediction server (http://www.cbs.dtu.dk/services/ProP/).

2.5. Western Blot Analysis

1. 2× Reducing sample buffer: 0.5 M Tris–HCl, 10% (v/v) glycerol, 10% (w/v) SDS, 0.01% (w/v) bromophenol blue. Store in 5 ml aliquots at –20°C. Add β mercaptoethanol (2%, v/v) prior to use as required.
2. Blocking solution: Tris-buffered saline (TBS) solution containing 5% (w/v) skim milk and 0.2% (v/v) Tween 20. Prepare fresh prior to use. May be stored overnight at 4°C.

3. Anti-caldesmon antibodies: Mouse monoclonal anti-caldesmon clones C21 (Cald-21) and Cald-5 (both from Sigma), used at 1:500 dilution.
4. Anti-mouse HRP-conjugated secondary antibodies (Silenus, Australia) used at 1:10,000 dilution.
5. ECL Plus chemiluminescence detection kit (GE Healthcare).

2.6. Immunohistochemistry

1. 10× High-salt TBS stock: 3 M NaCl, 0.05 M Tris–HCl, pH 7.6.
2. 10× Citrate buffer stock: 0.1 M Trisodium citrate, pH 6.0.
3. Blocking buffer A: 12% Normal rabbit serum, 6% foetal calf serum, 2% normal human serum, 0.1% Tween 20 in 1× high-salt TBS. Blocking buffer B: 15% Normal horse serum, 0.1% Tween 20 in 1× high-salt TBS. Blocking buffer C: 10% Normal horse serum, 0.1% Tween 20 in 1× high-salt TBS.
4. Histosol and graded ethanol. Wax pen.
5. Hydrogen peroxide solution (6%) in 100% methanol.
6. Shaking platform, microwave oven.
7. Primary antibodies: Affinity-purified sheep anti-PC6 antibody (home-made) (10). Anti-caldesmon antibodies: Mouse monoclonal anti-caldesmon clone C21 (Sigma).
8. Secondary antibodies: Biotinylated rabbit anti-sheep (BA 6000, Vector), biotinylated horse anti-mouse (BA 2000, Vector).
9. Negative control Ig's: Non-immune sheep IgG (home–made) (10)], normal mouse IgG (X0931; DAKO).
10. Vectastain Elite ABC Kit (PK6100; Vector).
11. DAKO Liquid DAB Chromogen System (KO 3466, Dako-Cytomation).
12. Harris haematoxylin (HHS32; Sigma-Aldrich, use 1:10 dilution in distilled H_2O).
13. DPX mounting solution (360294H; BDH).

3. Methods

3.1. Culture and Decidualization of Primary Human Endometrial Stromal Cells

1. Add 39–65 μl collagenase and 50 μl DNase to 1 ml PBS to make the collagenase/DNase enzyme mix.
2. Rinse the tissues thoroughly with 25 ml of warm PBS buffer, add 1 ml of the freshly prepared collagenase/DNase enzyme

mix, then cut the tissues with scissors into the smallest possible pieces.

3. Add another 1 ml of PBS into the tissues and incubate in a shaking water bath at 37°C for 20 min. Break up any clumps by pipetting up and down with a 1-ml pipette and continue the digest for a further 20 min at the same temperature and speed. Digestion is complete when glands are visible only as white stringy sediment (*see* **Note 2**).
4. Pour digested tissue into a 50-ml falcon tube containing 25 ml of DMEM/F12 to stop the digestion. Filter the digest first through 44-μm and then 11-μm nylon mesh.
5. Pellet the cells by centrifugation for 7 min at 1,450 rpm. Resuspend the pellet in 2 ml of DMEM/F12 medium containing 10% CS-FCS, 2 mM L-glutamine, 100 μg/ml streptomycin, and 100 IU/ml penicillin. Plate the cells in T25 or T75 culture flasks depending on tissue size. Due to the harsh digestion conditions, cells should be washed gently with sterile PBS after 40 min to remove contaminating red blood cells and epithelial cells. Remove residual PBS and replace with 15 ml fresh culture medium.
6. Grow the cells at 37°C until confluence (typically ~2–4 days). Passage the cells into T75 cm^2 flasks (~1×10^6 per flask) and grow to ~80% confluence.
7. Replace the medium with DMEM/F12 containing 2% CS-FCS, 2 mM L-glutamine, 100 μg/ml streptomycin, and 100 IU/ml penicillin, and incubate for 24 h.
8. Add cAMP (to a final concentration of 500 μM) to the cells and culture at 37°C for 4 days.

3.2. Harvest of Decidualized Cells and PC6 Cleavage Assay

1. Decidualized cells are harvested by pipetting 2 ml of solubilization buffer into the culture flask, and the cells are dislodged by gentle agitation with a Teflon cell scraper. The lysate is transferred to a fresh tube, and the culture flask washed with an additional 2 ml of solubilization buffer. The second wash is combined with the first, and any cell debris is pelleted by centrifugation at $10{,}000\times g$ for 10 min at room temperature.
2. The clarified lysate is concentrated to ~10 μl using a centrifugal concentrator (Pall Life Sciences; 3 kDa molecular weight cut-off), and the retained proteins are diluted 100× in a 1:1 mix of DMEM/Ham's F12 at pH 7.4 (*see* **Note 3**). The final lysate volume is 1 ml.
3. The lysate is transferred in 2× equal 500 μl volumes to a 96-well microtitre plate, and 10 U of recombinant human PC6-A (rhPC6A) are added to one well. To monitor PC6 activity, 100 μl pERTKR-MCA fluorogenic peptide substrate is added to each well.

4. The cleavage reaction is allowed to proceed for 2 h at 37°C. The reaction is monitored by measuring hydrolysis of the fluorogenic substrate every 60 s using a Wallac Victor 2 spectrophotometer with excitation/emission wavelengths of 355 nm/460 nm, respectively. An example of the assay result is provided in **Fig. 18.2**.
5. Following digestion, each sample is transferred to a Nanosep 3-kDa spin concentrator; the sample is concentrated to ~10 μl, and then 1 ml of solubilization buffer is added. Proteins are reduced and alkylated by incubation for 90 min in the presence of 5 mM TBP and 10 mM acrylamide monomer (*see* **Note 4**). The reaction is quenched by the addition of DTT to 10 mM.
6. Proteins are precipitated by addition of 10 volumes of acetone and incubation at room temperature for 90 min (*see* **Note 5**). Precipitated protein is pelleted at 10,000×*g* for 15 min at room temperature, the supernatant is removed (*see* **Note 6**), and the protein pellet is air-dried completely.
7. The protein pellet is then dissolved in ~100 μl of CyDye labelling buffer. The pH of the sample is checked using a B212 hand-held pH meter, and if required the pH is adjusted using solubilization buffer (*see* **Note** 7). The pH should be 8.1 at room temperature, which will increase to 8.5 when the sample is chilled to 4°C for subsequent labelling. The sample is then snap-frozen and stored at –80°C in tightly sealed, screw-capped centrifuge tubes until required.

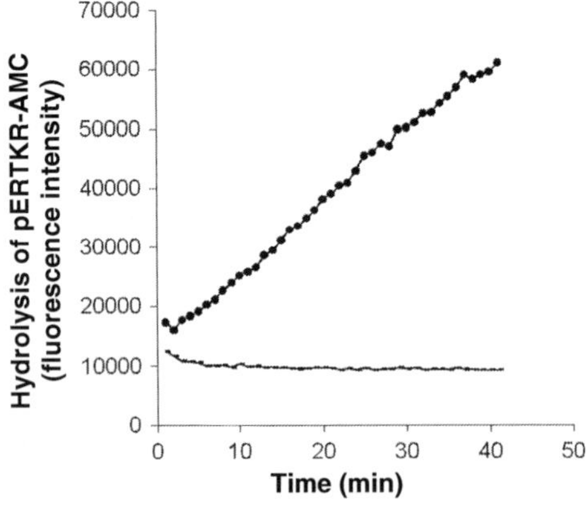

Fig. 18.2. Cleavage of pERTKR-AMC by rhPC6 in decidual cell lysates. Cell lysates were incubated in the presence (•) or the absence (-) of 10 U of rhPC6 and assayed directly for the hydrolysis of 100 μM of PC fluorogenic substrate, pERTKR-AMC, over 50 min.

3.3. DIGE Labelling and 2D PAGE Separation of Proteins

1. Protein lysates are thawed to room temperature, the supernatants clarified by centrifugation at 10,000×g for 10 min at room temperature, then transferred into fresh 1.7-ml microcentrifuge tubes. Protein assay is carried out using a Bradford microassay, and all samples are adjusted to an identical protein concentration (between 5 and 10 mg/ml).
2. Proteins are then labelled using CyDye minimal dyes, as recommended by the manufacturer. All steps are performed on ice and are protected from exposure to light. Typically we use minimal dyes at 200 pmol/μl concentration with no apparent effect on labelling efficiency (*see* **Note 8**).
3. On completion, each labelled protein sample (i.e., Cy2, Cy3, and Cy5 labelled) is combined and then precipitated in 10 volumes of acetone as described above, in the dark. The air-dried pellets are resuspended in 100 μl of IEF buffer.
4. The conductivity of each sample is measured using a hand-held conductivity meter. Conductivity should ideally be ~200 μS (*see* **Note 9**); however, we have routinely had success with samples up to ~600 μS and much higher (depending on the sample). If conductivity is above 600 μS, the sample is concentrated using 3-kDa Nanosep centrifugal concentrators and diluted in IEF buffer until reaching an appropriate conductivity.
5. DTT to 10 mM and ampholytes to 0.5% (appropriate to the pH gradient to be used for IEF) are added, and the labelled proteins are separated in the first dimension using commercially available immobilized pH gradient IPG strips and standard IEF procedures. IPG strips are cup loaded at the cathodic (alkaline) end of the strip. We typically use an IPGPhor II instrument, with the following parameters applied to all pH ranges (except alkaline pH 6-11 IPG strips): maximum 60 μA current per strip, 100 V for 90 min, 300 V for 90 min, 500 V for 3 h, gradient to 1,000 V over 4 h, gradient to 8,000 V over 3 h, and finally constant 8,000 V overnight until reaching 60,000 V h (*see* **Note 10**). On completion of IEF, the IPG strips are transferred to a clean tray, tightly sealed, and stored at –20°C for 1 h to overnight.
6. Second-dimension SDS-PAGE is carried out in 8–16% Tris–glycine polyacrylamide gradient gels, poured in-house between low-fluorescence glass plates. Separation is carried out overnight at 50 V using standard Laemmli buffers in a Dodeca gel apparatus, which provides excellent resolution and reproducibility. A circulating cooling system is used to keep the tank buffer at a constant 25°C.

7. Separation is judged complete when the dye front has exited the end of the gel. Each gel is then scanned using an FLA 5100 multiwavelength fluorimager and specialty dual-wavelength filter sets to detect CyDye fluorescence (*see* **Note 10**). The gels are then placed in sealed plastic bags with a small amount of Laemmli SDS-PAGE running buffer and stored at 4°C in the dark. We recommend as short a time as possible for storage, prior to excision of specific protein spots of interest for identification by mass spectrometry.
8. Each image is analysed using Progenesis PG240 SameSpots software. Careful image alignment is paramount for good expression data; we manually inspect and adjust each image as required to ensure optimal alignment. Spot detection and expression analysis are carried out automatically by the software. Typically we allow ~1–2 days (depending on experiment size) to manually evaluate all detected protein spots and accept/reject expression changes as real.
9. Following cleavage, a *bona fide* PC6 substrate should produce three (or more) protein products; at least two of these should be observed by the 2D-DiGE method described, depending on their size. These will reflect the following:
   i. Cleavage of the full-length pro-protein, leading to a decrease in the amount of the pro-protein present following PC6 treatment.
   ii. The cleaved protein with the pro-sequence removed, present at an increased level following PC6 treatment.
   iii. The much smaller pro-sequence removed from the pre-protein, present at an increased level following PC6 treatment.

   To identify/prioritize suspected cleavage products, we generally assess expression changes on the following criteria (in order of importance):
   i. Statistical significance; assessed for log-transformed, normalized spot volumes, and typically $p \leq 0.05$ (or more stringent, depending on experimental design);
   ii. %CV in normalized spot volume, which can help to identify those proteins whose p values may achieve significance but that exhibit unacceptably high experimental variability;
   iii. Fold difference between treatments of greater than twofold (however, we usually include lower fold change values as we find that fold change considered in isolation is unreliable).
10. Protein spots of interest are then excised directly from the gel using a ProPicII robotic spot picker, based on

X–Y coordinates exported from PG240 SameSpots (*see* **Note 11**). Gel plugs are deposited into 96-well, V-bottom polypropylene microplates that have been pre-rinsed three times in 100% acetonitrile (ACN). All liquid is removed from the gel plugs, which are then sealed well with sealing tape and stored at –80°C prior to mass spectrometric identification.

3.4. Identification of Potential PC6 Substrates by Mass Spectrometry

1. Protein spots are typically identified using standard mass spectrometry techniques. Each gel plug (containing the protein spot of interest) is washed and dehydrated according to standard procedures, and then rehydrated in 3 μl of ice-cold recombinant porcine trypsin (20 ng/μl). The gel plug is incubated on ice for 20 min, inhibiting trypsin auto-digestion and allowing the trypsin to absorb into the gel plug, followed by addition of 20 μl of 25 mM ammonium carbonate solution. The plate is sealed and incubated overnight at 37°C in a humidified chamber.
2. Digested peptides are routinely desalted and concentrated using C18 ZipTips, as recommended by the manufacturer. When applying MALDI-TOF-MS/MS for protein identification, the peptides are eluted from ZipTips in MALDI-TOF matrix solution and spotted directly onto a MALDI target plate. To ensure consistent crystallization of the matrix, the target plate is incubated at a constant 22°C on a heated platform for ~30 min or until individual spots are completely dry. Alternately, LC-MS/MS may also be used to identify proteins. In this case, we elute proteins from ZipTips using a solution of 50% acetonitrile (v/v) and 0.1% formic acid (v/v). These samples may then be directly injected for online separation and MS/MS using an ion trap or other mass spectrometer.
3. The analysis of the data obtained depends on the instrumentation and software used to acquire and process spectra, but some important considerations include the following:
   1. Database search may be limited to human or other organism of interest;
   2. Carbonylamide-cysteine (CAM) and oxidation of methionine residues must be taken into account;
   3. Generally, no more than one missed cleavage event should be allowed;
   4. Number and intensity of peptides matched, with suitable validation of peptide matches;
   5. Coverage of the candidate protein sequence;
   6. Position on the 2D gel may be related back to the predicted M_r and pI of the protein; care must be taken,

however, as these will change following cleavage. It may not be possible to predict M_r or pI if cleavage occurs at an unknown PC recognition site; in addition, post-translational modifications may affect migration.

3.5. Validation of the Identified Proteins as Bona Fide Substrates of PC6

Following identification by mass spectrometry, it is important to confirm that the protein is indeed a PC substrate. We first examine the primary amino acid sequence of the identified protein for the presence of potential PC cleavage sites. This is followed by experimental validation using Western blotting techniques. It is anticipated that PC6 substrates will accumulate in their processed forms (lower molecular weight), while the precursor forms (higher molecular weight) will decrease following PC6 cleavage. To further confirm that the identified protein is a physiological substrate, we determine whether this protein and PC6 are co-expressed in vivo in the same cells or tissues of interest by immunolocalization. The following details the procedures involved, using the example of caldesmon as the PC6 substrate being validated.

3.5.1. Characterization of Putatively Identified PC6 Substrates by Bioinformatics

1. Proteins putatively identified by mass spectrometry as cleaved by PC6 should theoretically possess a predicted PC cleavage site. The primary amino acid sequence of each identified protein should therefore be uploaded into the publicly available ProP program. This program performs a general scan for identification of PC consensus sites.
2. The location of each predicted PC cleavage site is mapped to the amino acid sequence of the protein, in order for further validation to be carried out. This is important for prediction and interpretation of subsequent data. Subsequent validation strategies must be able to demonstrate both the cleavage of the full-length substrate and the formation of the cleaved product/s. Candidate proteins not identified to contain any PC cleavage sites are likely to be the "indirectly regulated" proteins rather than direct substrates.
3. In the case of caldesmon, which is identified in our study (9), several isoforms have been reported; it is therefore necessary to establish which isoform has been identified as a substrate of PC6. Mass spectra should be manually inspected and the identified peptides mapped against the peptide sequence of each isoform. While not always conclusive, the location of peptides in the amino acid sequence can provide evidence as to which substrate isoform has been identified. In the case of caldesmon, peptides unique to isoform 1 are observed in every case, providing unequivocal evidence that the protein identified is caldesmon isoform 1.

3.5.2. Demonstration of PC6-Induced Cleavage Using Western Blot

1. Decidualized HESCs are prepared as detailed above, and 10 μg aliquots of protein (treated with or without rhPC6-A) dissolved in a 1:1 volume of 2× reducing sample buffer (containing 2% (v/v) β-mercaptoethanol). The protein lysates are heated at 95°C for 5 min in a heating block and then separated in a 10% polyacrylamide SDS-PAGE gel using standard Laemmli running buffer. A typical run is performed at a constant 200 V and takes ~45 min on average.
2. Proteins are then blotted to polyvinylidene fluoride (PVDF) membrane using standard Western blotting protocols. Proteins are transferred at 90 V for 1 h, using coloured protein standards to demonstrate correct transfer. Following transfer, the PVDF membrane is blocked overnight at 4°C by incubation in blocking solution.
3. Primary antibodies against caldesmon (either Cald-5 or Cald-21, depending on the epitope being detected) are diluted 1:500 in blocking solution at room temperature. The membrane is incubated with the primary antibody for 1 h at room temperature, with gentle agitation. Cald-5 is specific for an epitope N-terminal to the predicted PC6 cleavage site; Cald-21 is specific for an epitope C-terminal to the cleavage site.
4. Each membrane is washed in blocking solution again and then incubated with secondary anti-mouse antibody as above. Following a second wash, cleaved and/or uncleaved caldesmon forms are detected using chemiluminescence (**Fig. 18.3**). All signals detected are normalized against β-actin as a protein loading control.

3.5.3. Demonstration of the Co-expression of PC6 and Its Substrate in Decidual Cells *In Vivo* in the Human Uterus

1. Five micrometre serial sections of formalin-fixed tissue blocks are cut and dried. The sections are deparaffinized and rehydrated through a series of graded ethanol (*see* **Note 12**).
2. To retrieve antigens, the slides are submerged in 0.01 M citrate buffer and microwaved for 5 min on high power (800 W microwave). The slides are then cooled slowly on ice or at 4°C (approx 30 min), rinsed twice with high-salt TBS, washed for 5 min with 0.6% Tween 20 in high-salt TBS, and rinsed twice more with high-salt TBS.
3. The endogenous peroxidase activity is blocked by incubating with 6% hydrogen peroxide in 100% methanol (1:1, v/v) for 5–10 min at room temperature. The slides are then rinsed twice with high-salt TBS, washed for 5 min with 0.6% Tween 20 in high-salt TBS, and rinsed twice more with high-salt TBS.
4. Non-specific binding is blocked with blocking buffer A (for PC6 staining) or B (for caldesmon staining) for 30 min at room temperature.

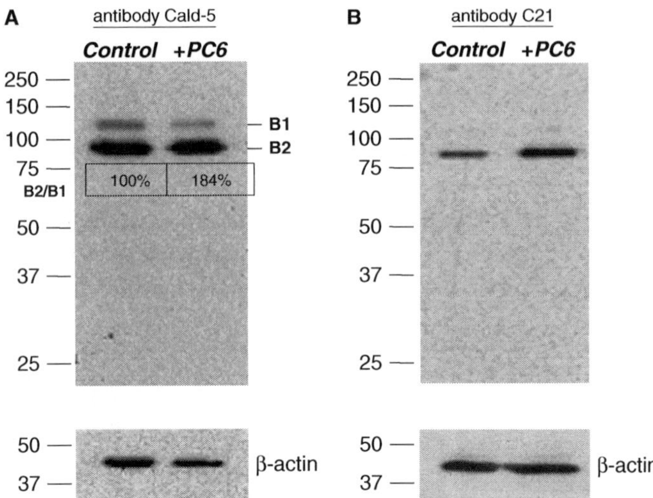

Fig. 18.3. Western blotting to confirm caldesmon as a PC6 substrate in decidual cells. (**a, b**) Western blotting of decidual cell lysates following treatment with (+PC6) or without (control) rhPC6 using two anti-caldesmon antibodies: an N-terminal antibody Cald-5 (**a**), and a C–terminal antibody C21 (**b**). The change in intensity of bands (B_1 and B_2) in (**a**) following rhPC6 treatment is shown by the ratio between the two bands (B_2/B_1). The protein loading was checked by probing for β-actin.

5. The sections are incubated with primary antibodies at 37°C for 30 min (for caldesmon antibody, 1:1,000 dilution in blocking buffer C) or 60 min (for PC6 antibody, 4 μg/ml diluted in blocking buffer A). Negative controls are included for each section using IgGs of non-immunized mouse (for caldesmon staining) or sheep (for PC6 staining) as the primary antibody.
6. The sections are washed three times (5 min each) with 0.6% Tween 20 in high-salt TBS and rinsed with high-salt TBS. The sections are then incubated with appropriate secondary antibodies for 30 min at room temperature. Use biotinylated horse anti-mouse IgG (1:200 dilution in blocking buffer C) for caldesmon staining and biotinylated rabbit anti-sheep IgG (1:200 dilution in blocking buffer A) for PC6 staining.
7. Signals were amplified with Vectastain Elite ABC Kit and visualized with diaminobenzidine. Sections were counter-stained with Harris haematoxylin, dehydrated, and mounted with DPX.
8. Both caldesmon and PC6 are strongly expressed in the decidual cells (**Fig. 18.4**), demonstrating co-expression of PC6 and its substrate caldesmon in the same cells. This co-expression is specific only to decidual cells, as the vascular smooth muscle cells of the blood vessel are strongly positive

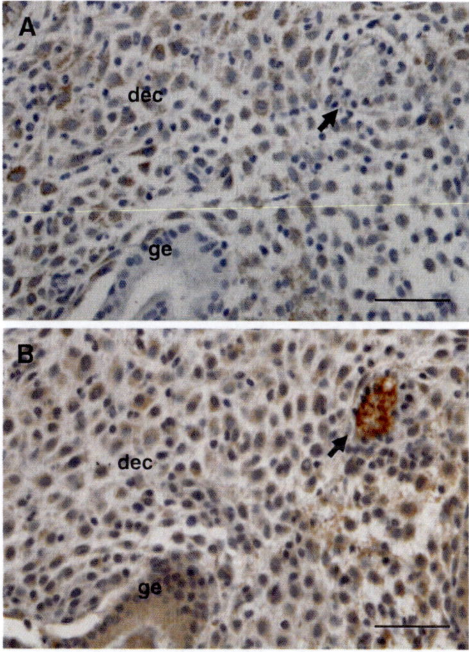

Fig. 18.4. Co-immunolocalization of caldesmon and PC6 in decidual cells in the human endometrium. Both caldesmon (**a**) and PC6 (**b**) are strongly expressed in the decidual cells (dec) in the secretory phase of the menstrual cycle. In contrast, the vascular smooth muscle cells of the blood vessel (highlighted by ↑) are strongly positive for caldesmon (**a**) but negative for PC6 (**b**), whereas the glandular epithelium (ge) is positive for PC6 (**b**) but negative for caldesmon (**a**). Bar = 50 μm.

for caldesmon but negative for PC6, whereas the glandular epithelium is positive for PC6 but negative for caldesmon (**Fig. 18.4**).

4. Notes

1. We typically use a single vial of trypsin per experiment; however, trypsin may also be diluted in a small volume of 0.01 M HCl, aliquoted, and stored at –80°C until used.
2. This protocol utilizes a harsh digestion step that breaks up most of the glands. It is therefore most useful for obtaining large numbers of stromal cells and less useful for isolating epithelial cells.
3. Dilution of the urea is absolutely required to ensure maximal PC6 activity during the cleavage reaction.

4. Acrylamide monomer is made from reagent-grade acrylamide powder as a 1 M stock, and 100 μl aliquots may be snap-frozen and stored at –20°C.
5. Incubation may also be performed overnight at –20°C. The sample must first be warmed to room temperature to ensure all urea crystals are dissolved before continuing.
6. Acetone must be completely removed; any residual acetone will affect subsequent separation by isoelectric focussing.
7. The pH of the sample is critical to ensure correct labelling by CyDye minimal dyes. In our experience, the pH may be anywhere between 8.0 and 9.0 but must be the same in each sample, otherwise labelling will not be equivalent between samples.
8. The key to achieve good labelling is to ensure that all factors are the same. Each sample must have an identical pH, protein concentration, and volume. In addition, it is extremely important to establish that the minimal Cy3 and Cy5 dyes do not lead to labelling bias. We routinely label identical aliquots of a pooled protein sample with both Cy3 and Cy5, and then compare the labelling. Any proteins showing differential labelling with one or the other dye must be eliminated from the final analysis.
9. Conductivity can vary depending on the sample type; however, it is imperative for good first-dimension separation by IEF that the conductivity is the same in all samples. Our rule of thumb is no more than ±10 μS difference between samples to be separated by IEF.
10. Total volt hours can vary between experiments; typically, we allow ~5 h at a constant 8,000 V to ensure complete focussing.
11. We routinely place multiple triangulation points around the periphery of each gel prior to scanning to facilitate easier software-mediated excision of protein spots using a robotic spot cutter. We use a red Artline permanent marker to place points, which fluoresce brightly in the red channel and do not rub off during storage. If robotic spot-cutting facilities are not available, we recommend Coomassie Blue G250 staining followed by careful manual matching.
12. We routinely rehydrate sections as follows:

Histosol	2 min
Histosol	2 min
100% Ethanol	2 min
100% Ethanol	2 min
70% Ethanol	2 min
Water	1 min

Acknowledgements

This work was supported by the National Health and Medical Research Council of Australia (project grant #611804 and fellowship #494808 to G.N.), CONRAD USA (to G.N.), and an Ovarian Cancer Research Fellowship (to AS).

References

1. Seidah, N. G., Day, R., Marcinkiewicz, M., and Chretien, M. (1998) Precursor convertases: An evolutionary ancient, cell-specific, combinatorial mechanism yielding diverse bioactive peptides and proteins *Ann NY Acad Sci* **839**, 9–24.
2. Khatib, A. M., Siegfried, G., Chretien, M., Metrakos, P., and Seidah, N. G. (2002) Proprotein convertases in tumor progression and malignancy: Novel targets in cancer therapy *Am J Pathol* **160**, 1921–35.
3. Taylor, N. A., Van De Ven, W. J., and Creemers, J. W. (2003) Curbing activation: Proprotein convertases in homeostasis and pathology *FASEB J* **17**, 1215–27.
4. Nakayama, K., Watanabe, T., Nakagawa, T., Kim, W. S., Nagahama, M., Hosaka, M., Hatsuzawa, K., Kondoh-Hashiba, K., and Murakami, K. (1992) Consensus sequence for precursor processing at mono-arginyl sites. Evidence for the involvement of a Kex2-like endoprotease in precursor cleavages at both dibasic and mono-arginyl sites *J Biol Chem* **267**, 16335–40.
5. Watanabe, T., Nakagawa, T., Ikemizu, J., Nagahama, M., Murakami, K., and Nakayama, K. (1992) Sequence requirements for precursor cleavage within the constitutive secretory pathway *J Biol Chem* **267**, 8270–4.
6. Seidah, N. G., Mayer, G., Zaid, A., Rousselet, E., Nassoury, N., Poirier, S., Essalmani, R., and Prat, A. (2008) The activation and physiological functions of the proprotein convertases *Int J Biochem Cell Biol* **40**, 1111–25.
7. Seidah, N. G., and Chretien, M. (1999) Proprotein and prohormone convertases: A family of subtilases generating diverse bioactive polypeptides *Brain Res* **848**, 45–62.
8. Bredemeyer, A. J., Lewis, R. M., Malone, J. P., Davis, A. E., Gross, J., Townsend, R. R., and Ley, T. J. (2004) A proteomic approach for the discovery of protease substrates *Proc Natl Acad Sci USA* **101**, 11785–90.
9. Kilpatrick, L. M., Stephens, A. N., Hardman, B. M., Salamonsen, L. A., Li, Y., Stanton, P. G., and Nie, G. (2009) Proteomic identification of caldesmon as a physiological substrate of proprotein convertase 6 in human uterine decidual cells essential for pregnancy establishment *J Proteome Res* **8**, 4983–92.
10. Nie, G., Li, Y., Wang, M., Liu, Y. X., Findlay, J. K., and Salamonsen, L. A. (2005) Inhibiting uterine PC6 blocks embryo implantation: An obligatory role for a proprotein convertase in fertility *Biol Reprod* **72**, 1029–36.
11. Okada, H., Nie, G., and Salamonsen, L. A. (2005) Requirement for proprotein convertase 5/6 during decidualization of human endometrial stromal cells in vitro *J Clin Endocrinol Metab* **90**, 1028–34.
12. Duckert, P., Brunak, S., and Blom, N. (2004) Prediction of proprotein convertase cleavage sites *Protein Eng Des Sel* **17**, 107–12.

Chapter 19

Neurophenotyping Genetically Modified Mice for Social Behavior

Ramona M. Rodriguiz, Jennifer S. Colvin, and William C. Wetsel

Abstract

Sociability in mice is a multidimensional adaptive and functional response. Due to its complexity, it is important that researchers use well-defined behavioral assays that are easily replicated with clearly defined ethograms. In the Mouse Behavioral and Neuroendocrine Analysis Core Facility at Duke University, we have developed a broad series of tests that examine different components of neonatal and adult social behaviors that include sociability, sexual behavior, aggressive and territorial responses, and maternal behaviors. While the purpose of this chapter is not to provide an exhaustive description of all mouse social tests available, we provide investigators with a description of basic procedures and considerations necessary to develop a successful social behavior testing program within their laboratories.

Key words: Social behavior, sociability, aggression, sexual behavior, resident-intruder, social dyadic, maternal behavior, mother–pup interactions, ultrasonic vocalizations.

1. Introduction

In the past, alterations in behavior have been attributed to changes in monoamines, acetylcholine, and amino acid neurotransmitters. Despite this fact, many chemical messengers affect behavior. Neuropeptides are one class of transmitters and they are proposed to control a variety of physiological and behavioral functions (1). To investigate these reputed functions, genes responsible for encoding peptide processing and degrading enzymes, neuropeptides themselves, and their receptor protein(s) have been disrupted in mice. In the Duke University Mouse Behavioral and Neuroendocrine Analysis Core Facility, we have identified a number of neuroendocrine, psychiatric, and neurological phenotypes

M. Mbikay, N.G. Seidah (eds.), *Proprotein Convertases*, Methods in Molecular Biology 768,
DOI 10.1007/978-1-61779-204-5_19, © Springer Science+Business Media, LLC 2011

in mice (2–18). Since the ultimate goal in making mutants is to understand human illness, it is important to identify the full range of behavioral abnormalities displayed by the mice. Primary disorders and comorbidities should be studied so that parallels can be drawn to conditions in humans. Although the Duke Facility can evaluate many different mouse behaviors, the present chapter focuses only upon social behavior because these responses are critical for reproduction, maternal care, and affiliation. Nevertheless, it should be emphasized that social behavior should be analyzed as a function of the species/strain, age and sex of the animal, and according to the context in which the test is administered.

2. Materials

2.1. Testing Area, Lighting, and Circadian Influences

The social tests that we describe consist of the resident-intruder, dyadic, social affiliation, sexual behavior, mother–pup interaction, and maternal behavior tests. These tests require a designated room or defined lab area where lighting, external noise, and activity are controlled. Typically, mice are acclimated to the test environment at least 12 h before evaluation. Most social tests are conducted during the dark cycle (at least 1 h into the dark cycle and less than 2 h before light onset) and under <5 lux illumination or under "red" light (6, 14, 15, 19). Mother–pup interactions can be observed at anytime.

2.2. Test Apparatus

For the resident-intruder test, a mouse clear-plastic "shoe-box" cage (29 × 18 × 13 cm) is used (**Fig. 19.1**). For the social

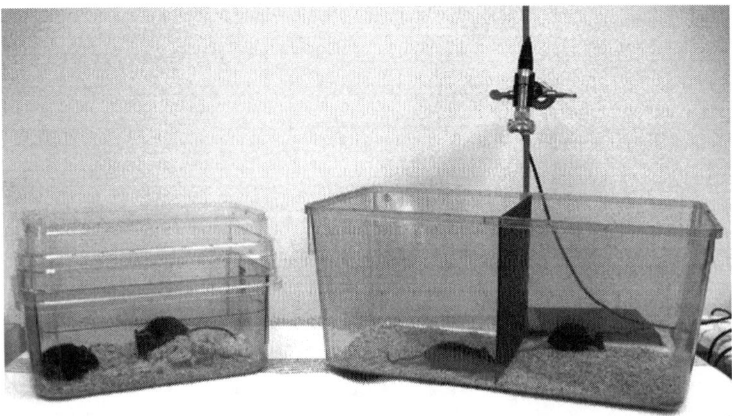

Fig. 19.1. Resident-intruder and dyadic test chambers. For resident-intruder (*left*) a mouse "shoe-box" cage is used with the wire cage top, food, and water bottle removed; the clear filter top keeps the mice from escaping during testing. For the dyadic test, a rat cage is divided by a solid partition that is removed after 5 min of acclimatization.

dyadic test, a rat clear-plastic "shoe-box" cage (48 × 26 × 20 cm) is divided in half by a solid partition (**Fig. 19.1**). The mother–pup interaction test is conducted in the clear-plastic home cage (**Fig. 19.2**). For sexual behavior, mice are housed individually in clear Plexiglas cages (20 × 39 × 19 cm). On the day of testing, the cage is placed on top of a stand with a mirror slanted at 45° to the cage to afford full view of the cage bottom (20) (**Fig. 19.3**). For social affiliation, a Plexiglas three-chambered apparatus (42 × 22 × 20) and two partner cages (10 cm in diameter and 11 cm high; OfficeMax – silver mesh pencil cups with fishing weights on the top of the cup) are used (**Fig. 19.4**). All social test chambers contain 1/8″ cob bedding (Anderson Inc., Maumee, OH).

2.3. Video Cameras

An infrared camera or one sensitive to low illumination is used. It should have >420 lines of resolution and view the test area in entirety. The camera is positioned at the side and/or top of the cage – depending upon how the behavior will be scored. Cameras should be interfaced both to a video monitor that permits video quality and clarity to be monitored and to a computer for live scoring (Noldus *Ethovision* or *Observer*, Leesburg, VA) or for use with behavioral recognition programs (Clever Sys *Social Scan*, Reston, VA; BiObserve, Bonn, Germany) (*see* **Note 1**).

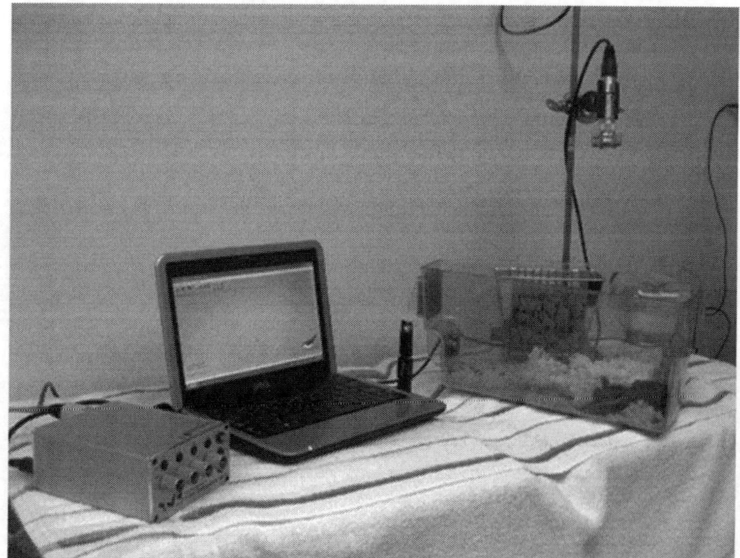

Fig. 19.2. Testing chamber for maternal behavior and monitoring of ultrasonic vocalizations. The ultrasonic microphone is positioned directly over the litter. The cage lid is removed 15 min before testing to permit the dam to acclimate to the testing conditions. This setup can be modified where a wire cage top remains on the cage to keep animals from escaping and allows long-term monitoring of maternal behaviors and pup ultrasonic vocalizations.

Fig. 19.3. Apparatus for testing sexual behavior. This test is conducted in a plexiglas chamber placed upon a stand with an angled mirror. Videotaping occurs from the side to produce a duel image which permits detailed assessment of social interaction and mating behavior. In this image, the male (*black*) target mouse is interacting with a (*white*) Swiss-Webster OVX estrogen–progesterone primed female. The posture and slightly elevated tail of the female indicates she is receptive to the investigations of the male.

Fig. 19.4. Social affiliation test apparatus. The apparatus is divided into three chambers. The mouse is placed into the center chamber and the time and amount of contacts or interactions it has with the non-social and social stimulus objects are recorded. The panels can also be removed and the mouse can be tested with the same protocol in a single large chamber.

2.4. Detection of Audible and Ultrasonic Vocalizations

While mice utter audible vocalizations, most rodents emit ultrasonic vocalizations. Microphones and software for recording vocalizations are commercially available (*see* **Note 2**). The microphone is positioned directly above the test chamber and not at an angle because echoes perturb recordings. Recordings from individual mice are best monitored in a sound attenuated chamber (**Fig. 19.5**).

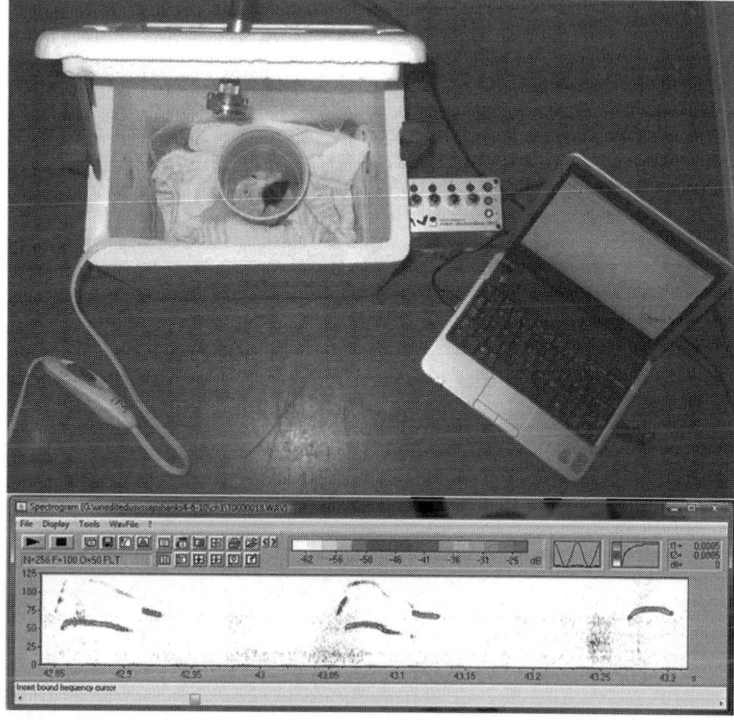

Fig. 19.5. Setup for monitoring of pup ultrasonic calls. Ultrasonic calls of individual pups can be assessed in sound-precluding test boxes, where the holder for the pup is maintained at a neutral nest temperature (32°C). Note the position of the ultrasonic microphone in the lid of the box, which is shut during testing. At the *bottom panel*, recordings of several calls from a 9-day-old pup are shown. The *y*-axis shows frequency of calls (kHz) and the *x*-axis shows the time scale in seconds. All three calls begin at 42.85, 43.08, and 43.27 s. The first two calls have two distinct components, with stable lower frequency elements (~50 kHz) and more dynamic multi-frequency elements (range 74–125 kHz). The final call has a single 75–80 kHz element. Calls were recorded and analyzed with Avisoft Bioacoustics (Berlin, Germany) software.

2.5. Test Partners

Partner mice are used for the resident-intruder, dyadic, social affiliation, and sexual behavior experiments. In the three former tests, male and female C3H/HeJ mice that are approximately the same size and age as the target mice are used. Note, C3H/HeJ mice give high levels of social investigation without initiating attack, but are susceptible to vision abnormalities as they age and must be screened for visual tracking of a moving object (6). For male sexual behavior, ovariectomized (OVX) Swiss-Webster females (Taconic Farms, Germantown, NY) are used; these mice are primed with estradiol benzoate and progesterone (Sigma-Aldrich, St. Louis, MO). For female sexual behavior, C57BL/6J or BALB/c males are used. To track mice, animals can be marked with ink, tattoo, or hair dye (leave the dye on the fur ~10 min, rinse with warm water, the dye marks are visible after several hours under infrared light). For most tests, partners are group-housed. To reduce reactivity and promote social investigation, partners may be handled a week before testing. Twenty-four hours before testing, partners are identified, tail-marked, weighed, and matched to target mice. Typically, partners are used once per test per day, except for tests of social memory or sexual behavior (*see* **Sections 3.2** and **3.4**). If the partner has not initiated attack behavior, been aggressed, or physically attacked, that partner can be reused every 48–72 h (*see* **Note 3**).

3. Methods

3.1. Defining Sociability in the Mouse

Sociability is a multidimensional behavior that requires detection and reactivity to stimuli, exploration, impulse control, emotionality, motivation, and problem-solving abilities. For this reason, social behaviors should be characterized into social domains.

3.1.1. The Social Ethogram

Table 19.1 depicts a social ethogram that lists 54 discrete behaviors that the target and partner animals may display. The ethogram classifies the behaviors into eight primary domains with three secondary domains. The primary domains include mild social investigation, intense social investigation, stationary reactivity, locomotor reactivity, threatening postures, aggression, submission, and withdrawal (*see* **Note 4**). Latency measures, misdirected behaviors, and solitary unclassified behaviors comprise secondary domains. Behaviors are scored for both target and partner animals, since interpretation of the target's social responses is contingent upon the behavioral initiations and responses of the partner (15; *see* **Note 5**). Given that emotional responses will change after the first agonistic encounter (6, 15), it is important that all behaviors be analyzed before and after the first display of threatening postures or aggression – regardless of which animal initiates the response. For example, reactivity prior to the first agonistic encounter is often reflective of the emotional state of the animal to social contact (6, 14, 15), whereas responses following the encounter are defensive and performed to reduce the probability of aggression. Behaviors may be scored as frequencies or total durations (*see* **Note 6**).

3.1.2. Analyses of Data from Social Behavior Tests

While each social paradigm has specialized measures, scoring behaviors are fairly universal across tests. The ethogram in **Table 19.1** will suffice for most testing.

1. Behaviors can be scored as individual events and reported as frequencies or scored as states where both frequencies of a given behavior and its duration are reported.

Table 19.1
Behavioral ethogram for social interaction[a]

Mild social interaction[b]
1. Approach
2. Stretch-attend
3. Sniff face
4. Sniff tail (tip or end)
Intense social interaction[c]
5. Sniff side
6. Sniff back
7. Sniff tail (base)
8. Nose (beneath tail)
9. Grooming face
10. Grooming head (back, sides, and neck)
11. Grooming body (not climb-groom, *see* 32)
12. Touch (but not pushing or shoving, *see* 31)
Stationary reactivity[d]
13. Eyes closed
14. Vocalization
15. Boxing
16. Holding
17. Immobility/freezing
18. Pushing
19. Kicking
Locomotor reactivity[e]
20. Kicking
21. Darting or escape behavior
22. Jumping
Threatening postures[f]
23. Feinting
24. Rushing
25. Mounting
26. Clawing
27. Pushing/shoving
28. Climb-grooming
29. Tail rattling
Aggression
30. Biting
31. Chasing
32. Lunging

Table 19.1 (continued)

33. Wrestling
34. Attacking
Withdrawal behaviors[g]
35. Turning away (from the other animal)
36. Walking away (from the other animal)
37. Avoiding (the other animal)
38. Withdrawal (walking backward from the other animal)
Submissive behaviors[h]
39. Submissive – mouse may lie on side, back, or assume "flat" position
40. Submissive – mouse will stand upright and/or remain rigid
Misdirected behaviors
41. Tunneling (front or side, crawling under abdomen or between rear legs of animal)
42. Following (walking behind the other mouse but within one body length or closer)
Latency measures
43. To first approach or investigation
44. To first reactivity to social investigation
45. To first agonistic exchange
46. To first aggressive exchange
Unclassified or solitary behaviors
47. Self-grooming
48. Laying down
49. Jumping
50. Tremor
51. Freezing
52. Vocalizations
53. Sleeping
54. Sifting (when cob bedding is present)

[a] *See* (6, 14, 15, 44) for detailed descriptions of behaviors.

[b] *Mild Social Interaction* refers to brief investigative contact with the nose and no physical contact.

[c] *Intense Social Interaction* denotes prolonged physical contact with the nose, mouth, or paws.

[d] *Stationary Reactivity* involves reactive responses to social contact that are performed within a small space.

[e] *Locomotor Reactivity* includes reactive responses to social contact that are performed while the animal is disengaging or leaving the proximity of the other animal.

[f] *Threatening Postures* concern brief agonistic responses which do not include biting or attacking.

[g] *Withdrawal Behaviors* refer to slow disengagement from the other animal.

[h] *Submissive Behaviors* are typically observed after threatening postures or aggression.

2. Under most conditions, incidences of threatening postures and aggression are of greatest interest to the investigator. In addition to frequencies and durations, the latency to the first agonistic event (i.e., threatening posture or aggressive behavior) and latency to first attack or bite are scored.
3. Since social interactions change after the first agonistic event, behaviors should be reported as frequencies and durations before and after this occurrence.
4. If social interaction is not observed, then the videotapes need to be reviewed for evidence for stereotyped or perseverative behaviors or other competing/impeding responses (e.g., locomotor deficiencies, anosmia, blindness, anxiety, and fear).
5. If automated scoring programs are used and mice are videotaped from the top of the cage, tracings of the animal's activity and exploration can be assessed (**Fig. 19.6**). These data provide important information about patterns of exploration, escape, or perseverative responses.

3.2. Social Affiliation

Social affiliation is a simple and rapid test that examines basic social behaviors. The test evaluates the propensity of the animal to explore stimuli in three phases: non-social objects, a novel social stimulus, and a familiar social reference. The test is flexible, in that

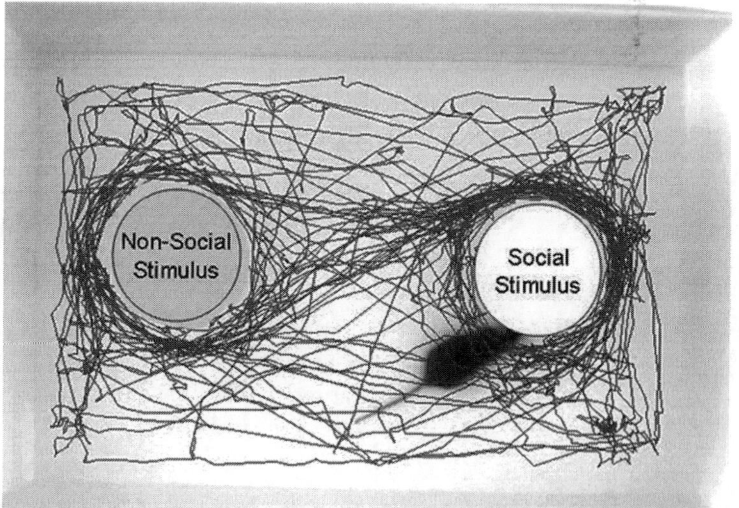

Fig. 19.6. Activity tracings from a social affiliation/preference test. Tracings provide important information about the exploration pattern of the mice. The mouse was tested in a single chamber apparatus to measure the ability of the animal to discriminate social and non-social stimuli. Tracings were produced with BiObserve (Bonn, Germany) software.

the types of social partners can be varied and, if desired, animals can be tested at different intervals to assess short-term, long-term, or remote social memory (*see* **Note 7**).

3.2.1. Preparation of Test Animals

One week prior to testing, partner mice should be trained to sit inside the social stimulus cage in the test arena for 5–10 min each day. Twenty-four hours before testing, target mice should be tail-marked and matched for weight and sex to a partner animal.

3.2.2. Social Affiliation Testing

1. First phase: non-social exploration. Two empty cages (non-social stimulus 1 or NS1 and non-social stimulus 2 or NS2) are placed in the center of the two outer chambers of the apparatus (**Fig. 19.4**). Often the cages are empty, but two identical non-social stimulus objects may be placed inside each cage (*see* **Note 8**).
2. Testing begins when the target mouse is placed in the center chamber. The responses of the target animal are videotaped for 10 min.
3. Second test phase: social/non-social exploration/ preference. At the end of the first test phase, the target is placed into a clean holding cage. One of the non-social stimulus cages (NS2) is removed and replaced with an identical cage containing a partner animal (social stimulus 1 or SS1). The target is reintroduced into the center chamber and its responses are videotaped for 10 min.
4. Third test phase: novel/familiar social preference. After the second test phase, the target mouse is removed to the holding cage. NS1 is removed and replaced with a novel test partner (SS2) where preference for the familiar SS1 and a novel SS2 partner is compared. Testing begins when the target mouse is reintroduced into the center chamber; behaviors are videotaped for 10 min. Intervals between the second and third test phases can be varied (*see* **Note 7**) to assess short- and long-term memory and can be repeated 10–14 days later for remote social memory.

3.2.3. Analyses of Social Affiliation Responses

1. Behaviors scored for frequency and duration during each test include contacts made with each test cage (orientation by the target to within 1 cm of the stimulus cage), time spent at each cage, and latency to make the first contact with each cage.
2. Total time spent exploring the two cages is recorded and a preference score is calculated: [(SS1 – NS1)/(SS1 + NS1)] or [(SS2 – SS1)/(SS2 + SS1)], where positive scores indicate preference for the novel stimulus, negative scores reflect preference for the familiar stimulus, and scores approximating "0" indicate no preference.

3. All incidences of responses toward the test cages should be scored, including freezing, lunging, tail rattling, feinting, and so forth. These behaviors provide important information for the interpretation of the final results.

3.3. Resident-Intruder and Social Dyadic Tests

Aggression refers to agonistic behaviors that support the initiation of an attack (21, 22). Aggression can be offensive or defensive; its categorization is affected by factors that regulate several levels of behavioral organization. A number of tests have been designed to examine agonistic responses in mice; the most notable are the resident-intruder test which is conducted in the target animal's home cage and the social dyadic exchange which occurs in a novel environment (6, 14, 15, 19) (**Fig. 19.1**).

3.3.1. Preparation of Test Animals

Test partners should be prepared as described (*see* **Section 2.5**). Target mice should be housed individually for at least 14–21 days before the onset of testing. Target mice are evaluated in the resident-intruder assay followed 7–10 days later by social dyadic testing. The cages should not be cleaned for at least 72 h prior to testing.

3.3.2. Resident-Intruder Test

1. Thirty minutes before testing, the wire cage top, food, and water bottle are removed (**Fig. 19.1**). Any excess amount of nesting material should be partially removed to reduce background noise and provide a "home base" for the target mouse. A filter or clear top should be placed on the cage, to reduce excessive rearing behaviors and the possibility of escape.
2. The target mouse should be videotaped for 5 min prior to introduction of the test partner to obtain baseline responses. If ultrasonic vocalizations are monitored, then the recording should commence with the onset of videotaping.
3. An age-sex-weight matched partner animal is introduced into an area of the target's home cage opposite the target animal, avoiding the nesting area.
4. Interactions are videotaped for 10 min. If an attack occurs, then this time should be recorded. If the attack is sustained for longer than 30 s without interruption, the test should be terminated and the partner removed. In the case of highly aggressive mice, tests may be reduced to 5 min in duration.
5. Once the partner is removed, the wire cage top, food, water, and filter top are replaced on the home cage and the animal is returned to the housing facility.

3.3.3. Social Dyadic Test

1. The target mouse is removed from the home cage and placed on one side of the test chamber and is separated by a solid

divider from its appropriate age-sex-weight matched partner (**Fig. 19.1**).

2. Target and partner mice are videotaped for 5 min. If ultrasonic vocalizations are monitored, then these recordings should begin with videotaping.
3. After the 5 min acclimatization period, the divider is removed and behavioral interactions are recorded for 10 min. If an attack is sustained for more than 30 s without interruption, the test should be terminated and the partner removed. In the case of highly aggressive mice, the test duration should be reduced to 5 min.
4. Upon completion of testing, the partner and the target animal are returned to their respective home cages (*see* **Note 9**).

3.3.4. Analyses of Social Behaviors in the Resident-Intruder and Dyadic Tests

Behaviors are scored and analyzed with the ethogram (*see* **Section 3.1.1**; **Table 19.1**).

3.4. Reproductive Behaviors

Reproductive behavior refers to mating behavior and partner preferences (7), as well as assessment of maternal behaviors, scent marking, and odorant communication (23). In males sexual behaviors can be observed ~10 days after the pubertal rise in testosterone, whereas in females they are observed ~1–2 weeks after vaginal opening (24).

3.4.1. Male and Female Sexual Behavior

1. Swiss-Webster females are OVX ~2 weeks before beginning the study.
2. Four days before testing, OVX females are treated with 10 μg estradiol benzoate dissolved in sesame oil (s.c.) for three consecutive days. On the fourth day, 500 μg progesterone in sesame oil (s.c.) is given 5–7 h before testing (*see* **Note 10**).
3. Target males are housed individually in polycarbonate cages (**Fig. 19.3**) with a small amount of cob bedding, nesting materials, food pellets, and water 4 days before testing. Chambers are placed in the test room 12 h before testing.
4. One hour before testing, the cob bedding, nesting materials, food, and water are removed. Testing commences 1 h following onset of the dark cycle.
5. The cage is placed on the mirror stand 10 min before testing. If ultrasonic calls are recorded, the microphone should be positioned directly above the cage. Video recordings should be from the side view (**Fig. 19.3**).
6. Testing starts when the female is introduced into the cage. All social interactions are videotaped for 30–60 min (*see* **Note 11**).

7. The frequencies and durations of social contacts between animals and the latencies, frequencies, and durations of mounting, intromission, and ejaculation are scored (7, 25). Receptivity of the female can be scored also (25) (**Table 19.2**).
8. For testing female sexual behavior, the procedures described above are the same, except the male partners are C57BL/6 or BALB/c mice. Males should be housed in the observation cages at least 4 days prior to testing, and the estrus cycle of females will have to be followed to determine when they will be most receptive (*see* **Note 12**). Alternatively, estrogen–progesterone primed OVX females can be used.

3.4.2. Maternal Behavior

Maternal behaviors are critical for the development and survival of offspring. Induction of maternal behaviors requires specific hormonal changes (26). Within the week after parturition, sensory cues emitted by the pups help to sustain maternal responses and modify these behaviors as the pups develop and become autonomous. Maternal behavior includes not only responses related to pup care but also preparatory behaviors before pup delivery and defense of the litter (27).

The primary variables used to measure maternal behavior include nest building; nursing, licking, and grooming the pups;

Table 19.2
Female receptivity[a]

Receptivity	Description
High	No attempts to avoid male before or during mounting or intromission. Male determines when contact is broken
Good	No attempt to avoid male before mounting, but may vocalize or hop during intromission. Male determines when contact is broken
Average	No attempt to avoid male before mounting, but may move away or kick male initially during mounting. Male determines when contact is broken
Poor	Female assumes defensive posture when male approaches and shows increased vocalization, kicking, and moves during intromission. Frequent attempts required by the male to mate successfully. Intromission may not occur. Female determines when contact is broken
Unreceptive	Female avoids or aggresses the male. She does not allow mounting; mating is incomplete

[a] *See* (25).

pup retrieval to the nest; and the presence of milk in the stomach of the pup. In addition, emission of ultrasonic vocalizations by the pups can be important mediators of maternal behavior (28, 29). Maternal behaviors are usually assessed in naturalistic settings in an undisturbed home cage and within the first 10 days after parturition (*see* **Note 13**).

1. Pregnant females are housed individually ~72 h before anticipated parturition (**Fig. 19.2**). Newborns should be undisturbed for 48–72 h (*see* **Note 13**).
2. Home cages with the dam and pups should be moved to the test area ~12 h before videotaping. If assessments are made across multiple days, the home cage should remain in the test area throughout testing.
3. Maternal behavior is best observed between postnatal days 3–10. If a single day is selected, then postnatal day 5 or 8 provides the best opportunity to observe behaviors and obtain ultrasonic calling from pups (*see* **Note 14**).
4. Food and water bottles can be left in place during videotaping; however, the bottle or food holder should not interfere with the observations (**Fig. 19.2**).
5. The microphone for ultrasonic vocalizations should be positioned directly above the litter (**Fig. 19.2**). A wire cage top can be positioned on the cage to contain the dam. Note, it is important that the dam is acclimated (10–30 min) to this change before testing. Dams should be observed before the change is made and their activity noted (e.g., nursing, eating, and sleeping); testing should not begin until the dam has resumed normal activity.
6. Assessment of maternal behaviors typically requires several recording sessions made each test day to provide enough data for analyses (*see* **Note 15**). Observational sampling can occur over 5–15 min every few hours or at more frequent intervals over 5 min once an hour, 10 s every 5 min, and so forth.
7. Ultrasonic vocalization recordings need to be synchronized with the videotapes to determine whether maternal responses precede or follow pup vocalizations.
8. Maternal behaviors are best analyzed as states, with frequencies and durations reported. Behaviors include maternal handling and manipulation of pups in the nest or within the home cage, licking or grooming pups, nursing, nest building, pup retrieval (if pups are outside the nest area), and duration spent on and off the nest. As dams may move the nest and pups, it is important to map these changes and synchronize them with the observed behaviors.

9. Duration and frequency of nursing bouts should be scored. Dams can engage in different types of nursing behaviors (high-arched back, low-arched back, and blanket postures; *see* 30). Variations in postures are related to eventual changes in the pup's adult stress responses and behaviors (30, 31).
10. If maternal behaviors are infrequent, maternal behaviors can be artificially induced by short-term pup separation. This procedure typically provokes maternal behaviors and ultrasonic vocalizations in the pups and provides a means to distinguish groups based upon genotype, drug treatments, or other manipulations. Typically maternal behaviors are monitored for 30 min before separation, at which time the entire litter is removed from the nest and placed into a holding cup maintained at 32°C. During separation (60–300 s), the dam's behaviors are videotaped. Ultrasonic vocalizations are monitored from the litter during the separation. After this time, the litter is returned to the home cage but to an area of the cage opposite that of the nest. This procedure provokes high rates of pup retrieval, grooming, and nest building by the dam (*see* **Note 16**).

3.5. Ultrasonic Vocalizations

Ultrasonic vocalizations are used by young and adult mice in social interactions or during emotional states. These calls are emitted at frequencies between 20 and 120 kHz and require special detectors that can record, analyze, and playback the calls (*see* **Section 2.4**; **Fig. 19.5**). Ultrasonic vocalizations can be monitored for individual animals in isolated or sound attenuated boxes or in freely interacting contexts of the home cage or test arena. When recording ultrasonic calls, positioning of the microphone is important since echoes of the emitted ultrasounds can affect the recordings.

Ultrasonic vocalizations are discrete utterances composed of single or multiple "syllables" that contain one or more notes (32–35) (**Fig. 19.5**). These syllables occur in sequence as a "phrase." When the phrases are repeated they are "motifs." The syllables of the calls have varying characteristics. For example, chirp-like syllables are emitted rapidly (20–200 ms in length); calls are rarely less than 20 ms in duration (32–35). Call frequencies for mice typically range from 35 to 110 kHz. More recently, alarm cries in rats have been reported at lower frequencies (36); hence, investigators should monitor all frequencies between 20 and 120 kHz (*see* **Note 17**).

Males are more likely to emit ultrasonic calls than females during social investigation, aggressive encounters, and sexual contexts and when presented with the urinary pheromones from the opposite sex (32, 35, 36). Male vocalizations are often at

70 kHz during sexual behavior since females are more receptive to males uttering 70 kHz calls (7, 37, 38). Parenthetically, lower frequency calls are also reported (7, 32). When vocalizing males fail to emit 70 kHz calls, estrus females may lose interest and in some circumstances aggress the male (7). By comparison, very few reports of adult female ultrasonic vocalizations are reported (39, 40). Despite the fact that neonatal mice call frequently during the mother–pup interactions (28, 29, 32), lactating females rarely utter ultrasonic calls in response to their offspring. Neonatal mice (0–11 postnatal days) typically emit calls between 60 and 100 kHz, which in turn elicits maternal behaviors such as pup retrieval, grooming, and nursing. Pups tend to elicit calls when exposed to clean bedding or dirty substrate from an unfamiliar cage. By contrast, detection of unfamiliar stimuli by the pups, such as the scent of an unfamiliar male, can provoke a marked suppression of pup calls (28, 29, 33, 34).

4. Notes

1. Video cameras (infrared or low-light) should be interfaced with any digital recorder (e.g., DVR and computer) or have a detachable USB device. A number of software programs are available that permit rapid conversion of videos to any media format for analyses (*see* www.any-video-converter.com).
2. Complete systems to assess ultrasonic vocalizations, including recording and analyses of the audio files, can be obtained from Avisoft Bioacoustics (Berlin, Germany) or Metris B.V. (Hoofddorp, the Netherlands). These systems provide the capability of recording a broad range of call frequencies (10–200 kHz) from a single microphone. If bat detectors are used, then a detector and microphone will be required for each frequency range (40–60, 60–80, and 80–100 kHz) sampled. Detectors can be interfaced to oscilloscopes where the number of calls per time period is identified and measured (41).
3. Multiple exposures of males to social dyadic testing can promote increased aggression or propensities to attack, even among C3H/HeJ male mice. Male C3H/HeJ partners that attack should be removed from further testing, unless submission or withdrawal in the target mouse is a goal of the study.
4. Not all social behaviors are readily observed in all paradigms. For example, during social affiliation test-

ing where physical contact between mice is precluded, incidences of attack are limited while tail rattling or lunging may be observed.

5. The bi-directionality of social interaction is an important consideration in the interpretation of results. Given that agonistic behaviors are natural responses of male mice defending a territory or of female mice defending a litter, it is important to know whether the "aggression" displayed by the target animal is offensive or defensive to the partner or whether it is unprovoked.
6. Behavioral scoring programs (Noldus *Observer*) or automated behavioral recognition programs (Clever Sys *Social Scan*) will output data as frequencies, durations, and average rates during a user-defined time period. Latencies to first occurrence of a behavior during the entire test or over a defined interval are provided in the outputs. If automated methods for scoring videotapes are not available, behaviors may be scored by hand at 5 s intervals. Here, the investigator uses a stop-watch and records the behaviors that are observed in the target and partner animals. Data are transcribed as the numbers of 5 s blocks in which each behavior occurs. Durations and latencies can be estimated within 5 s accuracy. Shorter intervals (2–3 s) can be used also to make more precise estimates of the latency and duration measures.
7. Approach and withdraw tendencies can be directly assessed using novel or familiar non-social stimulus objects. Additionally, test partners in the social stimulus cage can be varied so as to examine issues relating to social reference memory, recognition of an aggressive partner, and preferences for estrus or non-estrus females. Intervals between the second phase (non-social and social stimulus) and third phase (familiar social stimulus and non-familiar social stimulus) can be varied to permit examination of short-term (<1 h) or long-term (24–72 h) or remote social memory (>10 days).
8. Non-social stimulus objects should be plastic, glass, or a non-porous material that can be easily cleaned. Objects with distinct scents should be avoided.
9. Males should not be group-housed upon completion of resident-intruder and dyadic tests, as long-term isolation will promote territoriality and can contribute to excessive aggression and fighting. Additionally, since mice are usually not suitable for other behavioral tests following social isolation, resident-intruder and dyadic assessments should be the last tests run.

10. If the investigator plans to examine large numbers of mice over the course of the dark cycle, progesterone injections should be staggered so that testing falls within the 5–7 h window following progesterone treatment.
11. Sexually primed females may be used once during the test session. If OVX females are reused, they will need to be re-primed with estrogen and progesterone with a hiatus of 14–21 days between hormone treatments.
12. The length of the estrus cycle in mice is variable but consists of four stages: diestrus, proestrus, estrus, and metestrus (42, 43). It is advised that vaginal smears be collected in the morning of each day over an 8–10 day period to determine the cycling pattern. Note, collecting smears from naïve mice can disrupt the cycle and repeated collection over prolonged periods can prolong estrus cycles and can introduce infection obscuring the cycle (42, 43).
13. Mother–pup interactions are typically observed between postnatal days 3 and 11. Unnecessary disturbances of the home cage within the first 72 h after parturition should be avoided as this can reduce pup survivability by disrupting the establishment of mother–pup interactions necessary for the maintenance of maternal care. Postnatal days 0–3 can be examined if pregnant females are moved into observation cages 24–48 h before giving birth and remain undisturbed during observation (videotaping/ultrasonic vocalization recordings) over this time. After postnatal days 10–11, pups can thermoregulate and will become more autonomous as their eyes open. At this time pups will begin to leave the nest and expand their behavioral repertoire; ultrasonic vocalizations decrease. Observations during postnatal days 11–21 require an expanded behavioral ethogram (*see* **Table 19.1**).
14. Maternal behavioral should be examined according to longitudinal or cross-sectional designs. Longitudinal designs permit better control of baseline differences in maternal care and changes in behavior as the pups mature over time. However, repeated, short-term (60–300 s) manipulations of the litter can exert profound effects on dam behavior and on physiology of the offspring. Given these "handling" effects (31, 41), a cross-sectional design can be advantageous, where a litter is examined only at a single time point. The disadvantage of this approach is that many animals are required to examine various ages.
15. Activity of the dam and pups can vary across the circadian cycle. Hence, it may be necessary to conduct pilot studies where behavior is videotaped over several continuous

hours during the dark and light cycles. The videotapes can be scanned to determine the most appropriate times for testing.

16. Since removal of pups can produce marked changes in adult behavior and induce stress responses in these pups, this procedure should be limited to only a few instances during the neonatal and juvenile periods.
17. Many other features of the test environment can artificially produce ultrasonic sounds that may be confused with true vocalizations including, but not limited to, sound reverberations inside the cages, walking through the bedding, nails of mice scratching the cage floors, investigator's keys, ceiling fans, air vents, or other electronic devices. These sounds can be detected by performing an ultrasonic sweep of the test environment before testing and the features of this "noise" can be filtered. Since adult females rarely emit ultrasonic calls (*see* **Section 3.5**), potential noise created by bedding disturbances can be controlled by obtaining control recordings of a single adult female in the test environment over several minutes.

Acknowledgment

This work is partially supported by HD36015, MH019109, and MH082441.

References

1. Strand, F. L. (2003) Neuropeptides: General characteristics and neuropharmaceutical potential in treating CNS disorders *Prog Drug Res* **61**, 1–37.
2. Gainetdinov, R. R., Wetsel, W. C., Jones, S. R., Levin, E. D., Jaber, M., and Caron, M. G. (1999) Role of serotonin in the paradoxical calming effect of psychostimulants on hyperactivity *Science* **283**, 397–401.
3. Xu, F., Gainetdinov, R. R., Wetsel, W. C., Jones, S. R., Bohn, L. M., Miller, G. W., Wang, Y. M., and Caron, M. G. (2000) Supersensitivity to psychostimulants in mice lacking the noradrenaline transporter *Nature Neurosci* **3**, 465–71.
4. Ribar, T. J., Rodriguiz, R. M., Khiroug, L., Wetsel, W. C., Augustine, G. J., and Means, A. R. (2000) Cerebellar deficits in Ca^{2+}/Calmodulin kinase IV-deficient mice *J Neurosci* **20**(RC107), 1–5.
5. Nillni, E. A., Xie, W., Mulcahy, L., Sanchez, V. C., and Wetsel, W. C. (2002) Deficiencies in pro-thyrotropin-releasing hormone (pro-TRH) processing and abnormalities in thermoregulation in $Cpe^{fat/fat}$ mice *J Biol Chem* **277**, 48587–95.
6. Rodriguiz, R. M., Chu, R., Caron, M. G., and Wetsel, W. C. (2004) Aberrant responses in social interaction of dopamine transporter knockout mice *Behav Brain Res* **148**, 185–98.
7. Srinivasan, S., Bunch, D. O., Feng, Y., Rodriguiz, R. M., Li, M., Ravenell, R. L., Luo, G. X., Arimura, A., Fricker, L. D., Eddy, E. M., and Wetsel, W. C. (2004) Analysis of the infertility phenotype in male $Cpe^{fat/fat}$ mice *Endocrinology* **145**, 2023–34.
8. Pillai-Nair, N., Panicker, A. K., Rodriguiz, R. M., Gilmore, K. L., Demyanenko, G. P.,

Huang, J. Z., Wetsel, W. C., and Maness, P. F. (2005) Neural cell adhesion molecule-secreting transgenic mice display abnormalities in GABAergic interneurons and alterations in behaviour *J Neurosci* **25**, 4659–71.

9. Cawley, N. X., Zhou, J., Hill, J., Abebe, D., Romboz, S., Yanik, T., Rodriguiz, R. M., Wetsel, W. C., and Loh, Y. P. (2004) The carboxypeptidase E knockout mouse exhibits endocrinological and behavioral deficits *Endocrinology* **145**, 5807–19.
10. Sotnikova, T. D., Beaulieu, J.-M., Barak, L. S., Wetsel, W. C., Caron, M. G., and Gaintedinov, R. R. (2005) Dopamine-independent locomotor actions of amphetamines in a novel acute model of Parkinson's disease *PLoS Biol* **3**, 1–13.
11. Prado, V. F., Martins-Silva, C., de Castro, B. M., Lima, R. F., Barros, D. M., Amaral, A. J., Ramsey, A. J., Sotnikova, T. D., Ramirez, M. R., Kim, H.-G., Rossato, J. I., Koenen, J., Quan, H., Cota, V. R., Moraes, M. F., Gomez, M. V., Guatimosim, C., Wetsel, W. C., Kushmerick, C., Pereira, G. S., Gainetdinov, R. R., Izquierdo, I. A., Caron, M. G., and Prado, M. A. (2006) Mice deficient for the vesicular acetylcholine transporter are myasthenic and have deficits in social recognition *Neuron* **51**, 601–12.
12. Fukui, M., Rodriguiz, R. M., Zhou, J., Jiang, S. X., Phillips, L. E., Caron, M. G., and Wetsel, W. C. (2007) *Vmat2* heterozygous mutant mice display a depressive-like phenotype *J Neurosci* **27**, 10520–9.
13. Welch, J. M., Lu, J., Rodriguiz, R. M., Trotta, N. C., Peca, J., Ding, J.-D., Feliciano, C., Adams, J. P., Dudek, S. M., Weinberg, R. J., Calakos, N., Wetsel, W. C., and Feng, G. (2007) Cortico-striatal synaptic defects and OCD-like behaviours in *Sapap3*-mutant mice *Nature* **448**, 894–900.
14. Beaulieu, J.-M., Zhang, M., Rodriguiz, R. M., Sotnikova, T. D., Cools, M. J., Wetsel, W. C., Gainetdinov, R. R., and Caron, M. G. (2008) Role of GSK3β in behavioral abnormalities induced by serotonin deficiency *Proc Natl Acad Sci USA* **105**, 1333–8.
15. Nehrenberg, D. L., Gariépy, J.-L., Rodriguiz, R. M., Zhou, X., Lauder, J. M., Cyr, M., and Wetsel, W. C. (2009) An anxiety-like phenotype in mice selectively bred for aggression *Behav Brain Res* **201**, 179–91.
16. Allen, K. D., Griffin, T. M., Rodriguiz, R. M., Wetsel, W. C., Kraus, V. B., Huebner, J. L., Boyd, L. M., and Setton, L. A. (2009) Decreased physical function and increased pain sensitivity in mice deficient for type IX collagen *Arthritis Rheum* **60**, 2684–93.
17. Roberts, A. C., Diez-Garcia, J., Rodriguiz, R. M., López, I. P., Luján, R., Martínez-Turrilas, R., Picó, E., Henson, M. A., Bernardo, D. R., Jarrett, T. M., Clendeninn, D. J., López-Mascaraque, L., Feng, G., Lo, D. C., Wesseling, J. F., Wetsel, W. C., Philpot, B. D., and Pérez-Otaño, I. (2009) Down-regulation of NR3A-containing NMDARs is required for synapse maturation and memory consolidation *Neuron* **63**, 342–56.
18. Crooks, K. R., Kleven, D. T., Rodriguiz, R. M., Wetsel, W. C., and McNamara, J. O. (2010) TrkB signaling is required for development of the sensitizing and drug seeking effects of cocaine *Neuropharmacology* **58**, 1067–77.
19. Gariépy, J.-L., and Rodriguiz, R. M. (2002) Issues of establishment, consolidation, and reorganization in biobehavioral adaptation *Mind Brain* **3**, 53–77.
20. Rissman, E. F., Wersinger, S. R., Fugger, H. N., and Foster, T. C. (1999) Sex with knockout models: Behavioral studies of estrogen receptor α *Brain Res* **835**, 80–90.
21. Cairns, R. B., and Scholz, S. D. (1973) Fighting in mice: Dyadic escalation and what is learned *J Comp Physiol Psychol* **85**, 540–50.
22. Scott, J. P. (1958) *Aggression*, University of Chicago Press, Chicago, IL.
23. Arakawa, H., Blanchard, D. C., Arakawa, K., Dunlap, C., and Blanchard, R. J. (2008) Scent marking behavior as an odorant communication in mice *Neurosci Biobehav Rev* **32**, 1236–48.
24. Sisk, C. L., and Zehr, J. L. (2005) Pubertal hormones organize the adolescent brain and behaviour *Front Neuroendocrinol* **26**, 163–74.
25. McGill, T. E. (1962) Sexual behavior in three inbred strains of mice *Behaviour* **19**, 341–50.
26. Svare, B., and Gandelman, R. (1973) Postpartum aggression in mice: Experiential and environmental factors *Horm Behav* **4**, 323–34.
27. Weber, E. M., and Olsson, A. S. (2008) Maternal behaviour in *Mus musculus*: An ethological review *Appl Anim Behav Sci* **114**, 1–22.
28. Hofer, M. A., Brunelli, S. A., Masmela, J. R., and Shair, H. N. (1996) Maternal interactions prior to separation potentiate isolation-induced calling in rat pups *Behav Neurosci* **110**, 1158–67.
29. D'Amato, F. R. D., Scalera, E., Sarli, C., and Moles, A. (2005) Pups call, mothers rush: Does maternal responsiveness affect the amount of ultrasonic vocalizations in mouse pups *Behav Genetics* **35**, 103–12.

30. Francis, D. D., Diorio, J., Lui, D., and Meaney, M. J. (1999) Non-genomic transmission across generations of maternal behavior and stress response in the rat *Science* **286**, 1155–8.
31. Meaney, M. J., Diorio, J., Francis, D., Widdowson, J., LaPlante, P., Caldji, C., Sharma, S., Seckl, J. R., and Plotsky, P. M. (1996) Early environmental regulation of forebrain glucocorticoid receptor gene expression: Implications for adrenocortical responses to stress *Develop Neurosci* **18**, 49–72.
32. Sales, G., and Pye, D. (1974) *Ultrasonic Communication by Animals*, Chaucer Press, Richard Clay, Ltd., Bungay, Suffolk, pp. 149–201.
33. Branchi, I., Santucci, D., Vitale, A., and Alleva, E. (1998) Ultrasonic vocalizations by infant laboratory mice: A preliminary spectrographic characterization under different conditions *Dev Psychobiol* **33**, 249–56.
34. Hofer, M. A., Shair, H. N., and Brunelli, S. A. (2001) Ultrasonic vocalizations in rat and mouse pups *Curr Protocols Neurosci* **8**, 114.1–114.16.
35. Holy, T. E., and Guo, Z. (2005) Ultrasonic songs of male mice *PLoS Biol* **3**, 2177–86.
36. Litvin, Y., Blanchard, D. C., and Blanchard, R. J. (2007) Rat 22 kHz ultrasonic vocalizations as alarm cries *Behav Brain Res* **182**, 166–72.
37. Nyby, J. (1983) Ultrasonic vocalizations during sex behavior of male house mice (*Mus musculus*): A description *Behav Neural Biol* **39**, 128–34.
38. Pomerantz, S. M., Nunex, A. A., and Bean, N. J. (1983) Female behavior is affected by male ultrasonic vocalizations in house mice *Physiol Behav* **31**, 91–6.
39. D'Amato, F. R., and Moles, A. (2001) Ultrasonic vocalizations as an index of social memory in female mice *Behav Neurosci* **115**, 834–40.
40. Moles, A., and D'Amato, F. R. (2000) Ultrasonic vocalization by female mice in the presence of a conspecific carrying food cues *Anim Behav* **60**, 689–94.
41. Bell, R., and Smotherman, W. P. (1980) *Maternal Influences and Early Behavior*, SP Medical & Scientific Books, New York, NY.
42. Allen, E. (1922) The oestrous cycle in the mouse *Am J Anat* **30**, 297–371.
43. Champlin, A. K., and Dorr, D. L. (1973) Determining the stage of estrous cycle in the mouse by the appearance of the vagina *Biol Reprod* **8**, 491–4.
44. Gendreau, P. L., Petitto, J. M., Gariepy, J.-L., and Lewis, M. H. (1997) D1 dopamine receptor mediation of social and nonsocial emotional reactivity in mice: Effects of housing and strain difference in motor activity *Behav Neurosci* **111**, 424–34.

INDEX

Note: The letters 'f' and 't' refer to figures and tables respectively.

M. Mbikay, N.G. Seidah (eds.), *Proprotein Convertases*, Methods in Molecular Biology 768,
DOI 10.1007/978-1-61779-204-5, © Springer Science+Business Media, LLC 2011

M

N

P